T0305736

VEHICLE SAFETY COMMUNICATIONS

WILEY SERIES ON INFORMATION AND COMMUNICATION TECHNOLOGY

Series Editors: T. Russell Hsing and Vincent K. N. Lau

The Information and Communication Technology (ICT) book series focuses on creating useful connections between advanced communication theories, practical designs, and end-user applications in various next generation networks and broadband access systems, including fiber, cable, satellite, and wireless. The ICT book series examines the difficulties of applying various advanced communication technologies to practical systems such as WiFi, WiMax, B3G, etc., and considers how technologies are designed in conjunction with standards, theories, and applications.

The ICT book series also addresses application-oriented topics such as service management and creation and end-user devices, as well as the coupling between end devices and infrastructure.

T. Russell Hsing, PhD, is the Executive Director of Emerging Technologies and Services Research at Telcordia Technologies. He manages and leads the applied research and development of information and wireless sensor networking solutions for numerous applications and systems. Email: thsing@telcordia.com

Vincent K.N. Lau, PhD, is Associate Professor in the Department of Electrical Engineering at the Hong Kong University of Science and Technology. His current research interest is on delay-sensitive cross-layer optimization with imperfect system state information. Email: eeknlau@ee.ust.hk

Wireless Internet and Mobile Computing: Interoperability and Performance
Yu-Kwong Ricky Kwok and Vincent K. N. Lau

Digital Signal Processing Techniques and Applications in Radar Image Processing
Bu-Chin Wang

The Fabric of Mobile Services: Software Paradigms and Business Demands
Shoshana Loeb, Benjamin Falchuk, and Euthimios Panagos

Fundamentals of Wireless Communications Engineering Technologies
K. Daniel Wong

RF Circuit Design, Second Edition
Richard Chi-Hsi Li

Networks and Services: Carrier Ethernet, PBT, MPLS-TP, and VPLS
Mehmet Toy

Equitable Resource Allocation: Models, Algorithms, and Applications
Hanan Luss

Vehicle Safety Communications: Protocols, Security, and Privacy
Luca Delgrossi and Tao Zhang

VEHICLE SAFETY COMMUNICATIONS
Protocols, Security, and Privacy

Luca Delgrossi
Mercedes-Benz R&D North America, Inc.

Tao Zhang
Cisco Systems, Inc.

A JOHN WILEY & SONS, INC., PUBLICATION

Cover Illustration: © Daimler AG
Cover Design: John Wiley & Sons, Inc.

Published by John Wiley & Sons, Inc., Hoboken, New Jersey
Published simultaneously in Canada

For general information on our other products and services or for technical support, please
contact our Customer Care Department within the United States at (800) 762-2974, outside the
United States at (317) 572-3993 or fax (317) 572-4002.

Wiley also publishes its books in a variety of electronic formats. Some content that appears in
print may not be available in electronic formats. For more information about Wiley products,
visit our web site at www.wiley.com.

Library of Congress Cataloging-in-Publication Data:
Delgrossi, Luca.
  Vehicle safety communications : protocols, security, and privacy / Luca Delgrossi, Tao Zhang.
      pages cm. – (Information and communication technology series ; 103)
    Includes bibliographical references and index.
  ISBN 978-1-118-13272-2
  1. Vehicular ad hoc networks (Computer networks)–Safety measures.   2. Automobiles–
Safety appliances.   3. Automobiles–Collision avoidance systems.   I. Zhang, Tao,
1962-   II. Title.
  TE228.37.D45 2012
  629.2'76–dc23

                                                                    2012020858

Printed in the United States of America

10 9 8 7 6 5 4 3 2 1

CONTENTS

FOREWORD

Ralf G. Herrtwich

Over the past few decades, vehicles have advanced to assist their drivers in many ways. They brake automatically to avoid accidents, they maintain a perfectly safe distance from the car ahead, they avoid drifting out of lane, and they even evade pedestrians to prevent harm. All these features rely on sensors through which vehicles monitor themselves and their environment. But could these features not become infinitely better if vehicles participated in a little game of "I spy with my little eye . . . " and conveyed their own findings to the vehicles around them? Enter vehicle-to-vehicle communication.

In a world where everything is getting connected, the idea of communicating cars is somewhat obvious. However, the first manifestations of vehicle communications that are offered in most premium vehicles today keep humans in the loop: they allow passengers to browse the Internet or their e-mail or let drivers access their vehicle remotely, for example, to readjust the electrical charging procedure. Communicating directly from car to car, from machine to machine, is a whole new ball game, the rules of which are described in this book.

Both Luca Delgrossi and Tao Zhang are pioneers not only in defining but also in implementing vehicle-to-vehicle communication. From the early stages of Department of Transportation (DOT) research projects through Institute of Electrical and Electronics Engineers (IEEE) standardization to the current deployment tests, they were involved in many notable activities that shaped and refined what hopefully is going to become the *lingua franca* of the vehicle world. I congratulate them on their effort to share their knowledge through this book—because while our vehicles will eventually speak this new language, it is us engineers who have to learn it.

Where will this all lead? Improved vehicle and traffic safety, for sure. Vehicles can communicate dangerous situations to the ones behind them, leading to fewer surprises for drivers through advance warnings—about the end of a traffic jam behind the next curve, a patch of black ice in the next off-ramp, or a stalled vehicle in the right lane. Eventually, driver assistance systems will be able to take such information into account and combine it with their other sensors for a more automated response—reacting to an emergency braking ahead or to a vehicle taking the right of way. On top of this, much of the information collected from vehicles can be used for mobility improvements, and even vehicle efficiency can be increased through a more uniform movement of communicating vehicles. Eventually, vehicles could coordinate their maneuvers with messages, making vehicle-to-vehicle communication an element in the emerging field of autonomous driving.

Not that we are fully there yet, or will be in just a few years. The value of any networking technology grows with the number of entities that can communicate. Reaching a decent population of communicating vehicles on the road is not an easy task, especially because early adopters will be the ones reaping the fewest benefits initially. And there are issues such as security, which may create anxiety among users and thus hinder widespread deployment. It goes without saying that a book covering vehicle-to-vehicle communication in its entirety also addresses these issues.

Enjoy!

PROF. DR. RALF G. HERRTWICH
Director, Driver Assistance and Chassis Systems
Group Research and Advanced Engineering
Daimler AG

FOREWORD

Flavio Bonomi

It is my special privilege to be able to express a few introductory thoughts on this very timely and important book.

This book, written by two active participants in the research, standardization, and early field deployment of technologies for vehicle-to-vehicle and vehicle-to-infrastructure communications, and of related applications, will quickly become a reference in this field. It provides a clear, organized, and thorough guide into the technology, reaching out to the bleeding research edge, but also into the history, motivations, and applications.

The value of this book is particularly important for our industry, which is starting to understand and act upon the promises of a deep transformation in the field of transportation.

This book is an important testimony of a broad effort, involving academia and industry, led by passionate and visionary leaders, such as Luca Delgrossi and Tao Zhang, who worked for years to bring these technologies to maturity and, hopefully, to pervasive deployment.

Indeed, we believe their effort will find it appropriate manifestation, not only in the traditional road transportation, but also in adjacent sectors.

The technologies and applications described in this book will definitely play an important role in the evolution of smart transportation, smart cities, industrial automation, and, more in general, in the vast explosion of mobile connectivity broadly described as the "Internet of Things."

Luca, Tao, thanks for your many years of dreaming, research, and evangelization. And thanks for this beautiful work, which will introduce this valuable body of technology to many of the people who will help turn your dreams into reality!

FLAVIO BONOMI
Cisco Fellow, Vice President
Head of Advanced Architecture and Research
Cisco Systems

FOREWORD

Adam Drobot

The roles that automobiles play in our global society are as important as ever. Luca Delgrossi and Tao Zhang have written a remarkable book that addresses one of the most important issues in the minds of car designers, public regulators, and consumers: safety and what modern communications can bring to dramatically improve what is possible. This is a complex subject and I have to commend the authors for the lucid and structured approach that they have taken. First, they have brought together the facts about the current status of "safety" of automotive vehicles gathered from many points of view. Second, they have captured the spirit of technological advances and understood that reducing the number of automotive accidents and the consequences of such accidents is a never-ending quest, one that relies on practical adoption of what is available at a given point in time. It is also one in which new tools can be brought to bear as we exhaust the rate of improvement from existing approaches. The heart of these new tools is the exploitation of wireless communications and digital electronics. Third, they have exposed the importance of a multidisciplinary approach. It is one of the reasons that this book is so important: because it melds the fundamentals and subtleties from disparate technical communities. Imagine automotive engineering, safety, wireless engineering, and cyber-security all contributing to one body of work. The promise is a dramatic reduction in the number of automotive crashes; injuries to drivers, passengers, and pedestrians; and destruction of property.

We are at a point in time where the silicon revolution is forcing car designers and manufacturers to significantly reconsider and revise the features and control systems of automobiles. The same is true of the highway infrastructure that our automobiles use. Much of this is driven through the adoption by the public of powerful mobile devices that have ever-growing computing capability, access to stored and real time information, mobile communications, and much more powerful interfaces. They expect the capabilities that have entered their daily lives to show up in their automobiles. This may be for entertainment, convenience, looking after their automobiles, and certainly for safety. A little less visible is the large investment over time in digital sensors and actuators, artificial intelligence, and various subsystems, which are also being built into automobiles and into our highways, and which rely on the economics of replication and mass markets. So looking at an issue such as safety there are several progressions that we can expect. The first is from passive systems to active systems, to eventually autonomous systems capable of safely guiding automobiles with little or no driver intervention. The second is from stand-alone vehicles to ones that interact with other vehicles and with the surrounding infrastructure, to eventually deeply optimized systems that cooperatively share information and rely on collaborative decision making to improve safety and mobility and to reduce the impact on the environment. This is the world of the Internet and the cloud.

For the progressions described above to become reality there is deep technical homework to be done. The topics addressed in this book are the building blocks for completing that homework. They include the architecture of how information is moved and processed within a car, and how the car relies on external information. There is great value to the fact that the topics are dealt with at several levels and that aspects important to safety are clearly identified. The topics include the overarching approaches to vehicle connectivity and how connectivity can be used to satisfy various functions. An important component is the expository description of communication technologies and how they match up to requirements such as latency, jitter, and scalability. Important to all of this are illustrative calculations and simulations that allow the reader to understand why the details matter. One of the most important contributions is the in-depth exploration of security and privacy, and the underlying mechanisms of encryption and key/certificate management. Implementing systems that affect the fundamental controls of automobiles without getting this right is something that would be hard to imagine. At the same time it is important to understand what the issues are and to get to the heart of why satisfying the requirements is so difficult. Last, while we can idealize many of the concepts and analysis of what constitutes a system design for safety, there is nothing better than learning from experimentation and empirical common sense. The authors make it a point to capture what has been accomplished in the field. This is an important book for anyone dealing with research and engineering for connected vehicle systems, and who has a need to know what the underlying technologies are capable of.

In closing this Foreword I would like to congratulate Luca and Tao for writing this book. I know that they live in a world of deadlines and pressures that make it difficult to devote the time and have the discipline to write a book. Nevertheless, they have succeeded at providing us with an excellent work and have contributed to the codification and distribution of knowledge that others can build on. The hard work and late hours that they put in outside their daily duties is visible. The effort they have taken is much appreciated and should be commended.

ADAM DROBOT
Chairman
OpenTechWorks, Inc.
Dallas, Texas

PREFACE

We pay a high price every day in fatalities, injuries, and property damage caused by motor vehicle crashes. While many incremental steps have been taken to improve vehicle safety over the years, the number of vehicles and distance traveled continues to increase, making it more and more difficult to travel safely down the road. Therefore, it is crucial to create new safety systems that can significantly reduce the number of collisions and their severity. Wireless communications may be the cornerstone for these next-generation automotive safety systems.

Over the past decade, engineers have explored ways to use wireless communications to achieve another breakthrough in vehicle safety. Dedicated short-range communications (DSRC) allow for the acquisition of high-quality data that would otherwise be impossible to collect through onboard sensors, providing a rich complement to existing systems. Sharing these high-quality data among vehicles enables them to "see" the complete picture of their surroundings and perceive potential dangers. In 2012, the effectiveness of communications-based safety systems has already been demonstrated through early prototypes and field trials.

This book focuses on communications for vehicle safety. It illustrates the underlying philosophy, design principles, and protocols to build a full wireless communications system suitable for consumer vehicles. It describes unprecedented challenges as well as potential solutions for establishing trust among vehicles, securing data exchanges, and protecting privacy in consumer vehicle networks.

The design of vehicle safety communications systems presents a series of unique challenges. Typical vehicle speeds result in short time intervals for communications, requiring low latencies and fast channel setup. Unlike existing networks, intervehicle exchanges of safety-critical data are dominated by periodic broadcasts of small messages. Traditional mechanisms to improve data transmission reliability, such as packet acknowledgement and retransmission, are no longer effective, because vehicles are constantly moving and delayed packets will likely carry outdated information. Data transmissions occur in widely diverse environments, ranging from urban canyons to hilly terrains and rural areas, each with its unique impact on signal propagation and network performance. Furthermore, communications systems should be able to quickly adjust to highly dynamic vehicle movements and traffic densities.

Similarly, consumer vehicle networks impose unique security and privacy requirements. Vehicles must establish sufficient trust in the messages they receive within the very short time available for data exchanges. Protecting driver privacy introduces conflicting requirements with securing the communications, supporting vehicle safety applications, and detecting misbehaving or malicious entities. Nationwide consumer vehicle networks demand unprecedented system scalability and bring security and privacy management to a significantly higher level of complexity. Many solutions targeted at smaller networks have been found to be nonscalable, ineffective, or inefficient in this context.

Additional constraints are imposed by vehicle requirements. Onboard safety equipment must be built according to automotive grade criteria and be certified. Fixing problems or making changes to in-vehicle hardware or software can incur significant costs and inconveniences to consumers and manufacturers. Finally, the long lifetime of consumer vehicles imposes challenges to ensure backward compatibility between different generations of onboard communication and security systems.

Over the past decade, tremendous collaborative efforts have been devoted to developing vehicle safety communications technology by industry, academia, and government agencies. This book attempts to summarize the main results from these efforts and intends to provide a solid basis for further study. We tried to balance technical details and readability for a broad audience.

BOOK OUTLINE

The first three chapters are dedicated to *automotive safety*. They introduce the motivation for this work, the context, and the nature of vehicle safety applications. Chapter 1 presents road traffic statistics for the United States, Europe, Japan, and other parts of the world, showing the high price we have been paying in terms of human lives, injured people, and property damage, as well as demonstrating the real extent of the vehicle safety problem. Chapter 2 summarizes the evolution of automotive safety systems, from the introduction of passive features such as seat belts and air bags to active safety and the latest driver assistance systems. Chapter 3 describes vehicle architectures supporting these onboard safety systems, including electronic control units, sensors, and in-vehicle networks. It also discusses vehicle data as well as positioning and security.

Chapters 4 through 9 focus on *wireless communications for vehicle safety*. Chapter 4 introduces connected vehicles. It discusses vehicle communication modes and applications' needs, highlighting unique requirements for consumer vehicle networks. In addition, existing technologies are compared to evaluate suitability for vehicle safety communications. Chapter 5 describes allocated spectra for 5.9 GHz DSRC and the wireless access in vehicular environment (WAVE) standard protocol stack. Chapters 6 and 7 illustrate the physical and medium access control layer behaviors, respectively, of the Institute of Electrical and Electronics Engineers (IEEE) 802.11p standard. Chapter 8 presents a study to determine the optimal data rate for DSRC. Chapter 9 presents WAVE upper layer protocols, including the WAVE short message protocol (WSMP) and the IEEE 1609.4 standard for DSRC multichannel operations.

Chapters 10 through 12 illustrate representative *safety applications* that have been developed and demonstrated as part of recent collaborative efforts. Chapter 10 focuses on vehicle-to-infrastructure (V2I) applications, while Chapter 11 presents vehicle-to-vehicle (V2V) applications. Chapter 12 describes the state-of-the-art research on DSRC scalability and congestion control algorithms for consumer vehicle networks.

Chapters 13 through 21 are devoted to *security and privacy protection* for consumer vehicle networks. Chapter 13 identifies unique security and privacy threats and requirements in a large-scale consumer vehicle network. Chapter 14 describes the fundamental cryptographic mechanisms that are crucial to supporting security and privacy in vehicle communication networks. Chapter 15 focuses on how public key infrastructures (PKIs) can be extended to manage security credentials such as digital certificates for large-scale consumer vehicle networks, and discusses the issues that need to be addressed to make the use of digital certificates and the PKI privacy preserving. Chapters 16–18 present and analyze three classes of privacy-preserving digital certificate management methodologies: shared certificates, short-lived certificates, and group certificates. Chapter 19 presents ways to extend the solutions presented in Chapters 16–18 to protect driver privacy against potential breaches by the operators of security credential management systems. Chapter 20 is a brief comparison of the three privacy-preserving digital certificate management methodologies presented in the previous chapters. Chapter 21 presents the IEEE 1609.2 standard for supporting security over DSRC networks.

The last chapter of the book, Chapter 22, discusses the use of *fourth-generation cellular networks* to support selected vehicle safety communication applications.

LUCA DELGROSSI
TAO ZHANG

ACKNOWLEDGMENTS

Naming all the colleagues and friends who contributed to the research described in this book is a virtually impossible task. At the Crash Avoidance Metrics Partnership (CAMP), we found an ideal environment for fruitful collaborations. Credit has to be given, among others, to Mike Shulman, who serves as CAMP Program Manager. Farid Ahmed-Zaid, Hariharan Krishnan, Michael Maile, and Tom Schaffnit served as principal investigators in a series of national projects and led us through the implementation of always more-refined prototype systems. We gained a wealth of insight and knowledge through interactions with all CAMP engineers. At the Vehicle Infrastructure Integration Consortium (VIIC), we developed the vehicle onboard equipment and the end-to-end privacy-preserving security credential management system for the proof-of-concept trial and debated on policy implications for vehicle safety communications systems. By mentioning Ralph Robinson, Dave Henry, and Tom Schaffnit, all of whom have served as presidents of the VIIC, we intend to acknowledge all VIIC engineers and policy experts. BMW, Chrysler, Ford, General Motors, Honda, Hyundai-Kia, Mercedes-Benz, Nissan, Toyota, and Volkswagen-Audi joined forces at CAMP and VIIC to conduct research in a precompetitive environment.

The Mercedes-Benz team in Palo Alto, California, includes several DSRC pioneers who contributed to this work. Qi Chen and Daniel Jiang developed, in collaboration with Felix Schmidt-Eisenlohr of the Karlsruhe Institute of Telematics, the network simulator 2 (ns-2) that was used to derive many of the results presented in this book. They made their software freely available for networking researchers. Michael Maile led the team that developed the Cooperative Intersection Collision Avoidance System for Violations (CICAS-V) and is one of the world's experts in V2I systems. Craig Robinson was the lead developer of the Integrated Safety system publicly showcased in 2008. Gordon Peredo, Graham Brown, and Kyla Tirey have years of experience with V2V DSRC systems. They built tens of fully functional prototype systems and public demonstrations with passenger cars and commercial vehicles. Mike Peredo developed software for the roadside infrastructure and took many pictures presented in this book. Tessa Tielert achieved significant breakthroughs with her work on DSRC congestion control and scalability.

The results on privacy-preserving security for vehicular communications are based on close collaborations with many colleagues at Telcordia Technologies, including Stanley Pietrowicz, Hyong Shim, Giovanni Di Crescenzo, and Eric van den Berg. We had fruitful collaborations with many industry partners and automotive suppliers, but special thanks go to Roger Berg, Sue Graham, and the team at DENSO International America, who joined from the very beginning and developed platforms and systems that we are still using today.

Andrew Moran and Yvonne Peredo researched and verified data and facts presented in the first part of the book. John Kenney provided precious comments on an early version of the manuscript. Emma Asiyo and Greg Stevens were instrumental with their encouragement. Finally, we would like to thank our editorial staff: Diana Gialo and Kristen Parrish of Wiley and Stephanie Sakson of Toppan Best-Set Premedia Ltd. for their excellent support.

L.D.
T.Z.

TRAFFIC SAFETY

1.1 TRAFFIC SAFETY FACTS

Six million crashes involving over 10 million motor vehicles occur on average every year in the United States. In 2009, an estimated 5,505,000 motor vehicle crashes occurred, leading to 33,808 fatalities and 2,217,000 injured people, averaging 93 deaths every day or one every 16 minutes [NHTS11]. Vehicular accidents are the leading cause of death for people between the ages of 3 and 34 in the United States [NHTS09]. These figures account only for police-reported crashes and therefore the actual number of motor vehicle crashes is likely even higher.

A significant percentage of accidents occur at road intersections. In 2007, there were an estimated 2,392,061 intersection crashes, accounting for 39.7% of all crashes in the United States [FHWA09]. Of these accidents, 8061 were fatal and 1,711,000 caused injuries. It has been estimated that, on average, 250,000 accidents every year involve vehicles running a red light and colliding with another vehicle crossing the intersection from a lateral direction [NHTS07].

A recent study estimates the costs of crashes for metropolitan areas of different sizes and populations in the United States [Kitt10]. According to this study, the average annual costs of crashes per person in small, large, and very large metropolitan areas are $1946, $1579, and $1392, respectively. In addition to lost lives, motor vehicle crashes place a heavy economic burden on the

Vehicle Safety Communications: Protocols, Security, and Privacy, First Edition. Luca Delgrossi and Tao Zhang.
© 2012 John Wiley & Sons, Inc. Published 2012 by John Wiley & Sons, Inc.

society, including increased costs of medical care, disability, insurance, and property damage. In 2000, the annual economic cost to society due to motor vehicle crashes was estimated at around $230 billion in the United States, roughly equivalent to 2.3% of the country's gross domestic product (GDP) in the same year [NHTS02].

Motor vehicle crashes significantly affect traffic mobility as well. The U.S. Federal Highway Administration (FHWA) estimated that approximately 25% of traffic slowdowns are related to crashes and other traffic incidents. The estimated average annual costs of traffic congestion per person in small, large, and very large metropolitan areas in the United States are $214, $407, and $575, respectively [Kitt10].

1.1.1   Fatalities

Based on historical data published by the U.S. National Highway Traffic Safety Administration (NHTSA) and FHWA, motor vehicle accidents have been responsible for over 3,300,000 fatalities in the United States alone since 1899 [NHTS10]. As the automobile came into greater use, the fatalities increased sharply each year from 1899 to 1931. After remaining stable for a few decades, the annual death rate rose again until peaking at 53,543 in 1969 (Figure 1.1). Since then, the annual number of fatalities has held fairly steady, or even decreased somewhat, due to significant advances in automotive safety measures. With an increasingly mobile society, reducing traffic fatality has become a more difficult task to achieve.

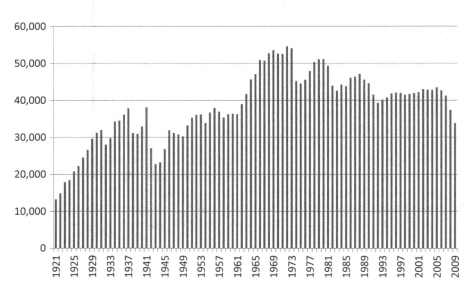

Figure 1.1. Total annual fatalities in the United States

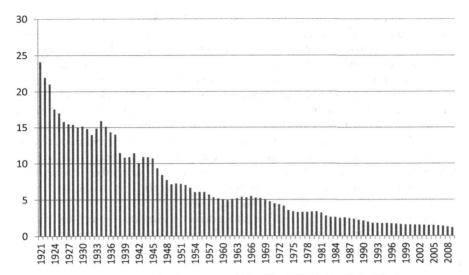

Figure 1.2. Annual fatality rate per 100 million VMT in the United States

The number of fatalities alone does not paint a complete picture of automotive safety. Since 1899, market penetration of automobiles has continued to increase significantly and the annual number of vehicle miles traveled (VMT) has exploded from 100 million in 1900 to over 3 trillion by 2007, according to FHWA statistics [FHWA07]. The number of fatalities per VMT has actually decreased. In 1921, the United States saw 24 fatalities per 100 million VMT, which were more than 21 times the record low 1.13 deaths per 100 million VMT in 2009.

Broader adoption of effective automotive safety systems, along with improved safety legislation and increased driver education efforts, has powered the reduction of fatalities and injuries despite the growing number of vehicles on the road and the distances traveled. As people continue to travel more, innovations become increasingly crucial to minimize traffic fatality (Figure 1.2).

1.1.2   Leading Causes of Crashes

According to NHTSA, the three most common causes of vehicle crashes are: control loss without prior vehicle action, lead vehicle stopped, and road edge departure without prior vehicle maneuver. In 2004, crashes under these circumstances accounted for an estimated 1 million lost functional years and $40 billion in direct economic costs in the United States [NHTS07].

Understanding the events that lead up to a motor vehicle crash is crucial to prevent future crashes. In 2008, the U.S. Congress authorized NHTSA to conduct a National Motor Vehicle Crash Causation Survey [NHTS08]. A representative sample of crashes from 2005 to 2007 was investigated. During the data collection process, the research team was granted timely permissions by local law enforcement and emergency responders to be on the crash scenes. Arriving on the scene before the crash was cleared by law enforcement gave the researchers access to relatively undisturbed information pertaining to the crashes and factors which led to these crashes. It allowed the researchers to discuss the circumstances of the crash with the drivers, passengers, and witnesses while the event was still fresh in their minds. The researchers were able to immediately and accurately reconcile the physical evidence with witness descriptions. Using this and other data, the researchers were able to assess the critical events that preceded the crash, the reasons for this event, and other factors that may have played contributing roles.

Ninety-five percent (95%) of the time, driver error was the critical reason for an accident. Driver errors can be classified into several categories: recognition, decision, performance, nonperformance, and other or unknown driver errors:

- Recognition errors accounted for 40.6% of all accidents due to driver error. Inadequate surveillance and driver distraction played a significant role in reorganization errors, accounting for 20.3% and 10.7% of driver error accidents, respectively.
- Decision errors accounted for 34% of all driver error accidents. The causes for decision errors were more numerous and varied than for recognition errors. Fast speeds were the most significant, being identified as a critical reason for 13.3% of crashes due to driver error.
- Performance errors constituted for 10.3% of all driver error crashes. The primary causes of performance errors are overcompensation and poor directional control. Noticeably, fatigued drivers were twice as likely as nonfatigued drivers to make performance errors.
- Miscellaneous nonperformance errors accounted for 7.1% of all driver error crashes. These included sleeping or having medical emergencies such as heart attacks while driving.
- Unknown driver errors accounted for the remaining 7.9% of all driver error crashes.

To prevent vehicle crashes, it is also important to understand prominent precrash events. The study has found that 36.2% of all accidents occurred while a vehicle was turning at or crossing an intersection. Traveling off the edge of the road is the second most frequent precrash event, accounting for 22.2% of all crashes. Traveling over the lane line constituted the critical precrash event for 10.8% of all collisions. A stopped vehicle served as the critical precrash event in 12.2% of all cases. Prevention and mitigation of these common causes of accidents therefore take top priority in safety research.

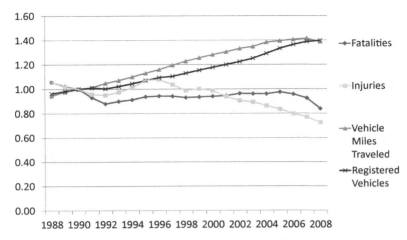

Figure 1.3. Traffic safety statistics in the United States (1988–2008)

1.1.3 Current Trends

Figure 1.3 shows traffic safety statistics in the United States between 1988 and 2008, including the number of registered vehicles, VMT, injuries, and fatalities. In this chart, each value is expressed as relative to the correspondent value for year 1990. Fatalities and injuries, although declined in recent years, have remained at high levels and the declines have been slow. This raises a concern that we are reaching the point where existing vehicle safety systems are not going to sustain the same rates of reduction in fatalities and injuries as they have in the past. The continuous rise in the number of vehicles on the road and in VMT calls for continuing innovations in vehicle traffic safety technologies.

1.2 EUROPEAN UNION

Countries in the European Union have been following a similar trend of increasing automotive safety as shown in Figure 1.4.

Germany has seen a significant long-term decline in fatalities, with a 79% reduction from 21,332 fatalities in 1970 to only 4477 fatalities in 2008. In addition, the annual number of crashes that caused injuries decreased from 414,362 to 320,614 in the same time period, an improvement of 23%. Remarkably, these declines in fatalities and injuries have been accomplished while the number of vehicles on the road nearly tripled [IRTA10]. These improvements were made possible by a combination of advances in safety technology, a highly developed road infrastructure, an advanced legal framework, and a highly sophisticated penalty point system. Stringent laws concerning

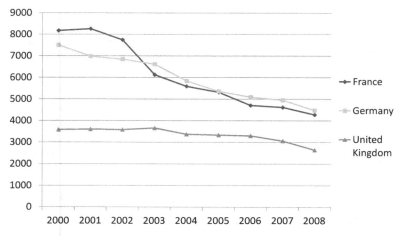

Figure 1.4. Annual fatalities in Germany, United Kingdom, and France (2000–2008)

intoxicated driving, speeding, and seat belt usage have all contributed to the long-term reductions in accidents and fatalities as well.

Traffic fatalities have also declined significantly in the United Kingdom. Between 1970 and 2008, the annual number of fatalities declined by 66% and the annual number of injury crashes declined by 35%, while the average distance traveled increased by 10% [IRTA10]. These percentages represent a decline from 7771 fatalities in 1970 to 2645 fatalities in 2008 and from 272,765 injury crashes in 1970 to 176,723 in 2008. The United Kingdom's traffic fatality rate is currently the lowest in the European Union, with 4.3 fatalities per 100,000 people [IRTA10]. As with Germany, the United Kingdom's improved traffic safety has largely been achieved through advances in safety technology, investments in road infrastructures, and enforcement efforts designed to curb excessive speeding and intoxicated driving. The United Kingdom has likewise placed a significant emphasis on educational programs to raise awareness of high-risk driving behaviors and the sanctions imposed for such behaviors.

France has also seen a significant long-term decline in the overall traffic fatality rate. Between 1970 and 2008, the number of fatalities decreased by 74% (from 16,445 in 1970 to 4275 in 2008) and the number of injury crashes by 68% (from 235,109 in 1970 to 74,487 in 2008) while the number of vehicles on the road tripled. The numbers are even more impressive when you consider the decline in fatalities per billion vehicle-kilometers, which fell from 90.36 in 1970 to a mere 8.1 in 2008, for a total improvement of 91% over that time period [IRTA10]. Further improvements continue to be made. Since 2002, France has implemented a focused road safety policy which includes effective measures regarding speed management, intoxicated driving, seat belt use, and strengthening of the demerit point system, all of which continue to impact traffic safety positively.

As in the United States, the reductions in traffic fatalities and injuries in the European Union countries have slowed down over the recent years (Figure 1.4), which suggests similar diminishing returns achievable through traditional vehicle safety technologies and calls for new thinking and innovation in vehicle safety technologies.

1.3 JAPAN

During the 1960s, the rapid increase in automobile traffic outpaced road constructions in Japan. The resulting increase in motor vehicle accidents became a public concern, prompting the government to take measures to reduce vehicular crashes. In 1970, following enactment of the Traffic Safety Policies Law, the Central Committee on Traffic Safety Measures was established and the first Fundamental Safety Program was formulated. Since 1971, the Central Committee on Traffic Safety Measures has continued to produce 5-year Fundamental Traffic Safety Programs which set forth the fundamental principles and goals for comprehensive and long-term measures for the safety of land, maritime, and air transport based on the Traffic Safety Policies Law.

A cornerstone of Japan's efforts to improve traffic safety has been a significant investment in road infrastructure enhancement. Safer roads have been achieved through improvements in expressways, bypasses, beltways, intersections, road lighting, road signs, and traffic signals. Safety measures were also enacted for pedestrians, including installation of sidewalks, development of shared pedestrian and bicycle paths, and addition of pedestrian overpasses and underpasses. As a result, pedestrian fatalities have decreased sharply, from 2794 in 1996 to 1943 in 2007, an improvement of approximately 31% [IATS08].

Japan's traffic fatalities have reduced significantly since the adoption of the first Fundamental Traffic Safety Program. Between 1970 and 2008, the annual number of fatalities decreased by 72% even though the number of injury crashes increased by 7% [IRTA10]. The annual number of fatalities in proportion to distance traveled decreased over that same time span by a remarkable 91% [IATS08]. The declining fatality rate has been sustained in recent years, despite a threefold increase in the numbers of vehicles and VMT. The fatality rate continues to decline as advancements continue to be made in automotive safety, decreasing by approximately 42% between 2000 and 2008 [IRTA10]. This is particularly remarkable and difficult to sustain due to the very high population density in Japan.

1.4 DEVELOPING COUNTRIES

While developed countries have been benefiting from declining traffic fatality rates, this has not been the case in many developing countries such as China and India. Developing countries currently account for 90% of the disability-adjusted life years lost to traffic injuries and deaths worldwide. This problem continues to escalate especially in Asia. It is projected that by 2020, vehicular deaths will increase by 80% in developing countries [KoCr03]. This includes fatality rate increases of almost 92% in China and 147% in India. Injuries due to vehicular crashes are the root cause of a significant portion of medical care sought in developing countries, accounting for up to one-third of the acute patient cases in many hospitals and between 30% and 86% of trauma admissions [OdGZ97]. Besides the toll on human lives, the economic cost of vehicular crashes in developing countries has been estimated at around US$65 billion, a heavy burden on the economy and a financial drain on national health-care systems [PSSM04].

A significant reason that developing countries have not experienced the same reduction in fatality rates as developed nations is that their road infrastructures are unable to keep pace with the sharp increases in the number of vehicles on the roads. This results in unsafe driving conditions and massive traffic congestions. Poor traffic conditions contribute to the prevalent fatalities of vulnerable road users such as pedestrians, bicyclists, and people using carts, rickshaws, mopeds, and scooters. This is in contrast to developed countries, where drivers and passengers are the primary victims [PSSM04]. Vehicles in developing countries are also significantly more likely to be involved in fatal crashes, 200-fold more likely in some cases, than in more developed countries [AATJ10].

Therefore, developing innovative automotive safety technologies is of utmost importance for the world as a whole, not merely for developed countries.

To reduce fatalities and injuries despite the rising number of vehicles and VMT, we must continue to discover new ways to prevent motor vehicle crashes and mitigate their damages.

REFERENCES

[AATJ10] G. Jacobs, A. Aeron-Thomas, and A. Astrop: "Estimating Global Road Fatalities," Department for International Development (DFID), ISSN 0968-4107, Transport Research Laboratory, Report 445, 2000.

[FHWA07] Federal Highway Administration: "Highway Statistics 2007: Public Road Mileage, Lane Miles, and VMT 1900-2007," Table VMT-421, 2011.

[FHWA09] Federal Highway Administration: "The National Intersection Problem," FHWA-SA-10-005, 2009.

[IATS08] International Association of Traffic and Safety Sciences: "Statistics 2007: Road Accidents Japan," Traffic Bureau, National Police Agency, 2008.

[IRTA10] International Traffic Safety Data and Analysis Group (IRTAD): "Annual Report 2009," Organization for Economic Cooperation and Development (OECD) International Transport Forum (ITF), 2010.

[Kitt10] M. J. Kittelson: "The Economic Impact of Traffic Crashes," Georgia Institute of Technology, 2010.

[KoCr03] E. Kopits and M. Cropper: "Traffic Fatalities and Economic Growth," World Bank Development Research Group, Infrastructure and Environment, Policy Research Working Paper 3035, 2003.

[NHTS02] National Highway Traffic Safety Administration: "The Economic Impact of Motor Vehicle Crashes, 2000," DOT HS 809 446, 2002.

[NHTS07] National Highway Traffic Safety Administration: "Pre-Crash Scenario Typology for Crash Avoidance Research," DOT HS 810 767, 2007.

[NHTS08] National Highway Traffic Safety Administration: "Motor Vehicle Traffic Crashes as a Leading Cause of Death in the United States," DOT HS 810 936, 2008.

[NHTS09] National Highway Traffic Safety Administration: "Traffic Safety Facts 2008," DOT HS 811 170, 2009.

[NHTS10] National Highway Traffic Safety Administration: "An Analysis of the Significant Decline in Motor Vehicle Crashes in 2008," DOT HS 811 346, 2010.

[NHTS11] National Highway Traffic Safety Administration: "Traffic Safety Facts 2009," DOT HS 811 402, 2011.

[OdGZ97] W. Odero, P. Garner, and A. Zwi: "Road Traffic Injuries in Developing Countries: A Comprehensive Review of Epidemiological Studies," Tropical Medicine and International Health, vol. 2, pp. 445–460, 1997.

[PSSM04] M. Peden, R. Scurfield, D. Sleet, D. Mohan, A. Hyder, E. Jarawan, and C. Mathers: World Report on Road Traffic Injury Prevention, World Health Organization, United Nations, Geneva, Switzerland, 2004.

2

AUTOMOTIVE SAFETY
EVOLUTION

2.1 PASSIVE SAFETY

Passive safety features are built into vehicles to minimize driver and passenger harm during a crash. Groundbreaking passive safety features include seat belts and air bags. They have been playing a crucial role in reducing traffic fatalities and have become integral—and in many countries mandatory—features in modern vehicles. The National Highway Traffic Safety Administration (NHTSA) reports that 322,409 lives have been saved in the United States between 1975 and 2008 through the use of child restraints, seat belts, air bags, and motorcycle helmets alone [NHTS09a].

2.1.1 Safety Cage and the Birth of Passive Safety

The introduction of the first safety cage, invented by Mercedes-Benz engineer Béla Barényi shortly after World War II, marked the birth of passive safety. This safety cage consisted of a strong central passenger cell flexibly connected to deformable crash cells at the front and rear, the precursor of the modern rigid passenger cell with front and rear crumple zones. The crash cells, and later the crumple zones, were specifically designed to deform in an accident, thus absorbing the kinetic energy of a collision.

In 1951, Mercedes-Benz was granted a patent on the design of a rigid passenger cell enclosed by crumple zones at the front and rear. This concept was first put to use on production vehicles in 1959 and has since become an

Vehicle Safety Communications: Protocols, Security, and Privacy, First Edition. Luca Delgrossi and Tao Zhang.
© 2012 John Wiley & Sons, Inc. Published 2012 by John Wiley & Sons, Inc.

industry-wide standard. Barényi's groundbreaking work spurred a long list of innovations that led to a dramatic improvement of passenger safety.

2.1.2 Seat Belts

The first patent for an automotive seat belt was issued in 1885. The first three-point seat belt, however, was patented and developed to its modern form many years later, in 1951, by Volvo engineer Nils Bohlin. By1965, all 50 states in the United States had passed laws requiring seat belts in the front seats of automobiles. The first federal seat belt law took effect a few years later on January 1, 1968, and required that all vehicles (except buses) be fitted with seat belts in all designated seating positions. By 1975, most of the developed world had followed suit, with laws requiring automakers to include seat belts for every seat in a car.

Initially, seat belt usage was not compulsory. In 1984, New York became the first state to pass a law which required vehicle occupants to wear seat belts. Currently, in the United States, 30 states have primary laws (i.e., a police officer may stop and ticket a driver for not wearing a seat belt), 19 states have secondary laws (i.e., a police officer may only stop or cite a driver for seat belt violation if the driver committed another primary violation, such as speeding or running a red light), and only New Hampshire does not have a law requiring seat belt usage for adults.

Due to increased legislation and safety awareness, observed national seat belt usage rates in front seats has climbed from 14% in 1983 to 83% in 2008 [IRTA10]. This has contributed significantly to an overall decline in vehicular fatalities in the United States and throughout the developed world. According to NHTSA, seat belts were responsible for saving 255,115 lives from 1975 to 2008 [NHTS09a], making it the single most effective passive safety system ever conceived, when compared with others whose impact can be similarly quantified.

2.1.3 Air Bags

The air bag is another passive safety device responsible for saving thousands of lives every year. By inflating rapidly upon collision, the air bag acts as a restraint system, preventing occupants from striking interior objects. The first air bag was designed in the early 1950s. However, it was not until the 1970s that air bags began to be featured in passenger cars in the United States.

According to NHTSA, air bags have prevented the deaths of over 27,000 people from 1995 to 2008 [NHTS09a]. In terms of quantifiable effectiveness in reducing fatalities, only the seat belt surpasses the air bag in the number of lives saved. In conjunction with each other, it is estimated that air bags and seat belts reduce one's fatality risk by as much as 61% [NHTS09b].

Many modern air bags utilize sensors to gather information such as the weight and seating position of the occupant and whether seat belts are in use.

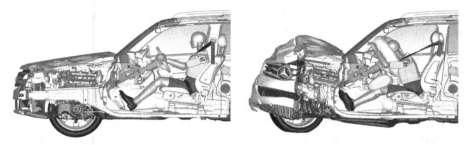

Figure 2.1. Passive safety systems. Courtesy of Daimler AG

In the event of a collision, this information can help determine the optimal force for air bag deployment. This adaptive force of deployment further increases the air bag's effectiveness by reducing the likelihood of injury from the air bag itself.

Figure 2.1 shows passive safety systems in action. The passenger cell acts as a high-strength safety cage that protects the occupants in the event of an offset or collision. This protection is further enhanced by an integrated air bag and seat-belt system.

2.2 ACTIVE SAFETY

While passive safety systems have proven to be invaluable assets in automotive safety, they seem to have reached their potentials. Over time it has become evident that further significant advances in vehicle safety could be achieved through electronic systems and active safety systems. Unlike passive safety systems that reduce harm during a collision, active safety systems aim to prevent vehicle crashes from occurring in the first place and to minimize the damage when collisions become unavoidable.

2.2.1 Antilock Braking System

The antilock braking system (ABS) prevents the wheels on a motor vehicle from ceasing to rotate during heavy braking, which helps the driver control steering by preventing a skid and allowing the wheels to maintain traction with the road surface. More specifically, ABS automatically changes the brake fluid pressure at each wheel to maintain optimal brake performance while not locking up the wheels. An electronic control unit (ECU) regulates the brake fluid pressure in response to changing road conditions or impending wheel lockup. This reduces the required stopping distance and improves driver's control during emergency braking on wet and slippery roads.

A typical ABS includes four wheel speed sensors, hydraulic valves within the brake hydraulics system, and an ECU. The ECU monitors the rotation speed of each wheel at all times, looking for signs of an impending wheel lock, such as when a wheel rotates significantly slower than the others. When a potentially dangerous slowdown of rotation is detected, valves are actuated to reduce hydraulic pressure to the brake at the affected wheel, thus reducing the braking force on that wheel and causing it to turn faster. ABS also works in the reverse situation, in which a wheel may be rotating too fast. In this case, brake hydraulic pressure to that wheel is increased so that the braking force is reapplied and the wheel is slowed down.

ABS is a critical component of many other active safety systems in automobiles. For instance, electronic stability control (ESC) systems add steering and gyroscopic elements to ABS to assist the driver in steering maneuvers during dangerous braking situations.

2.2.2   Electronic Stability Control

ESC is a computerized technology that improves a vehicle's stability by minimizing skids. When ESC detects loss of steering control, it automatically applies the brakes to individual wheels to help steer the vehicle where the driver intends to go, effectively utilizing ABS on individual wheels. ESC is a highly effective active safety system which could prevent one-third of all fatal accidents and reduce rollover risk by as much as 84%, according to the Insurance Institute for Highway Safety (IIHS) [IIHS06] and NHTSA [NHTS07]. It is such an important technology that NHTSA has mandated that all new vehicles in the United States be equipped with ESC by 2012 [NHTS07]. Also, the combined effects of ABS and ESC result in an estimated 30% reduction in fatal run-off-the-road crashes [NHTS09d].

Figure 2.2 illustrates an example of how ESC works. Without ESC on a curved road (left), the vehicle's front wheels move outwards. When ESC is engaged (right), it supports the driver's steering correction through brake intervention mainly at the inner rear wheel.

Just as ABS forms the foundation of the ESC system, ESC provides a foundation for new advances in active safety systems. The computing technologies required for ESC facilitate the development of further active and passive safety systems in the car, which in turn creates opportunities to address additional causes of vehicular crashes.

2.2.3   Brake Assist

In 1992, research conducted by Mercedes-Benz using their driving simulator in Berlin, Germany, revealed that more than 90% of drivers fail to brake with sufficient force during emergency situations. This discovery stimulated the development of brake assist systems (BASs) to assist the driver during braking maneuvers.

Brake assist technology was initially developed to detect circumstances in which emergency braking is required by measuring the speed with which the

 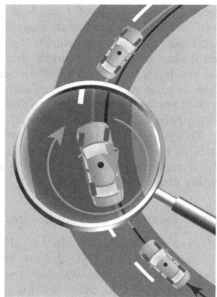

Figure 2.2. Electronic stability control. Courtesy of Daimler AG

brake pedal is pressed. When a panic braking condition is detected, BASs automatically boost braking power to mitigate the driver's tendency to brake without enough force, thereby reducing the required stopping distance.

In 1996, Mercedes-Benz became the first automotive manufacturer to introduce the BAS, making it standard equipment on all its models in 1998. Other manufacturers, including Volvo and BMW, soon followed suit. Volvo's Collision Warning with Auto Brake (CWAB) uses radar signals and a camera sensor to detect when a collision is likely and precharges the brakes so that full braking is applied as soon as the driver activates the brakes. The system uses a flashing light and a warning sound to alert the driver, and applies the brakes automatically when a collision becomes unavoidable if the driver has not heeded the warning and activated the brakes himself.

BASs continue to evolve. Today's vehicles often integrate collision avoidance systems with adaptive cruise control (ACC) systems to significantly reduce rear-end collisions.

2.3 ADVANCED DRIVER ASSISTANCE SYSTEMS

The majority of active safety systems use onboard sensors and actuators to assist the driver or take autonomous actions. Most advanced driver assistance systems provide visual and auditory warnings to the driver in the presence of potential hazards. Some systems can detect potential hazards and also actively avoid them. Furthermore, the combination of different advanced driver assistance systems can result in even more powerful solutions. For instance, ACC can work together with precrash features to create a comprehensive safety system for preventing collisions and mitigating the severity of unavoidable impacts.

Advanced driver assistance systems able to gather information about current road conditions can effectively mitigate the risk of distracted driving, which is becoming an increasingly common problem accounting for 16% of fatal crashes in 2008 and 2009 [NHTS09c] [NHTS10].

2.3.1 Adaptive Cruise Control

ACC, also known as stop-and-go cruise control and other manufacturer-specific names, enhances both comfort and safety. Like traditional cruise control, the ACC system maintains a speed set by the driver. However, ACC also uses forward-sensing radars (or laser scanners as a lower cost alternative) to detect the speed of the vehicle ahead and automatically adjust its speed to maintain a safe following distance. In the absence of obstacles or slower traffic ahead, ACC will operate like traditional cruise control, maintaining the speed set by the driver.

The driver may also specify a following distance. If the ACC system detects a slower-moving vehicle ahead, it will reduce throttle to maintain the following distance specified by the driver and return to the preset speed when the front vehicle either speeds up or moves to a different lane.

ACC systems are capable of slowing the vehicle all the way to a standstill and accelerating it back up to the preset desired speed if the traffic situation permits. Figure 2.3 shows distance keeping as displayed in the Mercedes-Benz S-Class instrument cluster.

Figure 2.3. Adaptive cruise control. Courtesy of Daimler AG

2.3.2 Blind Spot Assist

A variety of blind spot detection and assist systems have been implemented in production vehicles since 2005. Methods of detecting vehicles in blind spots evolved from side mirrors to sensor-based systems able to issue visual and audible warnings to the driver when the vehicle is about to enter an occupied lane. Some existing active blind spot assist systems cause the steering wheel or driver's seat to vibrate when the vehicle is getting too close to another vehicle. Others even take limited steering control to help avoid steering into another car.

In 2009, Volvo introduced a Blind Spot Information System (BLIS), which uses two door-mounted cameras to detect vehicles in blind spots and then visually alert the driver when switching lanes. The Mercedes-Benz Active Blind Spot Assist system uses short-range radar sensors that monitor the areas to the sides and rear of the car on both the left and the right. When another car is in the blind spot zone, the system displays a red warning symbol in the correspondent side mirror. If the driver ignores the warning and activates the turn indicator, the red warning symbol starts to flash and the system emits a sound inside the vehicle. If the driver attempts to change lanes despite the warning, the system activates ESC to apply the brakes to individual wheels to gently pull the vehicle back into its intended lane of travel.

2.3.3 Attention Assist

Drowsiness is a significant cause in road accidents. Approximately 16.5% of fatal crashes involve a drowsy driver, according to the American Automobile Association Foundation [AAAF10]. A field of development in active safety systems is to reduce drowsiness-related accidents with drowsiness detection and attention assist technologies. Some driver alert systems, like those employed by Volvo, make use of cameras to monitor lane-drifting patterns. Other driver alert systems, like those introduced by Toyota, can detect if the driver is becoming sleepy by monitoring the eyelids. A visual or audible alert is then issued to the driver. Depending on the system, the warning may be repeated until drowsy driving behavior stops.

The Mercedes-Benz Attention Assist system, introduced in 2010, uses sensors to monitor steering patterns, braking, acceleration, weather, and road conditions to determine a driver's alertness. If the system detects driver drowsiness or danger of the driver falling asleep at the wheel, it sounds warning chimes and flashes a coffee cup light icon.

2.3.4 Precrash Systems

Precrash systems are activated immediately before a crash to reduce the severity of the impact. They use onboard sensors to detect when a frontal or rear-end collision is unavoidable and whether brakes ought to be applied. When a crash is determined to be inevitable, precrash systems initiate measures to minimize occupant injury from the impending collision. Depending on the system, various actions may be taken, including precharging the brakes to maximize braking force as soon as the driver applies the brake, inflating seats for extra support, moving seats into an upright position, positioning head rests to minimize whiplash, rolling up the windows, and tensioning seat belts. All of these protective measures take place within less than 2 seconds immediately prior to the collision. Precrash systems of various types and capabilities are available from major automotive original equipment manufacturers (OEMs), and their deployments are expected to increase as their effectiveness continues to be demonstrated.

2.4 COOPERATIVE SAFETY

The many safety innovations introduced over the past 50 years mark the evolution from passive safety systems, which mitigate harmful effects during a crash, to active safety systems, which provide support during normal driving and immediately before a crash. These systems have made today's vehicles significantly safer and have been a key factor in maintaining and even reducing the number and severity of motor vehicle crashes.

The different phases in which safety systems can be effective with respect to the moment of the collision are represented in Figure 2.4. Systems such as ABS, ESC, and BAS provide driver support in crash avoidance when potential danger has been recognized. Precrash systems take measures to reduce the effects of unavoidable collisions in the short but crucial stage just prior to an

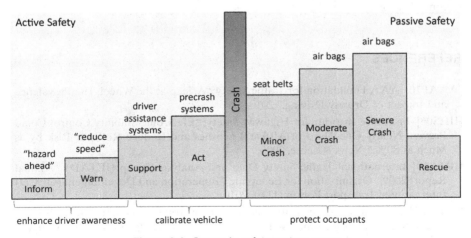

Figure 2.4. Stages in safety system usage

accident. Seat belts are instrumental during all types of crashes, but they are especially effective in minor to moderate crashes, in which they can prevent the driver and passengers from sustaining life-threatening injuries as a consequence of being ejected from the vehicle. The prompt deployment of air bags offers protection in moderate to severe crashes. Finally, remote assistance systems can aid in the rescue process after a crash has occurred.

Most passive safety systems have become commodities and are expected to be in all vehicles. In contrast, active safety systems are still high-end features and key differentiators between automotive OEMs competing to provide top-of-the-line safety to a growing customer base.

Automotive grade autonomous sensors are revolutionizing vehicular safety. Their functions are expected to evolve from assisting the driver to executing specific driving maneuvers and taking partial or even full control of the vehicle. Yet, as with any technology, autonomous sensors present some limitations. The ranges of today's sensors are comparable to the driver's vision and cannot be easily extended without compromising their accuracy and reliability. Furthermore, their performance suffers under certain conditions such as inclement weather and abnormal reflective surfaces in surrounding vehicles and road obstructions.

Today, automotive engineers are exploring wireless communication technologies as a natural extension to sensors-based vehicle safety systems. Preliminary results demonstrated the effectiveness of vehicle-to-vehicle (V2V) and vehicle-to-infrastructure (V2I) communications to support active vehicle safety. Communications provide vehicles with high-quality data that would be hard to obtain through other means.

The ability to exchange real-time safety-critical information among vehicles defines a new paradigm in automotive safety, where vehicles cooperate with each other and with the surrounding environment. Cooperation enabled by vehicle communications promises the next leap in vehicle safety and the potential to one day make traffic accidents a rarity.

REFERENCES

[AAAF10] AAA Foundation for Traffic Safety: "Asleep at the Wheel: The Prevalence and Impact of Drowsy Driving," 2010.

[IIHS06] Insurance Institute for Highway Safety: "Electronic Stability Control Could Prevent Nearly One-Third of All Fatal Crashes and Reduce Rollover Risk by as Much as 80%," News Release, 2006.

[IRTA10] International Traffic Safety Data and Analysis Group (IRTAD): "Annual Report 2009," Organization for Economic Cooperation and Development (OECD) International Transport Forum (ITF), 2010.

[NHTS07] National Highway Traffic Safety Administration: "Federal Motor Vehicle Safety Standards: Electronic Stability Control Systems; Controls and Displays," RIN: 2127AJ77, 2007.

[NHTS09a] National Highway Traffic Safety Administration: "Traffic Safety Facts 2008," DOT HS 811 162, 2009.

[NHTS09b] National Highway Traffic Safety Administration: "Fatalities in Frontal Crashes despite Seat Belts and Air Bags," DOT HS 811 202, 2009.

[NHTS09c] National Highway Traffic Safety Administration: "An Examination of Driver Distraction as Recorded in NHTSA Databases," DOT HS 811 216, 2009.

[NHTS09d] National Highway Traffic Safety Administration: "The Long-Term Effects of ABS in Passenger Cars and LTVs," DOT HS 811 182, 2009.

[NHTS10] National Highway Traffic Safety Administration: "Distracted Driving 2009," DOT HS 811 379, September 2010.

3

VEHICLE ARCHITECTURES

3.1 ELECTRONIC CONTROL UNITS

Modern vehicles are controlled electronically rather than purely mechanically. Most automakers use a few baseline vehicle architectures to build various vehicle models, which differ from each other largely in their electronic systems. In-vehicle devices and components are controlled by electronic control units (ECUs). An ECU is an embedded system that controls one or more electrical systems or subsystems in a vehicle. ECUs take in data from onboard sensors, perform calculations, and distribute instructions to the various in-vehicle electronic systems to maintain proper and efficient driving performance. ECUs govern practically every aspect of vehicle functions, from small tasks such as activating brake lights or opening windows to critical functions such as autonomous brake systems [NiPL08]. Each ECU typically works independently operating its own firmware, but complex tasks may require cooperation among multiple ECUs.

ECUs can be classified based on their roles. Modern vehicles may include ECUs for the power train, safety, comfort, infotainment, and telematics systems. Power train ECUs perform engine and transmission management. Safety ECUs activate and control in-vehicle safety systems such as air bags, adaptive cruise control, and brake assist systems. Comfort ECUs control electronic suspensions and temperature. Infotainment and telematics ECUs manage audio, video, and mobile communication systems that support infotainment and telematics [See06].

Vehicle Safety Communications: Protocols, Security, and Privacy, First Edition. Luca Delgrossi and Tao Zhang.
© 2012 John Wiley & Sons, Inc. Published 2012 by John Wiley & Sons, Inc.

Today, automotive electronics and control software is the main area of growth in automotive research and development, with 90% of all innovations driven by electronics and software [Shor11]. The value of the ECUs in each car has been steadily increasing since 2002. Currently, an estimated 40% of a vehicle's cost is determined by electronics and software, which has increased by 146% since the 1990s [Shor11].

Today's complex vehicle architectures typically incorporate 50–70 ECUs, while some high-end vehicle architectures feature up to 100 ECUs [NiPL08]. As an example, the 2010 Mercedes-Benz S-Class requires over 20 million lines of code, nearly as many as the new Airbus A380 if the plane's in-flight entertainment system is not taken into account [Char09].

The value of electronic components and ECUs on a vehicle is expected to continue to grow in the near future, as more ECUs and electronic systems are installed in each vehicle and in more vehicle models to support advanced vehicle technologies.

3.2 VEHICLE SENSORS

Autonomous sensors, such as radars and cameras, are versatile tools to aid and enable active safety systems in vehicles. Their presence is becoming more pronounced with the introduction and steadily increasing usage of systems such as lane tracking, object detection, and adaptive cruise control, among others. Sensor-based systems, although currently confined mostly to luxury vehicles, are increasingly being integrated into more vehicle models.

3.2.1 Radars

Radar sensors use radio waves to detect objects and determine their position and velocity. Vehicle developers have discovered many ways to use radars to detect and track the movements of other vehicles or road users. Vehicles equipped with radar-based safety systems typically contain a number of radar sensors arrayed in strategic positions throughout the vehicle.

Long-range radars with a 15°–20° field of view and up to approximately 250 m range are used to detect vehicles ahead to support adaptive cruise control systems. Less expensive radars operating at ranges around 150 m are used at lower vehicle speeds [Stev11]. Short-range radars with a wider field of view and up to about 50 or 60 m range are employed in various collision avoidance systems, including blind spot warning and forward collision warning.

3.2.2 Cameras

Cameras are used in many detection and warning systems for vehicle safety. They are often integrated with radars to provide more complete detection

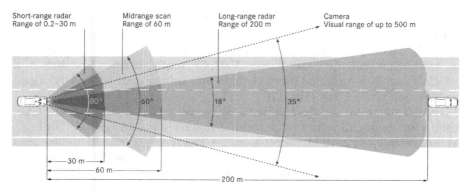

Figure 3.1. Onboard autonomous sensors. Courtesy of Daimler AG

capabilities. Radars' distance and velocity detection capabilities are complemented by cameras' superior angle of resolution, creating an enhanced coverage when both are implemented together.

Cameras can be integrated into monocular or stereo configurations. Stereo configurations are rare in production vehicles at this time. Mono cameras, which typically have a 50°–60° field of view depending on the lens and a range of approximately 100–200 m, are increasingly common in vehicles. They have proven especially useful in pedestrian detection, lane recognition, and lane keeping systems, along with speed limit detection programs and rearview cameras for parking assistance. Stereo cameras accomplish the same tasks as mono cameras, but may provide greater reliability. They work like human eyes, comparing data from both cameras to create a larger field of vision and more accurately identify objects and determine their distance and velocity.

Cameras are a low-cost option for some safety systems that can use either radars or cameras. However, these sensors are often used in conjunction with each other to provide more comprehensive and reliable information to vehicle safety systems. Vehicles which use both cameras and radars benefit from the best of both: superior distance and velocity detection from radars and more reliable object detection, especially of pedestrians and nonvehicle road features or obstructions, from cameras. Many vehicle safety systems, such as adaptive cruise control, brake assist, and forward collision warning, employ radars as primary sensors, with cameras enhancing accuracy and preventing false alarms.

Figure 3.1 shows a configuration of autonomous sensors, including a front camera as well as radars for short, medium, and long ranges.

3.3   ONBOARD COMMUNICATION NETWORKS

Communications among various vehicle subsystems are essential for the proper functioning of most components of modern vehicles. In addition, communications within each subsystem are needed to control actuators or receive feedback from sensors. Several onboard networks have been designed to fill these needs.

3.3.1 Controller Area Network

The dominant onboard network is currently the controller area network (CAN). CAN is a message-based vehicle bus standard that allows ECUs to communicate with each other within a vehicle. In a CAN, short messages are broadcast to the entire network. CANs were developed for the automotive industry to provide data transfers at a maximum of 1 Mbps with high immunity to electrical interference. CAN is also designed to self-diagnose and repair communication errors.

Communications over CAN are based on carrier sensing multiple access with collision detection (CSMA/CD) with arbitration on message priority (AMP). Each message consists of an 11-bit identifier, which designates the priority of the message, and data, which is transmitted onto the bus and sensed by all nodes. If the bus is free, any node may begin to transmit. If multiple nodes begin sending messages at the same time, the message which has the most dominant identifier will overwrite the others so that only the dominant message is received by all nodes. Priority-based bus arbitration is especially important in ensuring that time-sensitive safety systems are given priority over tasks of lesser urgency.

Low-speed CAN, specified in the International Organization for Standardization (ISO) 11519 standard [ISO94], offers 125 Kbps fault-tolerant data transfers. Higher speeds up to 1 Mbps are achievable with the second version of CAN, ISO 11898-1993 [ISO03], which is often referred to as Standard CAN Version 2.0A. An amendment to the second version, namely Standard CAN Version 2.0B or ISO 11898-1995 [ISO03], also achieves speeds up to 1 Mbps but allows for a larger number of unique message identifiers.

3.3.2 Local Interconnect Network

Local Interconnect Network (LIN) or Society of Automotive Engineers (SAE) J2602 [SAE05] is a low-cost communication bus designed for communications between smart sensors and actuators in automotive applications. It is designed as a cost-effective alternative for simpler systems when the higher bit rate CAN bus is not needed. Applications which typically use LIN are the instrument cluster, seats, mirrors, rain sensors, light sensors, door locks, and windows.

A LIN is comprised of a LIN master and one or many LIN slaves. In vehicles, the LIN bus is typically connected between sensors, actuators, and an ECU, which may act as a gateway to a CAN bus. LIN uses broadcast and the slaves synchronize to the master's clock to help keep the cost low. LIN has a number of advantages, including ease of use, availability of components, and cost-effective parts and implementation. These advantages make it a good complement to CAN whenever lower cost is priority and high speed or band-width is not essential.

3.3.3 FlexRay

FlexRay is a recently developed deterministic, fault-tolerant and high-speed bus system for vehicle communications [Flex10]. It was developed by the FlexRay Consortium as a next-generation vehicle communication bus. It is faster and more reliable than CAN, but also more expensive and requires a redesign of the network architecture. FlexRay offers two communication channels which have a data rate of 10 Mbps, a significant increase over other onboard network protocols [Voel08]. Differential signaling reduces the effects of external noise on the network, and dual-channel configurations offer enhanced fault tolerance and increased bandwidth [NIC09]. Despite the technological advantages of FlexRay, its implementation is likely to be limited to high-end systems in the near future, due to cost and the substantial challenges of onboard network architecture transitioning.

3.3.4 Media Oriented Systems Transport

Media Oriented Systems Transport (MOST) is a high-speed and low-cost fiber optical network for multimedia onboard communications. It is predominantly used to support high data volumes, especially for entertainment and information applications. The MOST bus uses a ring topology and synchronous data communication to transport audio, video, voice, and data signals via plastic optical fiber or electrical conductor physical layers. It is used in almost every vehicle brand worldwide. MOST devices can be used in audio and video devices, media interfaces, and sophisticated information and entertainment systems.

3.3.5 Onboard Diagnostics

Onboard networks such as CAN and LIN interconnect vehicle systems components and are accessible only to vehicle manufacturers and developers. They are not designed to enable access by mechanics or vehicle owners for vehicle customization, diagnostics, or repair. Instead, onboard diagnostics (OBD) systems have been devised to support diagnostic needs.

OBD systems with various self-diagnostic and reporting capabilities have been implemented in vehicles since the early 1980s. Early OBD systems would simply illuminate a malfunction indicator light if a problem was detected. Modern OBD systems use a standardized digital communications port to provide real-time data and standardized diagnostic trouble codes, allowing rapid identification and remediation of vehicle malfunctions. The new OBD-II standard, introduced in the mid-1990s, provides almost complete diagnostics for engine control and parts of the chassis, body, and accessory devices. While there are several OBD-II connection protocols, the command set is specified in the SAE J1979 standard [SAE07].

OBD systems have been proven useful for many applications. For example, OBD has been used for scanning, data logging, and analysis for ECU diagnostics. It has also been used for vehicle subsystems tuning, insurance risk determination, emission testing, and driver's supplementary vehicle instrumentation. The latter includes driver-installed instrumentation such as fuel economy analysis devices, trip computers, personal digital assistant interfaces, or navigation units.

3.4 VEHICLE DATA

Vehicle data transmitted over onboard networks are an excellent source of highly accurate and reliable information that can support a wide range of offboard applications as well as vehicle-to-vehicle (V2V) communication applications. For example, traffic management centers need accurate data on local traffic to be able to offer their services to the traveling public and improve mobility. The most useful traffic data can be collected or generated by the vehicles themselves as they travel along the road.

Vehicle data extend beyond the basic information on position, speed, and acceleration that can be estimated by other means such as using low-cost Global Positioning System (GPS) devices integrated in smart phones. Some examples of the rich information available on the vehicle include activation of headlights or fog lights indicating poor visibility, activation of windshield wipers for reliable assessments of local weather conditions, and sudden vertical accelerations for detection of potholes in the road surface. Probe data collection applications record relevant portions of vehicle data and send them to back-end servers able to aggregate, process, and analyze input from all probes on the road to generate accurate local traffic information in real time.

Applications typically access vehicle data through the vehicle's CAN. In the case of passenger vehicles, CAN messages and their formats are proprietary to individual original equipment manufacturers (OEMs) and may vary between vehicle models produced by the same manufacturer. Furthermore, OEMs may have different methods to define and calculate data values exchanged over their proprietary onboard networks. This makes it difficult to compare and aggregate data collected from multiple sources. In the United States, in order to overcome this problem, OEMs have developed the SAE J2735 standard [SAE09], a data dictionary which promotes interoperability between offboard applications by standardizing message sets, data frames, and data elements. The SAE J2735 standard is limited to specifying initial representative message structures and data elements and providing sufficient background information to allow application developers to properly interpret the data [SAE09].

In the case of heavy-duty commercial vehicles, by contrast, a common standard has been widely adopted for CAN data messages. This standard, SAE

J1939 [SAE10], is a higher-layer CAN-based protocol used for communications and diagnostics among vehicle components. The messages transmitted using J1939 typically consist of a 29-bit identifier which defines the message priority, an 8-byte data array, and information identifying which ECU sent the message. The maximum number of ECUs in a CAN J1939 control systems is 30, with a maximum CAN bus length of 40 m. Parameters embedded in the identifier are divided into a parameter group number and an 8-bit source address.

3.5   VEHICLE DATA SECURITY

In most modern vehicles, vehicle data travel along CAN buses. If an attacker has physical access to a vehicle's CAN bus, he can intercept, modify, and broadcast CAN messages. This provides the ability to inject false messages into a CAN bus.

Current vehicle architectures protect different onboard subsystems by using separated CAN buses for each of them. For example, CAN messages for vehicle safety systems are isolated from those used for vehicle entertainment systems. However, these separate CAN buses are usually connected by gateways that enable interactions among the different onboard subsystems. These gateways are susceptible to malicious hacking, allowing for false messages to be sent from one CAN bus to another.

Researchers at the University of Washington and the University of California, San Diego, showed that they could systematically control through the CAN bus a wide array of vehicle components, including engine, brakes, heating and cooling systems, lights, instrument panel, radio, and locks [KCRP10]. A range of techniques were used to gain control of and attack in-vehicle systems. These included sniffing messages broadcast over CAN buses through specialized software installed on a laptop PC. Packet sniffing is particularly effective in revealing how different in-vehicle components communicate with each other. Another technique they used is fuzzing, which consists of injecting random messages into the network to cause unintended behaviors and expose security vulnerabilities. In this experiment, fuzzing allowed the researchers to reset and control the functions of many ECUs. Finally, it was possible to reverse engineer the software code controlling onboard devices. For instance, attackers were able to reverse engineer functions of the telematics unit by using parts available through resellers. These techniques allowed researchers to gain control of a wide range of vehicle subsystems:

Radio: The researchers were able to completely control, and disable the user's control of, the radio. This allowed them to display arbitrary messages on the radio display, control the radio volume, and prevent the user from resetting the radio. It also allowed the researchers to produce clicks and chimes at arbitrary frequencies for various durations.

Instrument Cluster: The researchers were able to fully control the instrument cluster, allowing them to display arbitrary messages, falsify the fuel level and the speedometer reading, and adjust the illumination of instruments.

Body Controller: By reverse engineering and fuzzing packets, the researchers were able to control essentially all body control module (BCM) functions. They were able to lock and unlock doors, jam door locks, open the trunk, adjust interior and exterior lighting levels, honk the horn, disable windshield wipers, continuously shoot windshield fluid, and disable the key lock relay to lock the key in the ignition.

Engine: By fuzzing requests to the engine control module (ECM), the researchers were able to temporarily boost the engine revolutions per minute (RPM), disturb engine timing by resetting the learned crankshaft angle sensor error, disable all cylinders, and disable the engine such that it knocks excessively when restarted or cannot be restarted at all.

Brakes: By fuzzing the electronic brake control module, the researchers were able to lock the brakes.

Denial of Service: The researchers were able to disable the communications of individual components on the CAN bus at arbitrary times. Disabling communications with the ECM when the wheels are spinning reduces the car's reported speed immediately to 0 mph. Disabling communications with the BCM freezes the instrument panel cluster in its current state. The researchers were also able to prevent the car from turning on or off.

These vulnerabilities demonstrate the need for more advanced security measures for automobiles. As more vehicle functionality becomes software controlled, protecting vehicle data and communications over onboard networks becomes an essential task for vehicle designers. This will assume even greater importance for connected vehicles that will share data with each other to enhance safety and support mobility and convenience applications.

3.6 VEHICLE POSITIONING

Perhaps the most useful data for vehicular applications is an accurate estimate of the vehicle's current position and dynamics. This section discusses currently available solutions to determine global positioning.

3.6.1 Global Positioning System

Many vehicle applications depend upon a precise positioning system. In the United States, the GPS was created to fill this need. GPS relies on a constellation of Earth-orbiting satellites which communicate with GPS receivers to provide an accurate estimate of their locations on the globe. A GPS receiver uses high-frequency, low-power radio signals, synchronized clocks, and an almanac which contains the precise location of each satellite at any given moment to calculate how far it is from multiple satellites. Using this distance information, the receiver employs trilateration to determine its location on the globe, as long as there is an unobstructed line of sight (LoS) between the receiver and at least four GPS satellites. GPS data can be used to determine several valuable pieces of information, including how far a vehicle has traveled, how long it has been traveling, its current and average speeds, and its estimated time of arrival (TOA)) at the destination.

GPS can provide service to an unlimited number of users because the users interact passively with the system by simply receiving its signals. The system utilizes one-way TOA ranging, in which satellite transmissions are referenced to highly accurate atomic frequency standards. The satellites broadcast ranging codes and navigation data on two frequencies using the code division multiple access (CDMA) technique. These frequencies are designated L1 (1575.42 MHz) and L2 (1227.6 MHz) [KaHe06]. Each satellite transmits on these frequencies unique ranging codes which have low cross-correlation properties. These include a short code, called the coarse/acquisition (C/A) code, and a long code, referred to as the precision code. The navigation data transmitted by the satellites enable the receiver to determine the location of the satellite at the time of the signal transmission, while the ranging codes allow the receiver to determine the propagation time of the signal and therefore determine the satellite-to-user range. This technique requires the receiver to use an extremely precise clock [KaHe06]. Because such a precise clock is cost prohibitive in mainstream GPS receivers, at least four satellites are used to solve for distance and time, rather than just three that would be sufficient if the time was absolutely correct and synchronized between the receiver and the satellites. Using signals from four satellites reduces positioning error significantly while keeping GPS receivers in a reasonable price range.

GPS is a dual-use system regulated by the U.S. military for both civil and military uses. Its accuracy may be augmented through the use of differential GPS (DGPS) corrections and the Wide Area Augmentation System (WAAS). Currently, the GPS user range error (URE) is approximately 1 m, and the modernization procedures the system is undergoing are projected to improve the URE by four times [Gibb09].

In some situations, the vehicle's navigation system may lose its connections with a sufficient number of satellites. This occurs primarily due to an interruption in LoS between the receiver and the satellites, as when a vehicle enters a tunnel or ventures through an area with dense foliage overheads. Modern automobiles typically use a system of dead reckoning when GPS is temporarily not available. Using sensor information pertaining to wheel speed, steering wheel angle, yaw rate, and map matching, a vehicle navigation system may continue to give sufficiently accurate position estimates through several minutes of GPS outages. The accuracy of dead reckoning techniques is often sufficient for navigation purposes.

3.6.2 Galileo

Galileo is the European Union's answer to the need for a global navigation satellite system (GNSS). It operates on principles similar to GPS and is interoperable with GPS. Unlike GPS, however, Galileo is under civilian control and provides the same degree of accuracy to both civilian and military users. Galileo will be able to deliver real-time positioning accuracy down to the meter, guarantee service availability under all but the most extreme circumstances, and inform users within seconds of any satellite failure [ESA11]. The Galileo program is working toward full operational capability, which it is projected to reach by 2020. The European Space Agency's (ESA) first two navigation satellites, Galileo In-Orbit Validation Element-A (GIOVE-A) and GIOVE-B, were launched in 2005 and 2008, respectively, to reserve radio frequencies and test key technologies. The first two of four operational satellites designed to validate Galileo were launched in October 2011 [ESA11]. Two more satellites will be launched in 2012 to complete the in-orbit validation (IOV) phase, after which additional satellites will be launched to bring the system to initial operational capability (IOC) around 2015, whereas full operational capability (FOC) will be achieved by the end of the decade [ESA11]. The fully deployed Galileo system will consist of 30 satellites (27 operational and three active spares) in three orbit planes at a 56° angle to the equator, providing coverage even to the polar regions.

3.6.3 Global Navigation Satellite System

The Global Navigation Satellite System (GLONASS) was built in Russia. Operating on principles similar to GPS and Galileo, GLONASS currently consists of 22 GLONASS-M satellites and one experimental GLONASS-K satellite [Duma11]. Many of the satellites broadcast a second civil signal, making GLONASS the only GNSS system to do so [Gibb09]. The URE of GLONASS has been improved considerably in recent years to 1.8 m. This error range, although still high compared with GPS or Galileo, is well within the bounds specified in the GLONASS interface control document [Gibb09]. CDMA modulation was used in the GLONASS-K satellite and is expected to be implemented more extensively in the future, easing the difficulties of interoperability with GPS and Galileo [Duma11].

Today, devices able to simultaneously receive GPS and GLONASS data are already available in the market. Their accuracy and operational ranges benefit from the higher number of satellites that become available when the two systems are combined.

3.6.4   Positioning Accuracy

Researchers commonly refer to the following levels of positioning accuracy for vehicle applications:

- Road level accuracy sufficient to determine "which road" the vehicle is currently traveling on (precision of 5 m or better)
- Lane level accuracy sufficient to determine "which lane" the vehicle is currently traveling on (precision of 1.5 m or better)

Existing navigation applications require the less stringent road level accuracy. However, some safety applications require the more challenging lane level accuracy. For example, these safety applications could use lane level positioning to select the correct traffic control signal phase when the vehicle approaches signalized intersections with protected left or right turns where the turn phase differs from the phase for the straight crossing direction.

Several techniques such as DGPS corrections and the WAAS have been developed to achieve submeter accuracy with today's GPS receivers [KaHe06]. DGPS corrections provide sufficient information to reconcile positions resulting from estimation algorithms with a local base station's known fixed location, whereas WAAS relies on ground reference stations spaced approximately 500 mi apart and computes corrections on a regional basis.

REFERENCES

[Char09] R. Charette: "This Car Runs on Code," IEEE Spectrum Magazine, [online]. Available at: <http://spectrum.ieee.org/green-tech/advanced-cars/this-car-runs-on-code> (Accessed October 22, 2011), 2009.

[Duma11] P. Dumas: "GLONASS-K for Airborne Applications," Inside GNSS, [online]. Available at: <http://www.insidegnss.com/auto/julyaug11-Dumas.pdf> (Accessed October 28, 2011), 2011.

[ESA11] European Space Agency: "What is Galileo? Fact Sheet," [online]. Available at: <http://download.esa.int/docs/Galileo_IOV_Launch/Galileo_factsheet_20110801. pdf> (Accessed October 28, 2011), 2011.

[Flex10] FlexRay Consortium: "FlexRay Communications System Protocol Specification," Version 3.0.1, 2010.

[Gibb09] G. Gibbons: "GPS, GLONASS, Galileo, Compass: What GNSS Race? What Competition?" Inside GNSS, [online]. Available at: <http://www.insidegnss.com/node/1389> (Accessed October 28, 2011), 2009.

[ISO94] International Organization for Standardization (ISO): "Road vehicles—Low-Speed Serial Data Communication," ISO 11519, 1994.

[ISO03] International Organization for Standardization (ISO): "Road vehicles—Controller Area Network (CAN)," ISO 11898, 2003.

[KaHe06] E. Kaplan and C. Hegarty, eds.: Understanding GPS: Principles and Applications, 2nd edition, Artech House Inc., 2006.

[KCRP10] K. Koscher, A. Czeskis, F. Roesner, S. Patel, T. Kohno, S. Checkoway, D. McCoy, B. Kantor, D. Anderson, H. Shacham, and S. Savage: "Experimental Security Analysis of a Modern Automobile," IEEE Symposium on Security and Privacy, 2010.

[NIC09] National Instruments Corporation: "FlexRay Automotive Communication Bus Overview," [online]. Available at: <http://zone.ni.com/devzone/cda/tut/p/id/3352> (Accessed October 20, 2011), 2009.

[NiPL08] D. K. Nilsson, P. H. Phung, and U. E. Larson: "Vehicle ECU Classification based on Safety-Security Characteristics," Proceedings of Road Transport Information and Control (RTIC 2008), Manchester, UK, 2008.

[SAE05] Society of Automotive Engineers: "SAE J2602 LIN Network for Vehicle Applications Conformance Test," 2005.

[SAE07] Society of Automotive Engineers: "SAE J1979 E/E Diagnostic Test Modes," 2007.

[SAE09] Society of Automotive Engineers: "SAE J2735 Dedicated Short-Range Communications (DSRC) Message Set Dictionary," 2009.

[SAE10] Society of Automotive Engineers: "SAE J1939 Vehicle Application Layer," 2010.

[See06] W. See: "Vehicle ECU Classification and Software Architectural Implications," Technical Report, Feng Chia University, Taiwan, 2006.

[Shor11] R. Shorey: "Emerging Trends in Vehicular Communications: Communications in Electric Vehicles," IEEE New York Monitor, 2011.

[Stev11] R. Stevenson: "Long-Distance Car Radar," IEEE Spectrum Magazine, [online]. Available at: <http://spectrum.ieee.org/green-tech/advanced-cars/longdistance-car-radar/0> (Accessed November 4, 2011), 2011.

[Voel08] J. Voelcker: "Top 10 Tech Cars of 2008," IEEE Spectrum Magazine [online]. Available at: <http://spectrum.ieee.org/green-tech/advanced-cars/top-10-tech-cars-2008/3> (Accessed November 7, 2011), 2008.

CONNECTED VEHICLES

4.1 CONNECTED VEHICLE APPLICATIONS

A wide range of innovative applications will be enabled by wireless communications, allowing vehicles to connect with each other, the roadside network infrastructure, and the Internet. Connected vehicle applications can be classified into the following categories: hard safety, soft safety, mobility, and convenience applications.

4.1.1 Hard Safety Applications

Hard safety applications are targeted to avoiding imminent crashes and minimizing the damage when these crashes become unavoidable. Examples include emergency electronic brake light (EEBL) and the intersection movement assist (IMA).

EEBL broadcasts messages to inform neighboring vehicles of a hard braking maneuver. This application is particularly effective when obstacles are obstructing the driver's line of sight with the breaking vehicle.

IMA warns the driver when it determines that it is not safe to enter an intersection due to high collision probability with other vehicles in cross traffic. As such, IMA addresses one of the most challenging traffic situations: cross traffic is particularly dangerous because targets may become visible to the driver (and traditional automotive onboard autonomous sensors) when they are already too close to the vehicle.

Vehicle Safety Communications: Protocols, Security, and Privacy, First Edition. Luca Delgrossi and Tao Zhang.
© 2012 John Wiley & Sons, Inc. Published 2012 by John Wiley & Sons, Inc.

Hard safety applications impose the most stringent requirements on the communication system. The communication latency has to be minimized in order to offer the driver sufficient time to take action. Acceptable latencies for this class of applications typically have to be within 100 ms [KAEH11]. Furthermore, the communication system must provide high levels of reliability such as high message reception probabilities.

4.1.2 Soft Safety Applications

Soft safety applications are safety applications that are less time critical than hard safety applications. Examples include applications that warn the driver of weather, road, traffic, and other hazardous driving conditions. Such driving conditions include icy roads, construction zones, reduced visibility, potholes, and traffic jams. Soft safety applications increase driver safety but do not require immediate driver reaction because the hazards are not imminent. Typical actions a driver needs to take in response to soft safety application alerts would be to proceed with caution or take alternate routes to avoid the dangerous conditions ahead.

It is important to distinguish between hard and soft safety applications because they impose significantly different requirements on the communication system. Soft safety applications can usually tolerate communication latencies in the order of seconds or even longer, which are substantially longer than requirements for hard safety applications. Soft safety applications also have less stringent reliability requirements than hard safety applications. The longer time available for message dissemination makes it easier for soft safety applications to receive a sufficient number of messages over less reliable communication channels. Furthermore, failures to receive soft safety application messages typically cause less severe consequences than failures to receive hard safety applications.

4.1.3 Mobility and Convenience Applications

Mobility applications focus on improving traffic flow. Examples include navigation, road guidance, traffic information services, traffic assistance, and traffic coordination.

Convenience applications focus on making driving more enjoyable and providing greater convenience to the driver and passengers. These applications include point-of-interest notification, e-mail, social networking, media download, and application update.

Mobility and convenience applications can be delivered to drivers and passengers through consumer electronic devices, such as smart phones and other portable electronic devices, which drivers and passengers routinely bring into their vehicles. Such applications can also be provided through devices embedded into the vehicle. Better integration with the vehicle can provide additional advantages. For instance, the vehicle's display and sound system can offer a

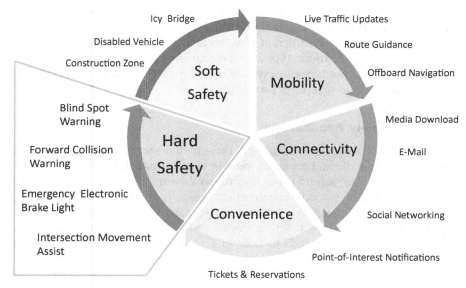

Figure 4.1. Connected vehicle applications

high-quality user interface designed to minimize driver distraction. Information from onboard sensors and vehicle subsystems can be integrated with information from external sources to enable more powerful applications.

Today, one prominent example of a system allowing consumers to stay connected on the road is Ford Sync. Ford Sync synchronizes with the mobile phones the driver and passengers bring into the vehicle to allow users to make hands-free phone calls, receive audible text messages, and control their mobile phones with voice-activated commands.

Mobility and convenience applications can tolerate even longer communication delays than soft safety applications. However, some mobility and convenience applications that download large files to vehicles will demand high communication bandwidths.

Figure 4.1 shows some connected vehicle applications categories, including a few examples for each category.

4.2 UNIQUENESS IN CONSUMER VEHICLE NETWORKS

Consumer vehicles wirelessly connected with each other and with the roadside infrastructure form a consumer vehicle network (CVN). CVNs will be the focus of this book. CVNs have many unique characteristics that have not been well addressed in the study of other types of networks.

Vehicle safety communications are often highly transient. When vehicles meet with each other, they often stay within communication range for very short time periods—often as short as a few seconds. This also applies to communications with the roadside infrastructure.

As consumer vehicles travel, they will meet many other vehicles that they have never met before. Vehicles do not know each others' addresses or security credentials a priori. Yet, they must communicate instantly to support hard safety applications.

The population of vehicles directly communicating with each other constantly changes over time. This makes it harder to optimize communication protocols and information dissemination strategies.

Fixing problems or making changes to the onboard communication hardware or software will typically require vehicles to be brought to service centers. Forcing drivers to bring their vehicles to service centers just for the maintenance of onboard communication systems will be time consuming, inconvenient, often costly to drivers, and should therefore be avoided whenever possible.

The long lifetime of consumer vehicles imposes challenges to ensure backward compatibility between different generations of onboard communication systems.

Most vehicles in a CVN belong to individual consumers. Onboard communication devices will not be managed by information technology organizations or experts. This makes it difficult to manage these devices and the applications, especially when vehicles lack frequent interactions with the roadside network infrastructure.

CVNs have to address unique security and privacy threats. Vehicles must establish sufficient trust in the messages they receive within the very short time available for communication. Also, a consumer vehicle communication system must be designed to adequately protect driver privacy. Designing a scalable and practically deployable privacy-preserving security system for national-scale CVN remains a tough challenge. Providing privacy introduces conflicting requirements with supporting some vehicle safety applications, securing vehicle communications, and detecting malicious vehicles that use vehicle communication capabilities to harm other vehicles or disturb the transportation system. A major form of security threats will be insider attacks. Adversaries can use vehicles with valid security credentials to attack the CVN. This makes it difficult to distinguish which vehicles are innocent and which are malicious.

CVNs are expected to be very large. There are approximately 256 million registered vehicles in the United States, according to statistics from the U.S. Department of Transportation (USDOT). Designing such a large-scale vehicle communication system is unprecedented. Many networking solutions proposed for smaller networks have been found to be not scalable, ineffective, or inefficient for CVNs.

4.3   VEHICLE COMMUNICATION MODES

Several vehicle communication modes can be envisioned as illustrated in Figure 4.2. Vehicles can directly exchange safety messages through local broadcasts (a). These messages can be further relayed across multiple hops to reach other vehicles not within the source vehicle's communication range (b). A roadside infrastructure network can broadcast messages to vehicles (c). Vehicles can communicate with application servers (d) or communicate with other vehicles through the infrastructure network (e).

4.3.1   Vehicle-to-Vehicle Local Broadcast

With vehicle-to-vehicle (V2V) local broadcast, a vehicle sends messages to all other vehicles within its communication range. These messages are not further relayed by other vehicles. This communication mode serves as the foundation for cooperative vehicle safety applications aimed at collision avoidance. For example, vehicles can use V2V local broadcast to inform neighboring cars of each other's current position, heading, and speed.

When a vehicle broadcasts a safety message, it typically does not know whether there are other vehicles nearby. It is generally difficult for a vehicle to know the network addresses of other vehicles a priori because the set of neighboring vehicles changes frequently. For this reason, short-range radios with native broadcasting capabilities are natural ways to support this communication mode.

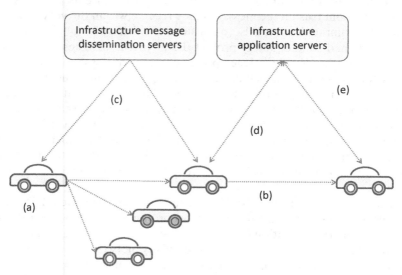

Figure 4.2. Vehicle safety communication modes. (a) V2V local broadcast; (b) V2V multihop forwarding; (c) I2V local broadcast; (d) V2I bidirectional communications; (e) indirect V2V

The transmission range of a short-range radio required to support vehicle safety applications depend heavily on several key factors such as vehicle speed and driver reaction time. Driver reaction time is the time it takes for drivers to react to road and traffic conditions after they observe a stimulus (alert). Extensive studies show that most alert drivers are capable of responding to urgent situations in less than 2.5 seconds when the stimulus is clearly identifiable and unambiguous [BiTD06] [GeMB00]. For example, consider two vehicles approaching each other at a speed of 120 km/h. The distance between these two vehicles reduces by approximately 33.3 m/s. The transmission range of the short-range radio must be at least 116.5 m, assuming that the driver will need 3 seconds to react to an alert and the vehicle will need at least 0.5 second to analyze incoming messages, assess collision risks, and generate the alert.

A fundamental issue in supporting V2V local broadcasting is how to achieve an acceptable performance when a large number of vehicles broadcast messages to each other. Broadcast is typically implemented over a radio channel shared by all users. The contention among users can increase rapidly with the number of users and the frequency of broadcasts, causing heavy loss of messages.

4.3.2 V2V Multihop Message Dissemination

With V2V multihop message dissemination, messages from one vehicle are relayed by other vehicles to reach vehicles that are not inside the source vehicle's communication range. When the number of hops is very low, this communication mode can be used to support hard safety applications. For instance, it could be used to relay EEBL messages to extend their distribution range. V2V multihop message dissemination could also be used for soft safety purposes such as the distribution of hazardous road and traffic information.

A fundamental issue in designing V2V multihop message dissemination protocols is how to balance performance and overheads. Important performance metrics include message dissemination delay and message delivery rate. Main overheads include the number of messages to be transmitted to implement a dissemination strategy, the amount of processing on each vehicle, and the implementation complexity.

A second fundamental issue is that the vehicles may not always form a connected V2V network. Message dissemination needs to work despite random temporary network partitions.

Many V2V multihop message dissemination protocols and mechanisms have been developed [BaBe03] [ChJR10] [JiCa02] [LiWa07] [STUH05] [TWBM08] [VaBe00] [ZhCh01]. They vary widely in terms of performance and overheads. They can be classified into the following categories:

- *Topology-Based Multicast*: These protocols establish and maintain a multicast topology, such as a multicast tree, over a connected V2V network for disseminating messages [RoPe99].

- *Geocast*: These protocols use location or geographical information to assist message dissemination by dividing vehicles into multicast groups or multicast zones and then distributing messages within each zone [CLPB02] [JiCa02] [KiSH08].
- *Enhanced Broadcast*: Enhanced broadcast seeks to benefit from the advantage of broadcast, such as low delay and ease of implementation, while reducing the number of vehicles that relay each message to reduce message overhead. A vehicle may decide whether to relay a message based on its distances to the source and destination, the number of duplicate messages it has received, and location and direction information.
- *Stochastic (Epidemic) Dissemination* [VaBe00] [ZhCh01]: Vehicles take advantage of their natural mobility to spread messages among themselves. Each vehicle decides whether it should relay a message based on a relay probability. For example, [ZhCh01] proposed a way for each vehicle to autonomously determine the relay probability based on random graph theories. Stochastic dissemination mechanisms also provide a foundation for supporting delay-tolerant networking to address random temporary partitions of a vehicle network.

Topology-based and geocast message dissemination mechanisms tend to incur excessive messaging and computing overheads for establishing, maintaining, and reorganizing multicast topologies, geocast zones, or groups over a large and highly dynamic vehicle network. Vehicles have to collect and maintain up-to-date information about the topologies or geocast zones, which further increase communication delays, required processing power on vehicles, and system complexity. Topology-based and geocast message dissemination mechanisms typically require the vehicles in the message dissemination region to form a connected network at the time of the message dissemination, which may often not be possible in a real-world situation.

Enhanced broadcast and stochastic dissemination mechanisms could strike better balances between message dissemination performance and overheads. The stochastic dissemination methods in [VaBe00] and [ZhCh01], for example, are shown to generate optimal dissemination delay and delivery rate while effectively controlling overheads.

4.3.3 Infrastructure-to-Vehicle Local Broadcast

In this communication mode, vehicles receive local broadcasts from the roadside infrastructure. For instance, infrastructure-to-vehicle (I2V) local broadcast can be used to send:

- traffic controller signal phase and timing information,
- dangerous road condition information,
- security credentials, and
- service advertisements.

I2V local broadcast can be implemented through short-range radio transceivers deployed along the roadside. It can also be implemented using cellular, satellite, or digital radio broadcast services to reach all the vehicles in a large geographical region. Satellite and digital radio broadcast services are widely available on vehicles today. They are commonly used to deliver real-time traffic and road conditions information to drivers and navigation devices.

4.3.4 Vehicle-to-Infrastructure Bidirectional Communications

Many mobility and convenience applications require the vehicle-to-infrastructure (V2I) bidirectional communications mode. Examples include navigation, Internet access for browsing or e-mail, electronic transactions for purchasing goods or services, and media download. Another important use of V2I bidirectional communications is for vehicles to communicate with security credential management systems. V2I communications can also be used to "broadcast" messages from a vehicle to other vehicles through infrastructure applications servers.

V2I communications can be supported using long-range or short-range radios. Cellular networking capabilities have been increasingly used in vehicle telematics devices. Short-range radio networks deployed along roadsides, at homes, or in public hotspots such as parking lots and vehicle service centers can be used by vehicles to access infrastructure-based services.

4.4 WIRELESS COMMUNICATIONS TECHNOLOGY FOR VEHICLES

Several existing wireless technologies can support vehicle communications. Short-range radios include Bluetooth, Wi-Fi, and dedicated short-range communications (DSRC). Long-range radios include cellular networks, satellite services, and digital radio broadcast networks.

Bluetooth is designed to support personal area networks to replace wired cables between devices nearby. It operates in an unlicensed 2.4 GHz industrial, scientific, and medical (ISM) frequency band. Bluetooth is increasingly used to pair mobile phones to vehicles. Such pairing enables hands-free calling from the vehicle, allows a vehicle's embedded display unit to be used to control mobile phones, and allows a mobile phone to use the vehicle's embedded sound systems. It also enables a vehicle to use the driver's or a passenger's mobile phone to communicate with the outside world to support a wide range of telematics applications such as making emergency calls when the driver loses her ability to do so in an accident, downloading digital maps, infotainment contents, travel information, or software updates to the device, and allowing the driver or passengers to access Internet-based or cloud-based applications. Bluetooth can also been used for some V2I communications when the vehicle is stationary or moving at very low speeds. These include, for example, allowing a vehicle to communicate with parking lot payment applications at a parking lot entrance. However, Bluetooth's limited communication range and high latency precludes its ability to support vehicle safety communications.

Wi-Fi networks, defined in the Institute of Electrical and Electronics Engineers (IEEE) 802.11 standards, can support V2V local broadcasts through their native ad hoc broadcast capabilities. However, modifications to the standards are necessary to reduce latency and achieve more reliable communications at high speeds in order to support vehicle safety communications. Wi-Fi networks can also support other forms of vehicle communications in their infrastructure mode. However, Wi-Fi hotspots have sparse coverage today and moving from one hotspot to another generally requires reconnection to the network. This could be tolerated by some nonsafety vehicle applications.

So far, IEEE 802.11 technology is the closest to meeting most requirements of hard safety applications. This led to the development of DSRC radio technology and standards. DSRC is essentially Wi-Fi in its ad hoc mode adapted to support communications among vehicles moving at high speeds. DSRC can be used to create large-scale CVNs able to support V2V and V2I communication modes.

Third-generation (3G) cellular networks offer communication latencies in the range of hundreds of milliseconds, thus meeting the requirements of some hard and soft safety applications. However, current 3G networks lack effective ways to support V2V local broadcasts, which are essential to most hard safety applications. Furthermore, sharing bandwidth with voice communications could cause unpredictable delays and could also overload the cellular system. Addressing this issue by using a separate cellular system for vehicle communications could be costly. Also, the long call setup latencies can introduce excessive delays for hard safety applications.

The predominate fourth-generation (4G) cellular network technology, Long-Term Evolution (LTE), offers significantly lower communication delay, higher capacity, and enhanced broadcasting capabilities compared with 3G cellular networks. LTE can also support very small radio cells. These include microcells covering up to 2 km, picocells covering up to 200 m, and femtocells covering tens of meters [3GPP11] [ChAG08] [LaZh11]. These small cells can be used to deliver even higher capacity and lower delay at hot spots such as homes, office building complexes, densely populated areas, and crowded road intersections. Therefore, LTE can support time-critical vehicle applications, including selected I2V local broadcast and V2V safety applications.

Table 4.1 provides a brief comparison of wireless communication technologies based on criteria such as latency, range, and ability to support vehicle communication modes.

The choice of communication technology to support connected vehicle applications depends on the applications' specific requirements. Multiple wireless technologies are likely to be used by vehicles to support the vast range of communication needs. Also, some applications can be supported with more than one wireless technology.

Table 4.1. Comparison of wireless communication technologies

	DSRC	Wi-Fi	Bluetooth	3G	4G LTE	SDARS
Range	Hundreds of meters	Hundreds of meters	Up to 100 m	Tens of kilometers	Tens of meters up to 100 km	Countrywide
End-to-end message delay	10 ms	10 ms	10 ms	From 50 to hundreds of milliseconds	Tens of milliseconds	10–20 seconds
Connection/call setup time	Not needed	3–5 seconds	3–4 seconds	Hundreds of milliseconds to seconds	~50 ms	Not applicable
V2V local broadcast	Yes	Yes	Impractical	With a server	With a server	No
V2V multihop messaging	Yes	Yes	Impractical	With a server	With a server	No
I2V local broadcast	Yes	Yes	Impractical	Not offered by all network operators	Not offered by all network operators	Yes
V2I bidirectional	Yes	Yes	Impractical	Yes	Yes	No

SDARS, satellite digital audio radio service.

Furthermore, wireless communication technologies have been advancing rapidly. The wireless technology implemented in a vehicle can become outdated while the vehicle is still on the road. New vehicles should be able to benefit from new wireless communication technologies and yet remain interoperable with older vehicles. These considerations call for innovative designs of vehicle communication systems that can take advantage of technology improvements while ensuring that crucial applications can remain interoperable among different generations of vehicles.

REFERENCES

[3GPP11] 3rd Generation Partnership Project: "3GPP TS 22.220 V10.8.0 (2011-12) Technical Specification Group Services and System Aspects; Service requirements for Home Node B (HNB) and Home eNode B (HeNB) (Release 10)," 2011.

[BaBe03] A. Bachir and A. Benslimane: "A Multicast Protocol in Ad-Hoc Networks Inter-Vehicle Geocast," IEEE Vehicular Technology Conference, 2003.

[BiTD06] S. Biswas, R. Tatchikou, and F. Dion: "Vehicle-to-Vehicle Wireless Communication Protocols for Enhancing Highway Traffic Safety," IEEE Communications Magazine, 2006.

[ChAG08] V. Chandrasekhar, J. G. Andrews, and A. Gatherer: "Femtocell Networks: A Survey," IEEE Communications Magazine, Vol. 46, 2008.

[ChJR10] R. Chen, W. Jin, and A. Regan: "Broadcasting Safety Information in Vehicular Networks: Issues and Approaches," IEEE Network Magazine, 2010.

[CLPB02] C. Cheng, H. Lemberg, S. Philip, E. Van Den Berg, and T. Zhang: "SLALoM: A Scalable Location Management Scheme for Large Mobile Ad-Hoc Networks," IEEE Wireless Communications and Networking Conference (WCNC'02), Orlando, Florida, 2002.

[GeMB00] D. V. McGehee, E. N. Mazzae, and G. H. S. Baldwin: "Driver Reaction Time in Crash Avoidance Research: Validation of a Driving Simulator Study on a Test Track," 16th Triennial Congress of the International Ergonomics Association and 44th Annual Meeting of the Human Factors and Ergonomics Society, Vol. 3, pp. 320–323, Santa Monica, CA, 2000.

[JiCa02] X. Jiang and T. Camp: "A Review of Geocasting Protocols for a Mobile Ad-Hoc Network," Proceedings of Grace Hopper Celebration (GHC), 2002.

[KAEH11] G. Karagiannis, O. Altintas, E. Ekici, G. Heijenk, B. Jarupan, K. Lin, and T. Weil: "Vehicular Networking: A Survey and Tutorial on Requirements, Architectures, Challenges, Standards and Solutions," IEEE Communications Surveys & Tutorials, 2011.

[KiSH08] M. Kihl, M. Sichitiu, and H. Joshi: "Design and Evaluation of Two Geocast Protocols for Vehicular Ad-Hoc Networks," Journal of Internet Engineering, vol. 2, no. 1, pp. 127–135, 2008.

[LaZh11] C. L. Lau and T. Zhang: "Planning and Control of LTE Femto Networks," 3rd International Workshop on Indoor and Outdoor Femto Cells (IOFC), Princeton, New Jersey, 2011.

[LiWa07] F. Li and Y. Wang: "Routing in Vehicular Ad Hoc Networks: A Survey," IEEE Vehicular Technology Magazine, 2007.

[RoPe99] E. Royer and C. Perkins: "Multicast Operation of the Ad-Hoc On-Demand Distance Vector Routing Protocol," 5th Annual ACM/IEEE International Conference on Mobile Computing and Networking (MobiComm), 1999.

[STUH05] M. Saito, J. Tsukamoto, T. Umedu, and T. Higashino: "Evaluation of Inter-Vehicle Ad-Hoc Communication Protocol," 19th IEEE International Conference on Advanced Information Networking and Applications, 2005.

[TWBM08] O. Tonguz, N. Wisitpongphan, F. Bai, P. Mudalige, and V. Sadekar: "Broadcasting in VANET," IEEE INFOCOM, 2008.

[VaBe00] A. Vahdat and D. Becker: "Epidemic Routing for Partially-Connected Ad Hoc Networks," Duke University Technical Report CS-2000-06, 2000.

[ZhCh01] T. Zhang and W. Chen: "Soft Mobile Ad-Hoc Networking," 2nd International Conference on Mobile Data Management (MDM), Hong Kong, 2001.

5

DEDICATED SHORT-RANGE COMMUNICATIONS

5.1 THE 5.9 GHZ SPECTRUM

In 1997, the Intelligent Transportation Society of America (ITSA) petitioned the U.S. Federal Communications Commission (FCC) to allocate 75 MHz of licensed spectrum at 5.9 GHz for the operations of intelligent transportation systems (ITS).

In 1998, the U.S. Congress enacted the Transportation Equity Act for the 21st Century [TEA98], which directed the FCC to consider the spectrum needs "for the operation of intelligent transportation systems, including spectrum for the dedicated short-range vehicle-to-wayside wireless standard." The FCC proposed to allocate 75 MHz of spectrum for use by dedicated short-range communications (DSRC) requiring "a short range, wireless link to transfer information between vehicles and roadside systems" [FCC98]. It was not until the next year that the FCC finally allocated the band: "By this action, we allocate 75 megahertz of spectrum at 5.850-5.925 GHz to the mobile service for use by Dedicated Short Range Communications ('DSRC') systems operating in the Intelligent Transportation System ('ITS') radio service. ITS services are expected to improve traveler safety, decrease traffic congestion, facilitate the reduction of air pollution, and help to conserve vital fossil fuels." [FCC99].

Today, in the United States, the 75-MHz spectrum between 5.850 and 5.925 GHz is referred to as 5.9 GHz DSRC.

It is often necessary to specify 5.9 GHz when referring to DSRC to differentiate the new spectrum from the older 900 MHz band of the same name,

Vehicle Safety Communications: Protocols, Security, and Privacy, First Edition. Luca Delgrossi and Tao Zhang.
© 2012 John Wiley & Sons, Inc. Published 2012 by John Wiley & Sons, Inc.

used for electronic toll collection. Throughout this book, the term DSRC will always be associated with the 5.9 GHz band.

5.1.1 DSRC Frequency Band Usage

The primary objective in allocating the new licensed DSRC band was to enable vehicle-to-vehicle (V2V) and vehicle-to-infrastructure (V2I) public safety applications to save lives and improve traffic flow.

In addition to these public safety applications, the FCC allocation permits private services to use a restricted portion of the band to spread deployment costs and to encourage quick adoption of DSRC technologies. To avoid conflicts with the primary public usage of the frequency band, private services are regulated so as not to interfere with or compromise public safety applications. The goal of the FCC was to prevent degradation of public safety applications, while still allowing nonsafety usage of the 5.9 GHz band.

5.1.2 DSRC Channels

The 75-MHz spectrum allocated in the United States at 5.9 GHz is structured into seven 10-MHz channels. The initial 5 MHz are used as a guard band protecting from adjacent frequencies (Figure 5.1). Each 10-MHz DSRC channel is identified by an even number from 172 to 184.

Channel 178, at the center of the spectrum (5.885–5.895 GHz), is the designated control channel (CCH) and is restricted to safety communications only. The FCC designated channel 172 (5.855–5.865 GHz) exclusively for V2V safety communications for accident avoidance and mitigation, and safety of life and property applications [FCC06]. The FCC also designated channel 184 (5.915–5.925 GHz) exclusively for high-power, long-distance communications to be used for public safety applications involving safety of life and property, including road intersection collision mitigation. Channel 172 is also known as high availability, low latency (HALL), whereas channel 184 is also known as high power, long range (HPLR). The remaining channels are designated as service channels (SCHs), available for both safety and nonsafety usage.

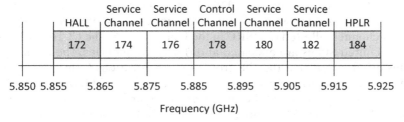

Figure 5.1. DSRC radio spectrum and channel allocation in the United States

The FCC allocation rules further specify that channels 174/176 and 180/182 may be combined to create two 20-MHz channels designated as 175 and 181, respectively.

5.1.3 DSRC Operations

The 5.9 GHz DSRC frequency band is a free but licensed spectrum. This should not be confused with other free usage bands in the United States, such as those allocated at 900 MHz, 2.4 GHz, and 5 GHz. The rules of use for these unlicensed bands, which are increasingly populated with Wi-Fi, Bluetooth, and other similar devices, place no restrictions on the communication technologies to be adopted and essentially provide guidelines for coexistence. The 5.9 GHz DSRC band, on the other hand, is restricted in both usage and communication technologies. FCC rulings require that all radios comply with the DSRC standards and regulate permitted usage over individual channels. In other words, unlike unlicensed bands, it is not permitted to adopt different radio technologies for 5.9 GHz, even if limited transmission powers are used.

From the perspective of vehicle safety communications, operating at 5.9 GHz has the advantage of avoiding bands crowded with smart mobile phones, personal navigation devices, and other non-DSRC devices. Operating in a dedicated environment where medium access is not subject to competition against other wireless stations significantly facilitates vehicle safety communications. For example, it requires no traffic prioritization to prevent nonsafety communications from interfering with time-critical safety communications.

Waves propagating in the 5.9 GHz band can offer communications with high data rates and low weather dependence. Also, 5.9 GHz DSRC is able to meet the communication range requirements imposed by most safety applications.

The high frequency spectrum nevertheless imposes some limitations. Since high-frequency signals cannot penetrate as far as low-frequency signals, solid objects such as walls tend to absorb waves at 5.9 GHz easily. This can significantly reduce achievable communication ranges in urban environments.

5.2   DSRC IN THE EUROPEAN UNION

The European Union (EU) also recognized the importance of a dedicated spectrum for ITS. The European DSRC spectrum is structured into five 10-MHz channels (including two for future expansions), as opposed to the seven 10-MHz channels allotted to DSRC in the United States. The CCHs for Europe and the United States are allocated at different frequencies. The main reason for this difference is that the Electronic Communications Committee wanted to introduce a guard zone between CCH (5.895–5.905 GHz) and SCH #1 (5.875–5.885 GHz) to avoid interference. This guard zone is in fact a second SCH (5.885–5.995 GHz) meant to be used only for low-priority, low-power messages. Under this scheme, the two most heavily used channels for safety applications will be the CCH and SCH #1 (Figure 5.2).

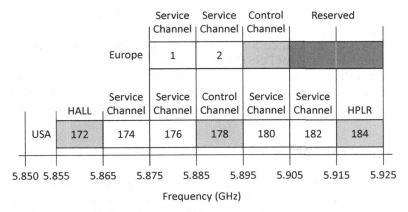

Figure 5.2. Overlapping EU and U.S. 5.9 GHz DSRC frequency bands

In 2009, the European Commission issued a standardization mandate to achieve ITS interoperability in the European Community [Ecom09]. This mandate (M/453) invited the European Telecommunications Standards Institute (ETSI) and the European Committee for Standardization (CEN) to prepare a coherent set of specifications and guidelines to support Europe-wide implementation and deployment of cooperative ITS.

5.3 DSRC IN JAPAN

In 1996, the Japanese Ministries of Transport, Construction, Posts and Telecommunications, International Trade and Industry, and the National Police Agency jointly released the Comprehensive Plan for ITS. This plan outlined nine areas of development and identified 20 distinct user services to be gradually realized by 2015.

In 2001, the Association of Radio Industries and Businesses (ARIB) developed the ARIB STD-T75 standard, designed specifically for DSRC [ARIB01]. As shown in Figure 5.3, the 5.770–5.850 GHz band is structured into seven downlink and seven uplink channels, each with a bandwidth of 4.4 MHz within its 5 MHz interval. The protocol specifies the interface between a Land Mobile

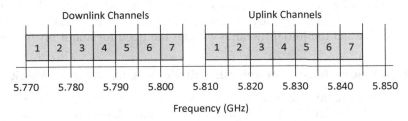

Figure 5.3. The 5.8 GHz DSRC frequency band in Japan

Station and a Base Station. The later published ARIB STD-T88 specifies DSRC application sublayer operations [ARIB04].

In contrast to the United States, V2V safety communications are not an officially intended usage of DSRC technology in Japan. Due to concerns raised in Japan that the 5.8 GHz band was too easily blocked by obstacles, the Ministry of Land, Infrastructure, Transport and Tourism (MLIT) opted to use a 10 MHz band in the 715–725 MHz range for V2V safety communications.

5.4   DSRC STANDARDS

Even before the official allocation of the 5.9 GHz spectrum, the U.S. Department of Transportation (USDOT) had reported to the U.S. Congress that the adoption of a communication standard for 5.9 GHz DSRC operations was "critical" for the development of ITS [DOT99]. The development of a set of standards for DSRC operations would provide a uniform platform to enable well-coordinated system deployment.

Initially, the Federal Highway Administration (FHWA) entered into cooperative agreements with five standards organizations, including the American Society for Testing and Materials (ASTM). FHWA decided that the DSRC standards should be "consensus standards" and should remain the property of the organization that developed them.

Subsequently, the FCC specified the licensing and service rules for ITS services based on 5.9 GHz DSRC and adopted the ASTM E2213-03 standard to ensure interoperability [ASTM03] [FCC03]. Furthermore, the FCC authorized nonexclusive geographic licensing of DSRC spectrum for use by roadside units (RSUs) and onboard units (OBUs) and established limits for transmission powers. It prohibited the use of commercial mobile radio services (CMRS) in the DSRC band.

5.4.1   Wireless Access in Vehicular Environments

In 2004, the DSRC standardization effort migrated to the Institute of Electrical and Electronics Engineers (IEEE). DSRC standards are based on IEEE 802.11a [IEEE99], with adjustments for low overhead operations in the 5.9 GHz spectrum. This made it possible not only to leverage the vast practical experience gained with IEEE 802.11 radios over the last decades, but also to use existing radio chipsets (mostly produced to operate in the nearby 5.8 GHz band) to conduct the initial research without the need for major investments.

5.4.2   Wireless Access in Vehicular Environments Protocol Stack

DSRC standards are defined as wireless access in vehicular environments (WAVE) [IEEE10c]. The WAVE protocol stack includes protocols to support communications for both safety and nonsafety applications in vehicular

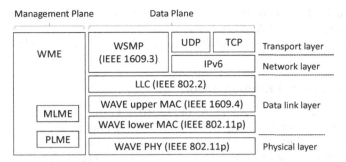

Figure 5.4. WAVE protocol stack

environments. It is structured into two parts: the data plane and the management plane.

Data plane protocols are responsible for over-the-air data exchange. As shown in Figure 5.4, DSRC applications may transmit data through the WAVE Short Message Protocol (WSMP) or traditional transport layer protocols such as the User Datagram Protocol (UDP) [Post80] and Transmission Control Protocol (TCP) [Post81], running on top of Internet Protocol version 6 (IPv6) [DeHi98].

WSMP, specified in the IEEE 1609.3 standard [IEEE10a], is used when Internet Protocol (IP) connectivity is either unavailable or not required. WSMP is used to support V2V and V2I safety applications. IP protocols are used for Internet connectivity.

The data link layer is divided into three sublayers: logical link control (LLC), and upper and lower media access control (MAC) layers. The LLC protocol defined by IEEE 802.2 [IEEE98] provides a standard interface to the lower layers. The behaviors at the upper MAC sublayer, including operations over multiple DSRC channels, are specified in the IEEE 1609.4 standard [IEEE10b].

The IEEE 802.11p protocol is the foundation upon which the WAVE protocol stack resides. IEEE 802.11p comprises the lower MAC layer and the physical (PHY) layer. It is meant for communications over a single DSRC channel. IEEE 802.11p specifies amendments to the IEEE 802.11 standard.

The WAVE management plane consists of the functions performed by the WAVE Management Entity (WME). These functions include handling of management frames, service advertisements, and IPv6 configurations. The WME also maintains a management information base (MIB) containing a DSRC station's configuration and status information. WME is supported by the MAC Layer Management Entity (MLME) and Physical Layer Management Entity (PLME).

In addition to the protocols illustrated in Figure 5.4, the WAVE protocol stack offers security functions, specified in the IEEE 1609.2 standard [IEEE06].

These security functions are orthogonal to the data plane and management plane defined by the WAVE architecture.

A complete description of WAVE standards can be found in [Kenn11].

5.4.3 International Harmonization

The DSRC bands allocated in Europe, Japan, and the United States differ in frequency and width, resulting in nonoverlapping channels. Furthermore, not every region is planning to adopt the same standards.

In 2009, the European Commission Directorate General for Information Society and Media and the USDOT signed the EU–U.S. Joint Declaration of Intent on Research Cooperation, which included a resolution to "strongly support development of global open standards which ensure interoperability through appropriate actions including, but not limited to, coordinating the activities of the standardization organizations" [EUUS09].

The USDOT Research and Innovative Technology Administration (RITA) and Japan's MLIT signed a similar Memorandum of Cooperation in the Field of Intelligent Transport Systems (ITS) "supporting development of globally open standards that ensure ITS interoperability."

REFERENCES

[ARIB01] Association of Radio Industries and Business: "Dedicated Short-Range Communication System," ARIB STD-T75, 2001.

[ARIB04] Association of Radio Industries and Business: "DSRC Application Sub-Layer," ARIB STD-T88, 2005.

[ASTM03] American Society for Testing and Materials: "ASTM E2213-03 Standard Specification for Telecommunications and Information Exchange Between Roadside and Vehicle Systems—5 GHz Band Dedicated Short-Range Communications (DSRC) Medium Access Control (MAC) and Physical Layer (PHY) Specifications," 2003.

[DeHi98] S. Deering and R. Hinden: "Internet Protocol, Version 6 (IPv6) Specifications," Request For Comments 2640, Internet Engineering Task Force, 1998.

[DOT99] United States Department of Transportation: "Intelligent Transportation Systems Critical Standards," 1999.

[Ecom09] European Commission Enterprise and Industry Directorate-General: "Standardization Mandate Addressed To CEN, CENELEC and ETSI in the Field of Information and Communication Technologies to Support the Interoperability of Co-Operative Systems for Intelligent Transport in the European Community," M/453 EN, 2009.

[EUUS09] European Commission Information Society and Media Directorate-General, USDOT Research and Innovative Technology Administration: "E. U.—U.S. Joint Declaration of Intent on Research Cooperation in Cooperative Systems," 2009.

[FCC98] Federal Communications Commission (FCC) Notice of Proposed Rule Making (NPRM): "Amendment of Parts 2 and 90 of the Commission's Rules to Allocate the 5.850–5.925 GHz Band to the Mobile Service for Dedicated Short-Range Communications of Intelligent Transportation Systems," FCC 98-119, 1998.

[FCC99] Federal Communications Commission (FCC) Report and Order 99-305: "Amendment of Parts 2 and 90 of the Commission's Rules to Allocate the 5.850–5.925 GHz Band to the Mobile Service for Dedicated Short-Range Communications of Intelligent Transportation Systems," FCC 99-305, 1999.

[FCC03] Federal Communications Commission (FCC) Report and Order 03-324: "Amendment of the Commission's Rules Regarding Dedicated Short-Range Communication Services in the 5.850–5.925 GHz Band (5.9 GHz Band)," FCC 03-324, 2003.

[FCC06] Federal Communications Commission (FCC) Memorandum Opinion and Order 06-110: "Amendment of the Commission's Rules Regarding Dedicated Short-Range Communication Services in the 5.850–5.925 GHz Band (5.9 GHz Band)," WT Docket 01-90, ET Docket 98-95, RM-9096, FCC 06-110, 2006.

[IEEE98] Institute of Electrical and Electronics Engineers (IEEE): "Telecommunications and Information Exchange between Systems — Local and Metropolitan Area Networks — Specific Requirements Part 2: Logical Link Control," IEEE 802.2, 1998.

[IEEE99] Institute of Electrical and Electronics Engineers (IEEE): "Telecommunications and Information Exchange between Systems — Local and Metropolitan Area Networks — Specific Requirements Part 11: Wireless LAN Medium Access Control (MAC) and Physical Layer (PHY) Specifications High-speed Physical Layer in the 5 GHz Band," IEEE 802.11a, 1999.

[IEEE06] Institute of Electrical and Electronics Engineers (IEEE): "Trial-Use Standard for Wireless Access in Vehicular Environments (WAVE)—Security Services for Applications and Management Messages," IEEE 1609.2, 2006.

[IEEE10a] Institute of Electrical and Electronics Engineers (IEEE): "Draft Standard for Wireless Access in Vehicular Environments (WAVE)—Networking Services," IEEE 1609.3/D5.0, 2010.

[IEEE10b] Institute of Electrical and Electronics Engineers (IEEE): "Draft Standard for Wireless Access in Vehicular Environments (WAVE)—Multi-Channel Operations," IEEE 1609.4/D6.0, 2010.

[IEEE10c] Institute of Electrical and Electronics Engineers (IEEE): "Telecommunications and Information Exchange between Systems — Local and Metropolitan Area Networks — Specific Requirements Part 11: Wireless LAN Medium Access Control (MAC) and Physical Layer (PHY) Spec.," IEEE 802.11p, 2010.

[Kenn11] J. Kenney: "Dedicated Short-Range Communications (DSRC) Standards in the United States," Proceedings of the IEEE, Vol. 99, no. 7, 2011.

[Post80] J. Postel: "User Datagram Protocol," Request For Comments 768, Internet Engineering Task Force, 1980.

[Post81] J. Postel: "Transmission Control Protocol," Request For Comments 793, Internet Engineering Task Force, 1981.

[TEA98] United States Congress: "Transportation Equity Act for the 21st Century," TEA-21, Public Law 105-178, 105th Congress, 1998.

6

WAVE PHYSICAL LAYER

6.1 PHYSICAL LAYER OPERATIONS

At the physical layer (PHY), the Institute of Electrical and Electronics Engineers (IEEE) 802.11 standard sets the specifications for frame transmissions and receptions over the air. This includes signal modulations, frame formats, and the way different elements of incoming frames need to be interpreted at the receivers.

This chapter focuses on PHY operations relevant to vehicle safety communications after a review of the most relevant IEEE 802.11 PHY mechanisms. We describe how the wireless access in vehicular environments (WAVE) PHY layer can be accurately modeled in software. We further describe a network simulator (ns) that we used to generate most of the results on WAVE radio performance presented in this book. Readers can find a comprehensive description of IEEE 802.11 standards and networks in the literature [Gast05].

6.1.1 Orthogonal Frequency Division Multiplexing

IEEE 802.11a radios use orthogonal frequency division multiplexing (OFDM) to transmit over the air. OFDM divides the spectrum into a number of narrowband subchannels, each carrying part of the information. Frequencies for each subchannel and transmission timings are chosen so that subchannel

Vehicle Safety Communications: Protocols, Security, and Privacy, First Edition. Luca Delgrossi and Tao Zhang.
© 2012 John Wiley & Sons, Inc. Published 2012 by John Wiley & Sons, Inc.

transmissions do not interfere with each other, hence the name "orthogonal frequency."

Each subchannel operates at low symbol rates, allowing interposition of guard intervals between subsequent symbols. These guard intervals make it possible to virtually eliminate intersymbol interference (ISI), resulting in high reliability at high data rates. OFDM is a simple yet efficient way to deal with signal distortion caused by multipath effects.

6.1.2 Modulation and Coding Rates

Data rates for IEEE 802.11 radios are determined by modulation schemes (variations in periodic waveforms) and coding rates adopted for data frame transmissions.

A modulation scheme is a method to vary periodic waveforms to convey information over the channel. Binary phase-shift keying (BPSK) modulation conveys 1 bit per periodic waveform per subcarrier. Since IEEE 802.11a radios concurrently transmit data over 48 subcarriers, BPSK modulation allows each OFDM symbol to carry 48 bits of information.

The coding rate represents the fraction of the total carried bits used for actual data bits. The rest are redundancy used to correct errors in the reception process. With a typical 20 MHz IEEE 802.11a channel, it takes 4 µs to send each periodic waveform, including the guard interval, to prevent intersymbol interference. The combination of BPSK modulation with 1/2 coding rate results in a 6 Mbps data rate for IEEE 802.11a radios.

As shown in Table 6.1, higher modulation and coding rate combinations result in higher data rates. However, higher data rates require cleaner signal at the receivers because frame reception is more prone to errors at higher data rates.

Table 6.1. IEEE 802.11a modulations and coding rates

Modulation	Coded Bits per Subcarrier	Coded Bits per OFDM Symbol	Coding Rate	Data Bits per OFDM Symbol	Data Rate for 20 MHz Channel (Mbps)
BPSK	1	48	1/2	24	6
BPSK	1	48	3/4	36	9
QPSK	2	96	1/2	48	12
QPSK	2	96	3/4	72	18
16-QAM	4	192	1/2	96	24
16-QAM	4	192	3/4	144	36
64-QAM	6	288	2/3	192	48
64-QAM	6	288	3/4	216	54

QPSK, quadrature phase-shift keying; QAM, quadrature amplitude modulation.

6.1.3 Frame Reception

6.1.3.1 Frame Preamble and Physical Layer Convergence Procedure Header An IEEE 802.11 sender may select any modulation and coding rate combinations defined in the standard to transmit a frame. A receiver needs to know whether the signal detected through the channel corresponds to a frame or just noise, the frame duration length, and what modulation and coding rate combination is being used by the sender, in order to successfully receive a frame.

Every IEEE 802.11 frame starts with a known bit sequence called the frame preamble. The purpose of the frame preamble is to notify receivers of the imminent arrival of a frame and assist them to lock on to the signal. The frame preamble is followed by the physical layer convergence procedure (PLCP) header, which contains details on the frame payload (frame body), including frame length, modulation, and coding rate.

The frame preamble and the Signal portion of the PLCP header are BPSK modulated in almost all IEEE 802.11 radio configurations. While the frame preamble has a zero coding rate, the Signal portion of the PLCP header is coded at 1/2 rate.

The frame payload follows the PLCP header and can be modulated according to any of the eight modulation and coding combinations illustrated in Table 6.1. The sender indicates which modulation and coding combination is to be used to demodulate the frame payload in the Rate data field of the PLCP header (Figure 6.1).

6.1.3.2 Frame Body When an IEEE 802.11 radio is listening for incoming frames, it continuously looks for the known pattern of the frame preamble by demodulating the received signal using BPSK. When the frame preamble pattern is detected, the receiver attempts to decode the PLCP header. Upon success, the receiver demodulates incoming waveforms according to the frame modulation, coding, and duration indicated in the PLCP header. The resulting raw bits are passed to the media access control (MAC) layer, where a cyclic

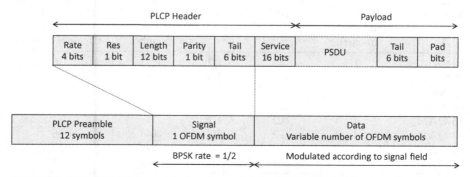

Figure 6.1. OFDM PLCP preamble, header, and data. PSDU, physical layer service data unit.

redundancy check (CRC) determines whether the frame was successfully received.

While a radio is transmitting over the channel, it is not able to receive any incoming frames. Also, the radio would not be able to receive any incoming frames if it missed the frame preamble and PLCP header for any reason, because it would have no information on how to receive the frame body.

Similarly, when a radio is already receiving a frame, it is not able to receive another incoming frame, because it would treat the preamble of the new incoming frame as part of the frame it is already demodulating. Furthermore, if the new incoming frame has a strong enough signal, it will collide with the earlier frame and prevent the earlier frame's successful reception.

Also, since the frame body can be transmitted with a different modulation and coding rate combination, a receiver might be able to successfully receive and decode the frame preamble and PLCP header, yet fail to receive the frame body, depending on signal quality and interference.

6.1.3.3 Frame Body Capture IEEE 802.11 specifies a robust process for searching for and decoding the frame preamble and PLCP header. If a new frame arrives when a radio is still receiving the frame preamble and PLCP header of an earlier frame, the radio may choose to lock onto the new frame if it has sufficiently higher power than the earlier frame. Furthermore, it is sometimes possible to capture a new incoming frame during the body reception of an earlier one. This mechanism, called frame body capture, is not part of the IEEE 802.11 standard but is implemented in some radio chipsets as an optional feature.

With frame body capture, the PHY continuously monitors the received signal strength. When a sudden sharp rise (e.g., greater than 10 dB) is detected, the receiver assumes the arrival of a new frame with a stronger signal. It then abandons the previous frame and attempts to decode the frame preamble and PLCP header of the new incoming frame. Upon success, it starts reception of the new frame.

The frame body capture mechanism is intended for typical Wi-Fi usage environments, such as homes and offices. Incoming frames associated with stronger signals are preferred over others because they are more likely originated by nearby nodes. Frame body capture can be useful for vehicle safety communications as well. A vehicle may prefer messages from closer vehicles over messages from vehicles that are further away, because nearby vehicles will more likely impose immediate risks than faraway vehicles. Frame body capture techniques can also be used to reduce the negative impact of hidden terminal effects.

6.2 PHY AMENDMENTS

The IEEE 802.11p PHY is an extension to the IEEE 802.11a PHY. It specifies minimal changes required for operations at 5.9 GHz. This allows us to build

IEEE 802.11p chipsets without the high costs associated with the development of new radio chipsets. Furthermore, since IEEE 802.11a radios operate at 5.8 GHz, it is possible to build IEEE 802.11p radios with existing IEEE 802.11a chipsets reconfigured to operate in the nearby 5.9 GHz band.

6.2.1 Channel Width

The IEEE 802.11p PHY is based on the same OFDM scheme as IEEE 802.11a, but it operates over 10 MHz channels instead of the 20 MHz channels used by IEEE 802.11a devices. Ten megahertz (10 MHz) channels can be implemented by doubling the OFDM timing parameters and halving the frequency parameters used for 20 MHz IEEE 802.11a transmissions. This also entails doubling the size of the guard intervals between transmitted symbols.

Downscaling channel width is motivated by a recent study demonstrating that the guard interval used for transmitting over a 20 MHz channel is not long enough to offset the worst-case excess delay spread introduced by multipath effects in vehicular environments [CHCS08]. That is, the guard interval is not long enough to prevent intersymbol interference within one radio's own transmissions.

The excess delay spread varies with the environment. In a suburban environment, empirical measurements show that 90% of the excess delays introduced by multipath effects are lower than 0.6 µs. This value increases to 1.4 µs on highways and reaches 1.5 µs in rural environments. Since the OFDM guard interval needs to be longer than the excess delay spread, the 0.8 µs guard interval defined for IEEE 802.11a over 20 MHz channels would not be sufficient in a vehicle communications environment. For this reason, IEEE 802.11p uses a channel width of 10 MHz with 1.6 µs guard intervals.

[SCHB07] indicate 8.5 MHz as the theoretical optimal channel width for WAVE, offering highest protection against excess delay spread. However, for ease of implementation with existing IEEE 802.11a chipsets, using 10 MHz wide channels in WAVE is a reasonable choice.

6.2.2 Spectrum Masks

Spectrum masks are used to limit excessive radiation from transmitting radios at frequencies beyond the intended bandwidth. Spectrum masks aim to protect adjacent channels from interference.

IEEE 802.11p amendments specify four masks for the 5.9 GHz DSRC spectrum, corresponding to class A, B, C, and D operations. Class C is expected to be adopted for vehicle safety communications. Table 6.2 indicates spectrum mask requirements for each class at different offsets from channel center. For each 10 MHz channel, the transmitted spectrum shall have a 0 dBr bandwidth not exceeding 9 MHz [FCC04].

The spectrum masks defined for WAVE are more stringent than the ones for current IEEE 802.11 radios.

Table 6.2. WAVE spectrum mask requirements

Class	Limit at 5 MHz Offset (dBr)	Limit at 5.5 MHz Offset (dBr)	Limit at 10 MHz Offset (dBr)	Limit at 15 MHz Offset (dBr)
A	−10	−20	−28	−40
B	−16	−20	−28	−40
C	−26	−32	−40	−50
D	−35	−45	−55	−65

6.2.3 Improved Receiver Performance

Vehicles in adjacent lanes have been shown to interfere with each other if they are operating in two adjacent channels [RBKL07]. For example, vehicle V_1 transmitting over channel 176 could cause interference, preventing vehicle V_2 in the adjacent lane (2.5 m apart) from receiving messages sent over channel 178 by another vehicle V_3.

Cross-channel interference is a well-known issue in wireless communications. The most effective and proper solution to this problem is through channel management policies. The definition of these policies is outside of the scope of IEEE 802.11 protocols. Nevertheless, given the even higher relevance of this issue in vehicular environments, IEEE 802.11p recommends improved receiver performance for adjacent channel rejections.

6.3 PHY LAYER MODELING

An essential tool for the study of vehicle safety communications is a simulator in software able to accurately model vehicle network operations. For our simulations, we relied on an overhauled version of the network simulator 2 (ns-2), one of the most popular tools adopted by researchers [KFKV10].

The first version of the ns was created in 1989 with support from the Defense Advanced Research Projects Agency (DARPA). Version 2 (ns-2) became available in 1996. Ns-2 is a discrete event simulation tool mainly built to simulate Transmission Control Protocol/Internet Protocol (TCP/IP), routing, and multicast protocols over wired and wireless networks.

The original implementation of IEEE 802.11 networks in ns-2 was strongly influenced by Sun Microsystems, University of California, Berkeley, and Carnegie Mellon University. This platform has been widely used by wireless networks researchers. Although this implementation served its main purpose well, researchers generally agree that it had limited accuracy in simulating radio behaviors [SLTH06]. For many years, this was not considered critical because it was used primarily for studying higher layer protocols rather than radio links. In the context of 5.9 GHz DSRC, however, it became essential to accurately simulate radio layer effects such as signal fading and interference due to hidden terminal effects.

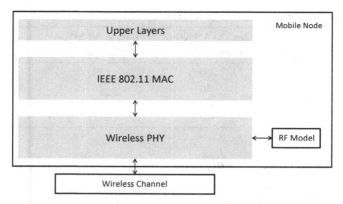

Figure 6.2. The ns-2 IEEE 802.11 simulation architecture

The desire to study vehicle safety communications at greater depth generated a series of efforts aimed at enhancing the ns-2 software modules that simulate IEEE 802.11 networks. In 2007, researchers from the University of Karlsruhe in Germany and Mercedes Benz Research & Development North America in Palo Alto, California, produced a new implementation based on a complete redesign of the PHY and MAC layer simulation models [CSJT07]. Over time, these new modules replaced the original modules and have now become part of the official ns-2 software distribution [NS-2HP].

In 2008, network simulator 3 (ns-3) was released. It features by default the new modules developed to accurately simulate IEEE 802.11. Today, this software is freely available from the ns-3 Web site [NS-3HP].

6.3.1 Network Simulator Architecture

In the ns-2 IEEE 802.11 simulation architecture (Figure 6.2), the over-the-air link is modeled by the wireless channel module.

The wireless channel module interconnects all the nodes in a simulation scenario and passes frames from each sender to all receivers. The wireless channel specifies the location and power associated with each transmission event, but it does not handle interference, collisions, or path loss calculations.

The wireless PHY module handles all physical layer operations, such as carrier sensing, signal-to-interference-noise ratio (SINR) tracking, and PLCP state management. Its design can support different MAC designs. Wireless PHY interacts with the radiofrequency (RF) model object, which computes received signal strength for each incoming transmission using one of the available RF propagation models and the node's position.

Wireless PHY takes each incoming frame from the wireless channel and requests the corresponding received power from the RF model. If the received power is greater than the carrier sensing threshold, the frame is simply passed above to 802.11 MAC, where further frame handling takes place. When wireless PHY receives a packet from 802.11 MAC, it passes it over to the wireless channel for transmission.

6.3.2 RF Model

The RF model module defines the RF signal propagation model. The ns-2 implementation currently offers four propagation models: free space, two-ray ground, shadowing, and Nakagami [Naka60].

Free space is the simplest model. It assumes that the radiated energy is spread uniformly and without loss over the surface of a sphere surrounding the antenna. In ns-2, the received power is calculated through the Friis equation (Eq. 6.1), which describes signal attenuation as proportional to the square of the distance:

$$P_r = \frac{P_t G_t G_r \lambda^2}{(4\pi d)^2},$$
(6.1)

where P_t is the transmission power, G_t and G_r are antenna gains, λ is the wavelength, and d is the distance between receiver and transmitter.

The two-ray ground model considers earth surface effects affecting signal propagation. In fact, beyond a certain crossover distance, the ground-reflected wave can partially cancel the line-of-sight wave and lead to faster degradation of the received power. This distance depends on the wavelength and respective heights of the transmitter and receiver. If a receiver is within the limits of the crossover distance, ns-2 calculates the received power through the Friis equation, just as with the free space model. For receivers beyond the crossover distance, however, the received power is calculated through the Rappaport equation (Eq. 6.2):

$$P_r = \frac{P_t G_t G_r (h_t h_r)^2}{d^4},$$
(6.2)

where h_t and h_r are the transmitter and receiver antenna heights, respectively.

The crossover distance, also known as Fresno distance, is calculated as follows:

$$d_c = \frac{4\pi h_t h_r}{\lambda}.$$
(6.3)

The main shortcoming of the free space and two-ray ground models is that they do not consider RF signal fluctuations. With these models, given a certain transmission power, distance, and heights of transmitter and receiver, the resulting received power is deterministic. Since reception and interference ranges become perfect circles, these RF models are often referred to as disc models. Deterministic RF models are usually sufficient for the study of sustained communications, because they provide the average signal quality over time. When studying shorter-term signal quality, however, RF signal fluctuations due to shadowing obstacles or multipath effects need to be taken into account.

The shadowing model is an improved RF model that accounts for received signal fluctuations caused by large-scale terrain features and obstacles in the vicinity of the antenna. The ns-2 implementation applies a log-normal distribution to the area-mean result calculated through the Friis equation. The probability density function for local power fluctuation is:

$$Prob(P_{r\_dBm}) = \frac{1}{\sqrt{2\pi}\sigma} e^{\left(-\frac{P^2_{r\_dBm}}{2\sigma^2}\right)}, \tag{6.4}$$

where P_{r_dBm} is the logarithmic representation of P_r in decibel-milliwatts and σ is the standard deviation.

Multipath fading is caused by reflected and scattered waves reaching the receiver with potentially large delay spreads. Multipath effects cause unpredictable fluctuations of the received signals. Signal fading introduced by multipath effects can be represented through probabilistic models such as the Nakagami distribution [Naka60]. The Nakagami distribution, related to the Γ distribution, has two parameters: m controls fading, whereas ω controls spread. The ns-2 implementation allows us to independently customize variance and area-mean for close, median, and far field propagation. The probability density function for local power fluctuation is:

$$Prob(P_r) = \left(\frac{m}{\Omega}\right)^m \left(\frac{P_r^{m-1}}{\Gamma(m)}\right) e^{-\frac{m}{\Omega}}, \tag{6.5}$$

where Ω is the local average of P_r, and m is the Nakagami fading parameter.

For moving stations, wave frequencies are subject to changes known as Doppler shifts that depend on the relative velocity between the source and the receiver. The Doppler shift frequency variation can be calculated as:

$$\Delta f = \frac{v}{\lambda}, \tag{6.6}$$

where v is the velocity of the transmitter relative to the receiver and λ is the wavelength.

The maximum Doppler shift for vehicle communications is about 2 KHz. Such a relatively small shift results in channel coherence times (i.e., the time intervals during which a channel maintains unvaried properties), which are usually longer than frame transmission durations. In other words, the amplitude and phase shifts imposed by the channel can be considered roughly constant during the transmission of a single frame. Based on these considerations, we can use a single received power value to represent the reception quality of the whole frame. In case of significantly longer frames, it would be necessary to use multiple values to represent the received powers associated with each individual frame.

6.3.3 Wireless PHY

Wireless PHY consists of two separate modules: the PHY power monitor and the PHY state manager. The former keeps track of received RF signals from transmissions, while the latter maintains PLCP states.

6.3.3.1 PHY Power Monitor   The PHY power monitor models the physical media dependent (PMD) sublayer. The PMD is the only entity that directly interacts with the analog RF signals. All information on received signals is processed and managed in this module.

The power monitor keeps track of noise and interference experienced by individual nodes. When the cumulative interference and noise level rise above the carrier sensing threshold, the power monitor notifies 802.11 MAC of physical carrier sensing status changes. A node's own transmissions are treated as carrier sensing busy.

6.3.3.2 PHY State Manager   The PHY state manager models the PLCP sublayer, keeping track of how wireless PHY switches among four operating modes. A state machine for the PHY state manager is represented in Figure 6.3.

The wireless PHY is in Searching state when it is neither in transmission nor reception of a frame. In this state, wireless PHY evaluates each transmission event notification from the wireless channel for potentially receivable

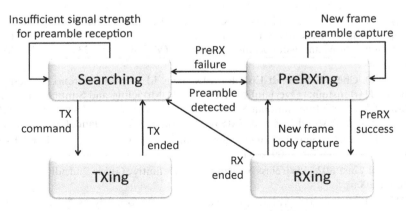

Figure 6.3. PHY state machine

frames. If a frame arrives with sufficient signal strength for frame preamble detection (i.e., SINR exceeds the BPSK threshold), wireless PHY moves into the PreRXing state.

The wireless PHY stays in PreRXing state for the duration of frame preamble and PLCP header reception. If the SINR stays above the BPSK threshold throughout this entire period, the wireless PHY moves into the RXing state.

While in the PreRXing state for one frame, if a new incoming frame has sufficient signal strength to prevent proper frame preamble and PLCP header reception for the current frame, the wireless PHY goes back to Searching. However, if the new frame has sufficiently higher signal strength for its own preamble to be detected, wireless PHY triggers frame preamble capture, staying in PreRXing state for the duration of the new frame.

In the RXing state, the wireless PHY handles the reception of the frame body. It monitors the SINR throughout frame body duration. If, at any time while in this state the SINR drops below the threshold required by the modulation and coding rate combination associated with the frame body, the wireless PHY marks the frame with an error flag.

When the frame body capture feature is enabled, it is possible for a later arriving frame to trigger the wireless PHY to move back to PreRXing to capture the new frame.

A transmit command issued from 802.11 MAC will move the wireless PHY into the TXing state for the duration of the frame transmission. The 802.11 MAC module should never issue a transmit command while the wireless PHY is in PreRXing, RXing, or TXing state. If a new frame comes in from the wireless channel when wireless PHY is in TXing state, it is ignored and only tracked by the power monitor as interference.

REFERENCES

[CHCS08] L. Cheng, B. E. Henty, R. Cooper, D. D. Stancil, and F. Bai: "Multi-Path Propagation Measurements for Vehicular Networks at 5.9 GHz," IEEE Wireless Communications and Networking Conference (WCNC), pp. 1239–1244, Las Vegas, NV, 2008.

[CSJT07] Q. Chen, F. Schmidt-Eisenlohr, D. Jiang, M. Torrent-Moreno, L. Delgrossi, and H. Hartenstein: "Overhaul of IEEE 802.11 Modeling and Simulation in ns-2," 10th ACM/IEEE International Symposium on Modeling, Analysis and Simulation of Wireless and Mobile Systems (MSWiM), Chania, Crete Island, Greece, 2007.

[FCC04] Federal Communications Commission:"Code of Federal Regulations CFR47," Sections 90.377 and 95.1509, 2004.

[Gast05] M. Gast:802.11 Wireless Networks:The Definitive Guide,2nd edition,O'Reilly Media, 2006.

[KFKV10] K. Fall and K. Varadahn: "The ns Manual (formerly ns Notes and Documentation)", The VINT Project. Available at: <http://www.isi.edu/nsnam/ns>, 2010.

[Naka60] M. Nakagami: "The m-distribution, a General Formula of Intensity Distribution of the Rapid Fading," Statistical Methods in Radio Wave Propagation, W. C. Hoffmann, Ed. Oxford, England, 1960.

[NS-2HP] Network simulator ns-2 home page: Available at: <http://isi.edu/nsnam/ns/>, 2011.

[NS-3HP] Network simulator ns-3 home page: Available at: <http://www.nsnam.org/>, 2011.

[RBKL07] V. Rai, F. Bai, J. Kenney, and K. Laberteaux: "Cross-Channel Interference Test Results: A report from the VSC-A project," IEEE 802.11 Task Group p report, DCN 11-07-2133, 2007.

[SCHB07] D. Stancil, L. Cheng, B. Henty, and F. Bai: "Performance of 802.11p Waveforms over the Vehicle-to-Vehicle Channel at 5.9 GHz," IEEE 802.11 Task Group p Report, 2007.

[SLTH06] F. Schmidt-Eisenlohr, J. Letamendia-Murua, M. Torrent-Moreno, and H. Hartenstein: "Bug Fixes on the IEEE 802.11 DCF Module of the Network Simulator ns-2.28," Department of Computer Science, University of Karlsruhe, Technical Report TR-2006-1, 2006.

7

WAVE MEDIA ACCESS CONTROL LAYER

7.1 MEDIA ACCESS CONTROL LAYER OPERATIONS

At the media access control (MAC) layer, the Institute of Electrical and Electronics Engineers (IEEE) 802.11 standard specifies channel access operations using carrier sensing multiple access with collision avoidance (CSMA/CA). This scheme includes a back-off algorithm and a handshake protocol designed to reduce frame collisions introduced by contention for media access and to address hidden terminal effects. The MAC layer filters out incoming frames that are not meant for the receiving node.

7.1.1 Carrier Sensing Multiple Access with Collision Avoidance

Using CSMA/CA, a station must sense channel status before sending a frame. As long as the medium is busy, the MAC layer must wait for the medium to become idle again. The station may only transmit the frame if the medium is continuously idle for a distributed coordination function interframe space (DIFS) interval plus an additional time randomly calculated by the MAC layer back-off algorithm. This random back-off delay is intended to reduce the likelihood of collisions when multiple stations are competing for medium access. If the channel is found busy during the DIFS interval or the back-off delay, the station should defer its transmission.

Vehicle Safety Communications: Protocols, Security, and Privacy, First Edition. Luca Delgrossi and Tao Zhang.
© 2012 John Wiley & Sons, Inc. Published 2012 by John Wiley & Sons, Inc.

The CSMA/CA scheme does not require any coordination among nodes, is simple to implement, and offers good performance over moderately loaded channels.

The IEEE 802.11 MAC layer adopts enhanced distributed channel access (EDCA) to offer four different traffic classes or access categories. EDCA permits to differentiate services and prioritize outgoing traffic at the MAC layer. Traffic prioritization is mainly provided through the adoption of arbitration interframe spacing (AIFS) intervals, which vary with the access category. The IEEE 802.11 MAC layer maintains separate ECDA queues, one for each access category. Frames assigned to these queues compete with each other for access to the medium.

7.1.2 Hidden Terminal Effects

The hidden terminal problem arises when two nodes cannot sense each other's signals directly, but can both sense the signals from a third node. Under these conditions, the two nodes cannot determine whether the other node is transmitting. When hidden terminal effects are present, the CSMA/CA scheme alone cannot effectively mitigate frame collisions over the channel.

Figure 7.1 illustrates an example of hidden terminal effect in a vehicular ad hoc network. In this example, both vehicle A and vehicle C can sense vehicle B. However, they cannot sense each other. Before transmitting a frame, vehicle A must sense the channel first to make sure no other vehicles are currently transmitting. Since vehicle A cannot sense vehicle C, it may assume that the channel is clear when, in fact, vehicle C is transmitting. Hidden terminals permit simultaneous attempts to access the medium, resulting in collisions over the channel.

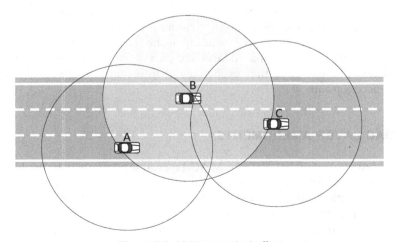

Figure 7.1. Hidden terminal effect

The IEEE 802.11 MAC layer may use a handshake protocol to reduce frame collisions caused by hidden terminals. Before transmitting a data frame, a node sends a Request to Send (RTS) frame indicating the intended destination. The destination of the RTS frame replies with a Clear to Send (CTS) frame. Any other nodes receiving the RTS or CTS frame should wait before attempting transmissions over the channel for a time duration indicated by the RTS and CTS frames.

In the previous example, vehicle A sends an RTS frame. Vehicle B replies with a CTS frame, indicating that the channel is clear. Vehicle C receives the CTS frame from B and therefore waits before attempting any further transmissions.

The RTS/CTS handshake only takes place before the transmission of packets exceeding a certain threshold, whereas smaller packets are sent immediately. It can be viewed as virtual carrier sensing that enables a node to sense another node when the two nodes cannot receive each other's signals directly. The RTS/CTS handshake is not used for broadcast and thus is unlikely to have an impact on most vehicle safety communications.

7.1.3 Basic Service Set

An IEEE 802.11 basic service set (BSS) is a group of radios configured to communicate with each other over the air-link. Transmissions from outsider BSSs are filtered out. An infrastructure BSS contains a station serving as the access point for the group. The IEEE 802.11 ad hoc mode, which does not require an access point, is called independent BSS (IBSS).

Each BSS is identified with a 48-bit basic service set identification (BSSID) at the MAC layer. For an infrastructure BSS, the BSSID will be the 48-bit MAC address of the access point. A special case of the BSSID is the wildcard BSSID, where all bits are set to "1."

Joining a BSS requires executing time-consuming operations such as scanning, authentication, and association. Scanning can be passive (listening to discover beacon frames) or active (sending probe request frames). Authentication may require the use of shared keys and encryption functions. The association process consists of further exchanges of information between the node and the access point.

7.2 MAC LAYER AMENDMENTS

The key purpose of IEEE 802.11p MAC layer amendments is to enable efficient communication setups while significantly reducing the associated overheads. Unlike physical layer (PHY) amendments, modifications to the MAC layer typically result in changes only to the radio software. This gives us relatively more freedom in the design of these amendments.

Traditional IEEE 802.11 MAC layer operations are too time consuming for wireless access in vehicular environments (WAVE). Many WAVE use cases cannot tolerate the delays introduced by operations such as scanning for BSS beacons, authentication, and association.

A key MAC amendment introduced to address this issue is the ability to send data frames outside the context of a BSS (OCB). This mechanism is available to WAVE radios that are not members in a BSS. These stations are allowed to transmit and receive data frames using a wildcard BSSID value without any previous setups. This means vehicles can immediately communicate with each other without any additional overheads as long as they operate in the same dedicated short-range communications (DSRC) channel and use the wildcard BSSID. A complete description of OCB communications in WAVE can be found in [Kenn11].

7.3 MAC LAYER MODELING

In the network simulator 2 (ns-2) implementation we have developed, MAC layer operations are modeled through six separate modules: Transmission, Reception, Channel State Manager, Back-Off Manager, Transmission Coordination, and Reception Coordination [CSJT07]. The designs of these modules reflect considerations and abstractions derived from the IEEE 802.11 standards. Figure 7.2 illustrates the modules used for modeling MAC layer operations in this ns-2 implementation as well as the interactions among them and with the PHY layer modules.

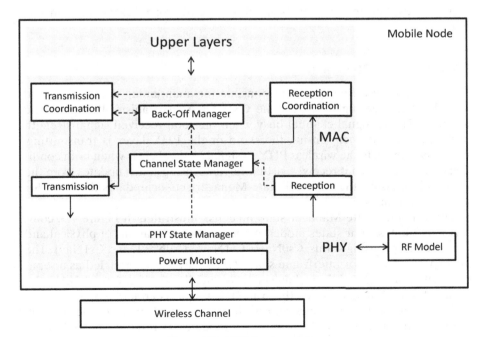

Figure 7.2. MAC layer modeling

7.3.1 Transmission

The Transmission module is the MAC layer interface to wireless PHY. It passes frames to the PHY layer for transmission. These frames include RTS and data frames from Transmission Coordination and acknowledgment (ACK) and CTS frames from Reception Coordination. The Transmission module can be in one of two states: idle and transmitting.

7.3.2 Reception

The Reception module completes the frame reception process initiated by wireless PHY. For each received frame, the Reception module performs a cyclic redundancy check (CRC) to verify that the frame was successfully received. The standard requires a node that just received a bad or unknown frame to wait for an extended interframe space (EIFS) interval. Since this is longer than the usual DIFS, the Reception module informs the Channel State Manager.

The Reception module applies address filtering on all successfully received frames and discards frames not intended for the node. Any frames that do not pass address filtering are examined before being discarded to see if they contain a network allocation vector (NAV) duration value. The NAV duration value tells a node how long it must delay its transmissions. When a NAV duration value is found, the Reception module passes it to the Channel State Manager.

The Reception module can be in either the idle or receiving state at any time.

7.3.3 Channel State Manager

The Channel State Manager is responsible for maintaining the physical and virtual carrier sensing status for the CSMA/CA mechanism. It expects the wireless PHY to signal channel busy when the total received signal strength rises above the carrier sensing threshold or the PHY layer is transmitting. Similarly, it expects the wireless PHY to signal channel clear when both conditions are gone. When it receives virtual carrier sensing status updates from the Reception module, the Channel State Manager sets or updates the NAV for the specified duration.

The Channel State Manager state machine, illustrated in Figure 7.3, comprises five states. Four states model the possible combinations of physical and virtual carrier sensing status: Cs0Nav0, Cs1Nav0, Cs0Nav1, and Cs1Nav1. The time spent waiting for interframe space (IFS) duration is modeled as a separate state. Since the IFS mechanism is essentially a self-enforced NAV, the Channel State Manager treats the Wait IFS state as channel busy.

The Channel State Manager returns the physical and virtual carrier sensing status in response to queries from other modules. It returns carrier sensing

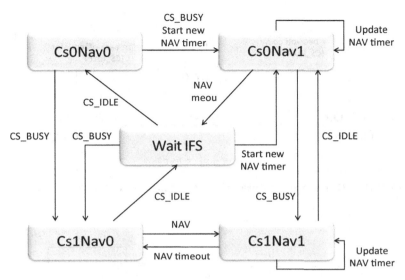

Figure 7.3. Channel State Manager state machine

idle when in CsONav0 and carrier sensing busy otherwise. The Channel State Manager actively signals the Back-Off Manager when it moves in or out of CsONav0 to indicate channel status changes. In turn, the Back-Off Manager resumes its back-off process or suspends an ongoing one.

7.3.4 Back-Off Manager

The Back-Off Manager maintains the back-off counter for the IEEE 802.11 MAC collision avoidance mechanism. Initially, the back-off counter is set to a random value, then decremented over time when the channel is idle. When the back-off counter reaches zero, the node transmits the packet.

The Back-Off Manager state machine, illustrated in Figure 7.4, comprises three states: No Back-Off, Back-Off Running, and Back-Off Pause. Each time the Channel State Manager signals carrier sensing idle, the Back-Off Manager decrements the back-off counter. When the Channel State Manager signals carrier sensing busy, decrementing the back-off counter is suspended until the next carrier sensing idle signal is received.

When the back-off counter reaches zero, the Back-Off Manager moves into the No Back-Off state and signals back-off zero to Transmission Coordination.

The Back-Off Manager assists the Transmission Coordination to run both the regular back-off and post-transmission back-off procedures, but is not aware of the difference between the two.

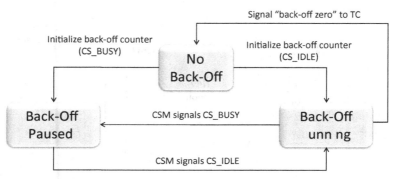

Figure 7.4. Back-Off Manager state machine. CSM, Channel State Manager; TC, Transmission Coordination

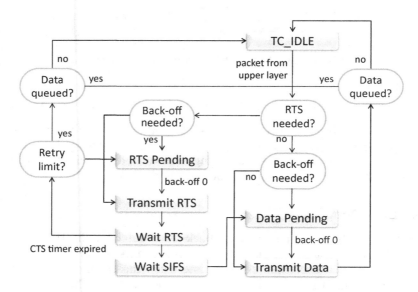

Figure 7.5. Transmission Coordination state machine

7.3.5 Transmission Coordination

The Transmission Coordination module manages medium access for packet transmission requests coming from the upper layer.

A simplified state machine for Transmission Coordination is represented in Figure 7.5. If the data frame to be transmitted is for broadcast or has a smaller size than the RTS threshold, only the states on the right side of the figure are activated before the data frame is sent. Otherwise, the RTS/CTS handshake becomes necessary and also, the states on the left side are activated.

When Transmission Coordination leaves the TC_IDLE state because of a packet transmission request coming from the upper layer, it first checks if an RTS frame should be generated. Then, it starts a back-off process at the Back-Off Manager (unless there is already one running) and moves into RTS Pending or Data Pending according to the RTS decision.

When Transmission Coordination is in RTS Pending or Data Pending, it instructs Transmission to send the RTS or data frame, respectively, as soon as it receives the signal indicating back-off zero from the Back-Off Manager. It is occasionally possible to bypass the RTS Pending or Data Pending state. For instance, if there is no back-off process running and the Channel State Manager reports carrier sensing idle, the Transmission Coordination can immediately transmit the RTS or Data frame. This is because the standard allows a radio to start transmissions right away if it has previously completed a post-transmission back-off cycle and the channel has been idle for more than DIFS.

When the transmission of an RTS frame completes, Transmission Coordination moves into the Wait CTS state and starts the CTS timer. If Reception Coordination does not signal the arrival of a CTS frame before this timer expires, it starts a new back-off process and moves back to the RTS Pending state. The Transmission Coordination repeats this process until a retry limit is reached. If the CTS frame arrives in due time, Transmission Coordination waits for SIFS duration before instructing Transmission to send the data frame.

After each unicast data frame transmission, Transmission Coordination moves into the Wait ACK state (not represented in Figure 7.5) and starts the ACK timer. If it does not receive an ACK reply indication from Reception Coordination before this timer expires, it starts a back-off process and moves back to RTS Pending or Data Pending state, respectively. Therefore, the attempted transmission of a data frame has three possible outcomes:

- it is a broadcast frame and it is transmitted once over the air,
- it is a unicast frame and the MAC layer has received the correspondent ACK, or
- the retry limit is reached.

In all these three cases, Transmission Coordination starts a post-transmission back-off process. Then, if it finds another packet in the send queue, it moves immediately into the RTS Pending or Data Pending state. Otherwise, it returns to the TC_IDLE state.

7.3.6 Reception Coordination

The Reception Coordination module receives from Reception filtered frames that are destined to the local node. It signals Transmission Coordination upon the arrival of CTS and ACK frames. It is responsible to generate CTS and ACK responses upon the arrival of RTS frames.

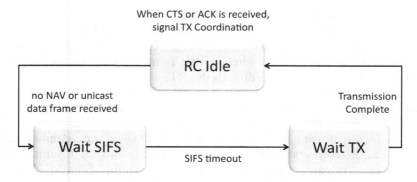

Figure 7.6. Reception Coordination state machine

As shown in Figure 7.6, the Reception Coordination state machine has three states: RC_IDLE, Wait SIFS, and Wait TX. The Reception Coordination module stays in the RC_IDLE state while waiting for control and data frames received from the Reception module.

When an RTS frame arrives, Reception Coordination queries the Channel State Manager to learn about the current NAV status. If the response indicates an active NAV, the RTS frame is discarded. Otherwise, Reception Coordination creates a CTS frame and moves into the Wait SIFS state with the SIFS timer set. When this timer expires, it immediately instructs Transmission to send the CTS frame. It then moves into the Wait TX state for the duration of the frame transmission before returning to TC_IDLE.

When a unicast data frame arrives, Reception Coordination creates an ACK frame, moves into the Wait SIFS state, and sets the SIFS timer. When this timer expires, it immediately instructs Transmission to send the ACK frame. Upon arrival of a CTS or ACK frame, Reception Coordination signals Transmission Coordination accordingly.

The Reception Coordination module passes data frames to the upper layer. It discards duplicate data frames possibly generated by the unicast retransmission mechanism.

7.4 OVERHAULED NS-2 IMPLEMENTATION

The need to model IEEE 802.11p PHY and MAC operations more accurately led to the development of an overhauled ns-2 implementation, which is now part of the official ns-2 code distribution [CJTD07]. Here, we show simulation results on frame reception rates and illustrate how the results obtained using the overhauled ns-2 differ from those obtained using the previous version ns-2.31.

Let us consider three scenarios with 133, 400, and 1000 cars, respectively. They represent moderate, dense, and extremely dense traffic. Let us assume that each vehicle broadcasts frames at 10 Hz using a transmission power level to achieve a theoretical 250-m communications range. Frame size is 250 bytes. Radiofrequency (RF) propagation is modeled with a Nakagami distribution.

The overhauled ns-2 implementation allows us to model greater details of frame transmission and reception operations [CSJT07]. The simulations are conducted with three different configurations: all capture features turned off (CapDisabled); preamble capture turned on (PreRXCap); and preamble and frame body captures both turned on (PreRX + DataCap). The results are compared with the original ns-2.31 implementation (Orig.NS-2.31). Figure 7.7 shows simulations results for average broadcast frame reception rates at various distances from the sender.

The original ns-2.31 implementation produces higher reception rates at longer distances but lower reception rates at shorter distances [CJTD07]. This is due to the less accurate MAC modeling in ns-2.31, which allows continuous

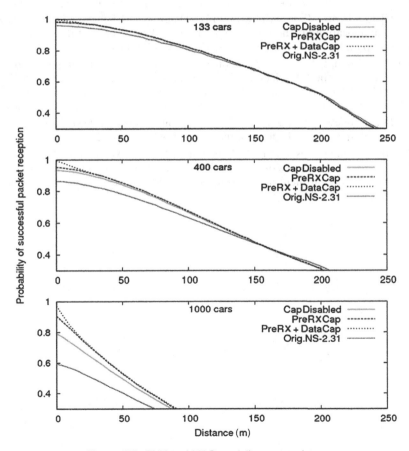

Figure 7.7. PHY and MAC modeling comparisons

incoming frames to extend collision state over long-time intervals [SLTH06]. Also, Figure 7.7 highlights the differences introduced by the preamble and frame body capture features. Both capture features significantly influence the reception rates at ranges close to the sender. In general, these differences are more pronounced for denser communication scenarios with increasing hidden terminal effects.

An experiment conducted with tens of real DSRC radios has confirmed that results obtained through the overhauled ns-2 implementation closely match field measurements [BaKW11].

REFERENCES

[BaKW11] G. Bansal, J. Kenney, and A. Weinfield: "Cross-Validation of DSRC Radio Testbed and NS-2 Simulation Platform for Vehicular Safety Communications," IEEE Vehicular Technology Conference (VTC Fall), San Francisco, CA, 2011.

[CJTD07] Q. Chen, D. Jiang, V. Taliwal, and L. Delgrossi: "IEEE 802.11 Based Vehicular Communication Simulation Design for NS-2," Proceedings of the International Workshop on Vehicular Ad Hoc Networks (VANET), Los Angeles, CA, 2007.

[CSJT07] Q. Chen, F. Schmidt-Eisenlohr, D. Jiang, M. Torrent-Moreno, L. Delgrossi, and H. Hartenstein: "Overhaul of IEEE 802.11 Modeling and Simulation in NS-2," 10th ACM/IEEE International Symposium on Modeling, Analysis and Simulation of Wireless and Mobile Systems (MSWiM), Chania, Crete Island, Greece, 2007.

[Kenn11] J. Kenney: "Dedicated Short-Range Communications (DSRC) Standards in the United States," Proceedings of the IEEE, Vol. 99, No. 7, 2011.

[SLTH06] F. Schmidt-Eisenlohr, J. Letamendia-Murua, M. Torrent-Moreno, and H. Hartenstein: "Bug Fixes on the IEEE 802.11 DCF module of the Network Simulator ns-2.28," Department of Computer Science, University of Karlsruhe, Technical Report TR-2006-1, 2007.

8

DSRC DATA RATES

Written in cooperation with Qi Chen and Daniel Jiang

8.1 INTRODUCTION

In traditional wireless networks, it is a common practice to adopt the lowest supported data rate for broadcast transmissions. This approach is justified by the fact that broadcasting represents an extremely small fraction of the overall traffic over these networks. The choice of the lowest data rate, which provides robustness at the expense of broadcast capacity, aims at maximizing the number of destinations experiencing successful frame receptions.

Vehicle safety communications, however, are characterized by a significantly different traffic pattern, where pervasive broadcasting dominates the traffic [JTMH06] [XMKS04]. Consequently, hidden terminal effects are amplified and it becomes essential to limit interference over the channel. The communication data rate is directly linked to interference. Higher data rates result in shorter frame durations generating lower levels of interference. However, to achieve comparable reception rates, higher data rates require higher transmission powers, leading to higher levels of interference. For these reasons, selecting an optimal data rate for vehicle safety communications is critical.

In this chapter, we present an analysis conducted to determine the optimal dedicated short-range communications (DSRC) date rate for supporting vehicle safety communications. The basic idea is to compare communication scenarios featuring different channel loads and observe how the network performance varies with the communication data rates. To facilitate comparisons among different scenarios, we introduce the notion of communication

Vehicle Safety Communications: Protocols, Security, and Privacy, First Edition. Luca Delgrossi and Tao Zhang.
© 2012 John Wiley & Sons, Inc. Published 2012 by John Wiley & Sons, Inc.

density (CD), a metric to estimate the overall channel load in the context of vehicle safety communications [JiCD07]. CD is used to create different traffic scenarios associated with comparable channel loads. The communications generated by these traffic scenarios are then simulated to determine the optimal data rate for DSRC. The results show that CD can effectively support the study of vehicle safety communications.

8.2 COMMUNICATION DENSITY

Wireless access in vehicular environments (WAVE) radios can measure the current DSRC channel load by means of the Institute of Electrical and Electronics Engineers (IEEE) 802.11 clear channel assessment (CCA) function [IEEE07]. CCA determines whether and when the medium is in busy or idle state. By periodically invoking CCA, a vehicle can calculate the fraction of time when the channel is busy, known as channel busy ratio (CBR), and use it as channel load estimate. However, representing channel load as a percentage of channel occupancy is not always convenient. For instance, when conducting simulation studies, it is difficult to determine channel occupancy a priori. For this reason, the concept of CD is introduced as an effective channel load metric for vehicle safety communications [JTMH06].

CD is defined as the number of carrier sensible events in the system per unit of time and space. The concept of CD was derived from a mathematical model created as a probabilistic analysis tool for IEEE 802.11 distributed coordination function (DCF) performance studies [TaJi05]. This model describes the broadcast performance of homogenous vehicles distributed on a linear highway, with signal attenuations modeled through Nakagami fading. Each vehicle generates messages according to the same Poisson distribution with rate λ. All messages have the same size M. The analysis shows that a vehicle's broadcast performance, expressed in terms of frame reception rates and media access delays, depends only on the message size M and the CD around the vehicle.

For practical purposes, CD can be approximated with the product of vehicle density, message rate, and transmission range. Although this product does not precisely match the number of carrier sensible events per unit of time and space, it can effectively serve the same purpose since these two quantities are generally proportional to each other. Simulation studies show that with Rayleigh fading, the number of carrier sensible events is about 2.82 times the CD computed in the above manner [JiCD07].

A key property of a good channel load metric is that different scenarios associated with the same value of the metric should produce consistent performance outcomes. Simulation results demonstrate that when vehicles broadcast with the same transmission power, different scenarios characterized by the same level of CD produce equivalent performances in terms of message reception rates and media access delays. This suggests that CD can be an

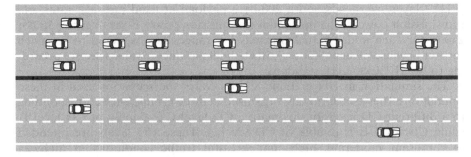

Figure 8.1. Sample highway scenario

effective metric for studying communication systems where traffic is dominated by pervasive broadcast.

The concept of CD is intuitive in homogenous scenarios where all vehicles use the same transmission power and broadcast messages at the same rate. For example, consider a road with vehicle density D where each vehicle is broadcasting at message rate λ and transmission range R. Vehicles along this road should experience similar broadcast performance as in a road scenario with vehicle density $2 \times D$, message rate $\lambda/2$, and transmission range R. The communication densities associated with groups of vehicles in the same area can be simply summed up to obtain the overall CD.

Figure 8.1 illustrates an example where vehicles travel on a highway in both directions. In one direction, vehicles are densely packed and move slowly. These vehicles could broadcast safety messages frequently and at very low power, given their close proximity to each other. In the opposite direction, a few sparse vehicles move at high speeds. This second group of vehicles could benefit from higher transmission powers and lower message rates. Communication densities associated with different groups of vehicles at this same location can be summed up and the resulting value (representing the overall CD) can be used to predict the broadcast performance of either group [TaJi05].

8.2.1 Simulation Study

One way of validating the concept of CD is through simulation studies. The idea is to compare different traffic scenarios associated with the same CD level and see whether they offer similar performance in terms of broadcast reception rates and channel access delays.

For these simulations, the vehicles are placed uniformly at random on a circular linear road with a 2-km perimeter. Distances between vehicles are calculated based on circle arc lengths. This can be viewed as a road around the equator of a greatly shrunken Earth. This arrangement eliminates boundary effects. Each vehicle broadcasts the media access control (MAC) and physical

(PHY) layer overheads and a 200-byte payload at fixed message rates and transmission powers. The maximum transmission power is set to either 100 or 250 m for each node. Actual transmission ranges are expected to be much shorter because of signal fading. All channel parameter values are set according to the IEEE 802.11p standard.

The simulation matrix in Table 8.1 illustrates the scenarios used in these simulations. These scenarios are organized in three groups, each corresponding to a different CD level. In this matrix, CD level C is double of CD level B, while CD level B is double of CD level A. These CD levels are intended to represent moderate, high, and stressful traffic conditions in highway environments.

Table 8.1 shows how each CD level is actually formed. For example, CD level A is equivalent to 100 cars distributed over a short highway segment, each transmitting 10 messages per second with a 250-m range. CD level C is equivalent to 400 cars, each transmitting 25 messages per second with a 100-m range.

For each CD level, five scenarios are specified. Of these scenarios, two are homogeneous (H1, H2) and three are heterogeneous (N1, N2, N3). In homogeneous scenarios, all vehicles share the same message rate and transmission power. Each heterogeneous scenario comprises two groups of vehicles intermixed with each other. Each group is characterized by different vehicle density, message rate, and transmission power.

For example, scenario H1 in CD level A produces (100 cars/km) × (250 m) × (10 messages/s) = 250,000 messages per second per kilometer. Scenario N1 in the same CD level produces $100 \times 250 \times 5 + 400 \times 100 \times 3.125 = 125,000 + 125,000 = 250,000$ messages per second per kilometer.

In heterogeneous scenarios, each group of vehicles contributes a different percentage of the overall CD level. For example, the group of vehicles with a 250-m transmission range contributes 50% of the total CD level in scenario N1, 80% in N2, and 20% in N3.

8.2.2 Broadcast Reception Rates

The broadcast reception rate represents the probability of successful reception of a broadcast message at a given distance from the message sender [TMJH04]. For scenarios sharing the same CD level, it is possible to directly compare their broadcast reception rates.

Figure 8.2 illustrates average broadcast reception rates for moderate channel load (CD level A). The x-axis shows the distance from the sender and the y-axis shows the average reception rate associated with a particular transmission power. The curves in this chart clearly form two sets of overlapping lines. The upper set of lines show average reception rates for 250-m range transmissions, while the lower set of lines show average reception rates for 100-m range transmissions. Average reception rates for CD levels B and C are shown in Figures 8.3 and 8.4, respectively.

Table 8.1. Communication density simulation matrix

Scenario	CD Level A (Moderate)			CD Level B (High)			CD Level C (Stressful)		
	Vehicle Density (cars/km)	TX Range (m)	Message Rate (Hz)	Vehicle Density (cars/km)	TX Range (m)	Message Rate (Hz)	Vehicle Density (cars/km)	TX Range (m)	Message Rate (Hz)
H1	100	250	10	200	250	10	200	250	20
H2	400	100	6.25	400	100	12.5	400	100	25
N1	100	250	5	100	250	10	200	250	10
	400	100	3.125	400	100	6.25	400	100	12.5
N2	100	250	8	200	250	8	200	250	16
	400	100	1.25	400	100	2.5	400	100	5
N3	100	250	2	200	250	2	200	250	4
	400	100	5	400	100	10	400	100	20

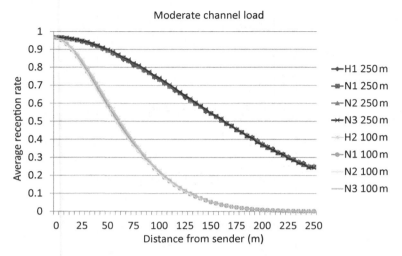

Figure 8.2. Average reception rates (CD level A)

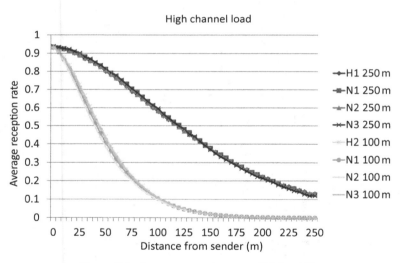

Figure 8.3. Average reception rates (CD level B)

As expected, at the same distance from the sender, higher CD levels result in lower average reception rates. The charts confirm that behaviors are consistent across groups of vehicles broadcasting at the same transmission power in scenarios characterized by the same CD level, regardless of how the scenarios are constructed. For example, with CD level A, a broadcast with 100-m transmission power has 60% average reception rate at 50 m from the sender both in homogenous and heterogeneous scenarios (Figure 8.2).

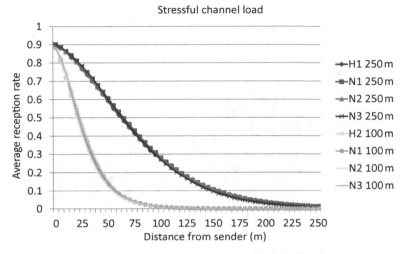

Figure 8.4. Average reception rates (CD level C)

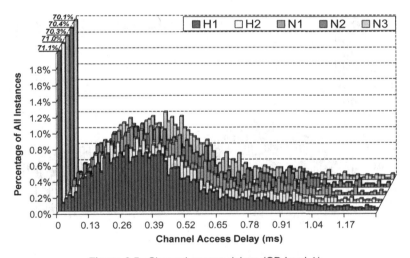

Figure 8.5. Channel access delays (CD level A)

8.2.3 Channel Access Delay

Channel access delay is defined as the time interval between the instant at which a frame is passed to the MAC layer from the upper layer protocol and the instant transmission over the air begins. Channel access delay is an important parameter for vehicle safety communications. According to the CD theory, channel access delay does not vary across scenarios associated with the same CD level [TaJi05]. This is confirmed by the simulations.

Figure 8.5 illustrates the channel access delay distribution for CD level A. Delays are grouped in 13-ms bins, the size of the MAC layer time slot

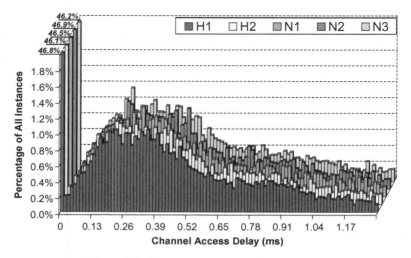

Figure 8.6. Channel access delays (CD level B)

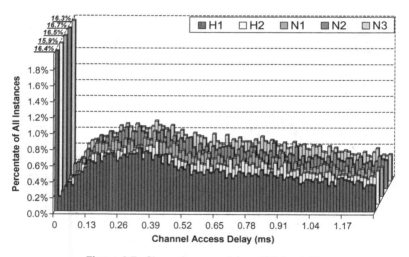

Figure 8.7. Channel access delays (CD level C)

parameter specified in the IEEE 802.11p standard. The *y*-axis shows the percentage of broadcasts that fall into each bin. Since channel access delay is not related to transmission power, it is possible to directly compare scenarios with the same CD level. Figure 8.5 shows that scenarios with CD level A present similar channel access delay distributions. Figures 8.6 and 8.7 confirm the result for CD levels B and C.

All the three figures show high percentages of immediate frame transmissions. They show how in congested environments the likelihood of activating

Table 8.2. Frame reception failure reasons

Number	Reason	Number	Reason
1	Signal too weak for frame detection	6	Already in PreRX for another frame
2	Signal too weak for preamble capture	7	Already in RXing for another frame
3	Signal too weak for frame body capture	8	Already in RXing with capture off
4	PreRX failed due to interference	9	Interrupted by transmit command
5	RXing failed due to interference	10	TXing

the back-off mechanism becomes much higher. At moderate channel load, over 71% of the frames are transmitted immediately. This percentage decreases to 46% at high channel load and to 16% at stressful channel load, where less than 1 in 5 frames is transmitted immediately without triggering the back-off algorithm.

8.2.4   Frames Reception Failures

Average reception rate and channel access delay are the most relevant broadcast performance metrics. However, it is interesting to examine the reasons for frame receptions failures with respect to the distance from the sender. The overhauled network simulator 2 (ns-2) can log frame reception failures and produce traces reporting the precise reasons why individual frames have been dropped during the reception process [JiCD07]. Such information provides deeper understanding of internal network behaviors and insight into potential optimization opportunities.

The 10 reasons for frame reception failure illustrated in Table 8.2 can be broadly grouped into a few main categories:

- *Transmission Busy (TXB)*: A vehicle is unable to receive an incoming frame because the frame arrives when the vehicle is already in transmission mode. This is expected to be a very rare case.
- *Prereception Busy (PXB)*: A vehicle is unable to successfully receive the preamble and physical layer convergence procedure (PLCP) header of an incoming frame. This could be the case when an incoming frame collided with a subsequent interference frame or an interference frame without enough overriding power to be captured.
- *Reception Busy (RXB)*: A vehicle is unable to successfully receive the frame body of an incoming frame. This could be the case when an incoming frame collided with a subsequent interference frame or an interference frame without enough overriding power to be captured.

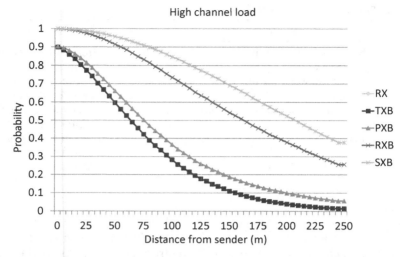

Figure 8.8. Frame nonreception probabilities for group N2 (CD level B)

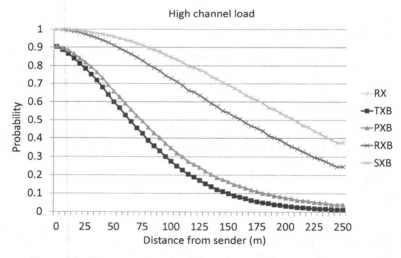

Figure 8.9. Frame nonreceptions distribution for group N3 (CD level B)

- *Searching (SXB)*: A vehicle in Searching state is unable to detect or receive an incoming frame preamble because of insufficient signal strength.
- *Insufficient Power (PWR)*: A vehicle is unable to receive an incoming frame body because of insufficient signal strength.

Figures 8.8 and 8.9 show the distributions of frame nonreception reasons for vehicle groups N2 and N3, respectively, at CD level B (high channel load). The transmission power used in these simulations corresponds to a 250-m maximum communication range. The lowest line in each chart represents the reception rate (RX). The difference between the RX curve and the TXB curve represents the percentage of frame nonreceptions caused by Transmission Busy. In Figures 8.8 and 8.9, these two curves are overlapping. This is due to the rare occurrence of nonreceptions due to frame arrivals when the radio is in transmission mode. The difference between the TXB and PXB curves represents the percentage of frame nonreceptions caused by Prereception Busy. The same applies to Reception Busy and Searching. The percentage of frame nonreceptions due to Insufficient Power is represented by the area above the SXB curve.

The results show that the overall network performance in terms of broadcast reception rates, channel access delays, and frame reception failures distribution is invariant across scenarios associated with the same level of CD, no matter how these traffic scenarios are constructed. This confirms that CD is an effective metric to measure channel load in vehicle safety communication networks.

The concept of CD significantly simplifies the study of vehicle safety communications, allowing us to focus on a small set of parameters and a limited number of scenarios [YEYR04]. CD makes it easier to study the design of distributed congestion control algorithms such as algorithms for adapting transmission powers and rates.

Next, we show how CD can be used to determine the optimal data rate for DSRC.

8.3 OPTIMAL DATA RATE

Data rate selection has been addressed in the literature by studying the number of hops and transmission power needed to meet a certain network performance [ToFe06]. Here, we show that it is possible to broadcast messages at different data rates, yet produce the same level of interference by adjusting the transmission power [ArRP05]. As a result, it becomes possible to fairly compare the performance associated with these transmissions and determine the optimal data rate [JiCD08].

8.3.1 Modulation and Coding Rates

Table 8.3 illustrates the modulation and coding rates specified in the IEEE 802.11p WAVE standard. Orthogonal frequency division multiplexing (OFDM) divides the 10-MHz channel into 52 orthogonal subcarriers, of which 48 are used to carry data. For each subcarrier, the following modulation schemes are available:

• Binary phase-shift keying (BPSK)
• Quadrature phase-shift keying (QPSK)

Table 8.3. IEEE 802.11p modulation, coding rate, and data rate

Modulation	Coded Bits per Periodic Wave Form	Coded Bits per OFDM Symbol	Coding Rate	Data Bits per OFDM Symbol	Data Rate (Mbps)	SINR Threshold (dB)
BPSK	1	48	1/2	24	3	5
BPSK	1	48	3/4	36	4.5	7
QPSK	2	96	1/2	48	6	8
QPSK	2	96	3/4	72	9	11
16-QAM	4	192	1/2	96	12	15
16-QAM	4	192	3/4	144	18	20
64-QAM	6	288	2/3	192	24	25
64-QAM	6	288	3/4	216	27	N/A

Table 8.4. IEEE 802.11p simulation parameters

Module	Parameter	Value
802.11 MAC	Header duration	0.000040
802.11 MAC	Symbol duration	0.000008
802.11 MAC	Contention window min	15
802.11 MAC	Contention window max	1023
802.11 MAC	time slot	0.000013
802.11 MAC	SIFS interval	0.000032
Wireless PHY	Frequency	5.9e + 9
Wireless PHY	PreRX duration	0.000040
Wireless PHY	noise floor	1.26e-13
Wireless PHY	Carrier sensing threshold	2.512e-13

- 16-point quadrature amplitude modulation (16-QAM)
- 64-point quadrature amplitude modulation (64-QAM)

As shown in Table 8.3, transmissions at high data rates are more efficient. However, they are also more susceptible to frame reception errors due to interference and noise. For this reason, a higher signal-to-interference-noise ratio (SINR) is required for higher data rates. The SINR threshold values shown in the table have been obtained through empirical experiments.

8.3.2 Simulation Study

The simulations described below were run with the overhauled ns-2 engine. Table 8.4 lists some of the parameters used for these simulations and their correspondent values, set accordingly to the IEEE 802.11p standard.

The SINR threshold values for frame reception used in the simulations are those listed in Table 8.3. Nakagami fading is applied in all simulations. Vehicles are randomly distributed along a 2000-m circular road (Figure 8.10).

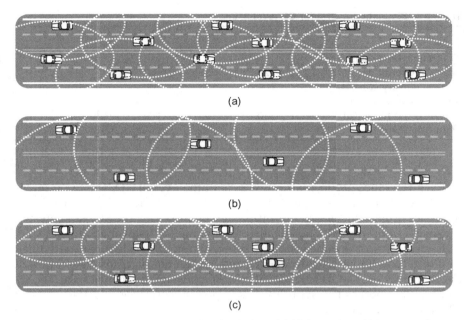

Figure 8.10. Three vehicle groups used in simulations. (a) High number of low-powered transmissions; (b) low number of high-powered transmissions; (c) mixture of high- and low-powered transmissions

Three groups of vehicles, namely reference group #1 (RG1), reference group #2 (RG2), and study group (SG), are intermixed on the road. All the vehicles in RG1 and RG2 are always configured to broadcast at 6 Mbps data rate. The transmission powers are set to reach the maximum ranges of 250 and 100 m, respectively, for groups RG1 and RG2. The SG consists of vehicles transmitting at different data rates and transmission powers depending on the simulation scenarios. The performances of vehicles in SG in different scenarios are compared to identify the optimal data rate in each vehicle safety communications scenario [JiCD08].

The reference groups RG1 and RG2 are introduced to make sure that the adjusted data rates and transmission powers used by SG vehicles combine to produce the same level of CD. This allows us to fairly compare different scenarios. This is accomplished by showing that the performance of vehicles in RG1 and RG2 match each other across different simulations.

8.3.3 Simulation Matrix

Each scenario is defined by two parameters: message size and overall CD. The message sizes used in these simulations are 125, 250, and 500 bytes. The CD levels used in these simulations are 250, 500, and 1000. This results in nine scenarios.

Table 8.5. Simulation matrix for CD level 250

	RG1	RG2	SG		
Communication density	50	50	150		
Share of overall density (%)	20	20	60		
Vehicle density (cars/km)	100	100	200		
Message frequency (Hz)	2	5	7.5	2.5	1.5
Transmission range (m)	250	100	100	300	500
Data rate (Mbps)	6	6	3, 4.5, 6, 9, 12		

Each reference group contributes 20% of the overall CD. The SG is intentionally configured to dominate the channel load. If the performance of the reference groups matches across different scenarios, then the SG data rate and transmission power must be equally contributing to the overall channel load.

Each scenario is divided into three subscenarios. Each subscenario is defined by a different SG message frequency and a different SG transmission power. For scenarios with a CD level of 250, the SG message frequencies are 7.5, 2.5, and 1.5 Hz, whereas the SG transmission powers correspond to 100, 300, and 500 m ranges, respectively.

For each subscenario, the first simulation is run to obtain a baseline performance using a 6 Mbps data rate.

Then, for each subscenario, four additional simulations are run in which SG vehicles transmit at different data rates (3, 4.5, 9, and 12 Mbps). In each case, the SG transmission power is adjusted to maintain a constant level of CD.

Table 8.5 shows the simulation matrix for the scenarios with an overall CD level of 250.

8.3.4 Simulation Results

The methodology adopted in this study is based on the premise that it is possible to adjust transmission powers so that the overall CD remains constant, whereas the data rates vary [JiCD08].

Figure 8.11 shows the reception rate of RG1 and RG2 across four subscenarios with an overall CD level of 250 and message size of 125 bytes. The transmission power corresponds to a theoretical range of 250 m for 6 Mbps data rate. The chart shows two sets of mostly overlapping lines, corresponding to RG1 and RG2, respectively. This indicates that the interference produced by the SG remains constant as data rates and transmission powers vary across different subscenarios.

At high CD levels and large message sizes, some of the reference groups' performance curves no longer completely overlap. This is due to channel saturation in these scenarios. For example, a 500-byte message takes about 700 μs to transmit at a 6 Mbps data rate. For a CD level of 1000, the number of carrier sensible events experienced at any point on the road per second is equivalent

OPTIMAL DATA RATE **89**

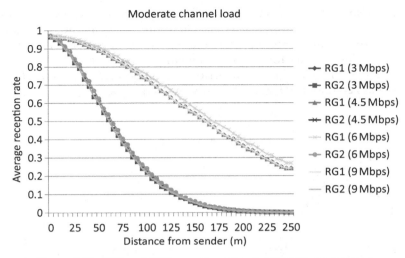

Figure 8.11. RG1 and RG2 reception rates at moderate channel load

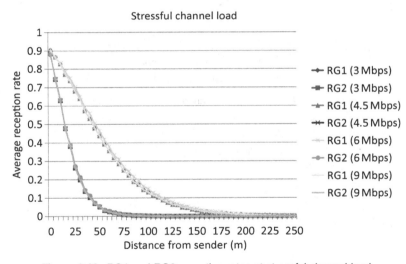

Figure 8.12. RG1 and RG2 reception rates at stressful channel load

to $1000 \times 2.82 = 2820$. In this case, the channel is stressed to the breaking point and it is therefore not possible to tune the transmission power of the SG to perfectly match the behaviors of the reference groups.

Figure 8.12 shows results from simulations under stressful channel load conditions. As in the previous scenario, the performance curves associated with RG1 and RG2 overlap. While not perfect, they are sufficient to provide a reasonable comparison among different data rates.

The matching performances of the reference groups allow us to compare the various data rates in a fair manner. Therefore, selecting the optimal data rate for DSRC translates into picking the top SG curve in each scenario.

The complete results for the 135 simulations can be found in [JiCD08]. Here, we present data rate comparisons at moderate, high, and stressful channel load conditions (Figures 8.13–15, respectively).

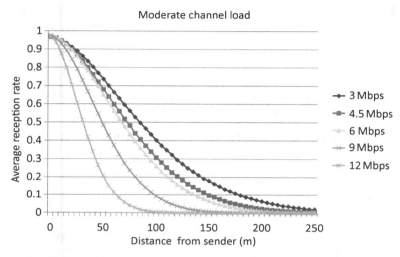

Figure 8.13. Data rates performance at moderate channel load

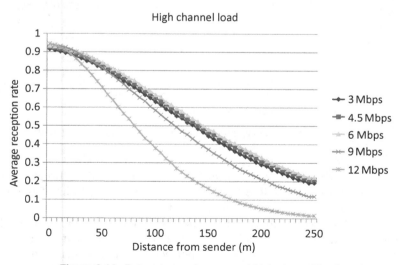

Figure 8.14. Data rates performance at high channel load

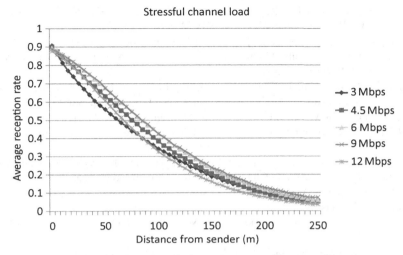

Figure 8.15. Data rates performance at stressful channel load

In almost all cases, the 6 Mbps data rate offers the best performance. Figure 8.14 illustrates the SG performance with a CD level of 500, message size of 250 bytes, and SG transmission power of 300 m at 6 Mbps. In this case, the curve representing transmissions at 6 Mbps is above all other curves.

Although 6 Mbps is consistently the top performer, there are cases where other data rates offer similar or even better performance. Low data rates work very well in moderately loaded channels. This is expected because when the channel is nearly empty, the impact of a longer duration in interference is minimized. Figure 8.13 shows a low stress scenario with a CD level of 250, message size of 125 bytes, and SG transmission power of 100 m. Similarly, high data rates become the top performers in very high stress scenarios, which can be explained with the same reason applied in reverse. Figure 8.15 shows a very high stress scenario with a CD level of 1000, message size of 500 bytes, and SG transmission power of 500 m.

The commonly assumed default 6 Mbps data rate turns out to be the best selection in most cases, except when the channel is either empty or saturated. This analysis confirms the long-held assumption on the default data rate for DSRC and provides further insight on why this choice makes sense. These findings allow researchers to bypass the need of evaluating different data rates in future DSRC studies, thus eliminating one dimension of complexity.

REFERENCES

[ArRP05] M. M. Artimy, W. Robertson, and W. J. Phillips: "Assignment of Dynamic Transmission Range based on Estimation of Vehicle Density," in Proc. of the Second ACM International Workshop on Vehicular Ad Hoc Networks (VANET), 2005.

[IEEE07] Institute of Electrical and Electronics Engineers (IEEE): "IEEE Standard 802.11-2007, Wireless LAN MAC and PHY Specifications," Section 7.3.2.22.2, 2007.

[JiCD07] D. Jiang, Q. Chen, and L. Delgrossi: "Communication Density: A Channel Load Metric for Vehicular Communications Research," Mobile Vehicular Networks (MoVeNet) Workshop, 2007.

[JiCD08] D. Jiang, Q. Chen, and L. Delgrossi: "Optimal Data Rate Selection for Vehicle Safety Communications," ACM MobiCom VANET Workshop, San Francisco, 2008.

[JTMH06] D. Jiang, V. Taliwal, A. Meier, W. Holfelder, and R. Herrtwich: "Design of 5.9 GHz DSRC-based Vehicular Safety Communications," IEEE Wireless Communications Magazine, vol. 13, no. 5, 2006.

[TaJi05] V. Taliwal, D. Jiang: "Mathematical Analysis of IEEE 802.11 Broadcast Performance in a Probabilistic Channel," DaimlerChrysler Technical Paper, 2005.

[TMJH04] M. Torrent-Moreno, D. Jiang, and H. Hartenstein: "Broadcast Reception Rates and Effects of Priority Access in 802.11 based Vehicular Ad-Hoc Networks," First ACM International Workshop on Vehicular Ad Hoc Networks (VANET), 2004.

[ToFe06] O. K. Tonguz and G. Ferrari: Ad Hoc Wireless Networks: A Communication-Theoretic Perspective, John Wiley & Sons Ltd., 2006.

[XMKS04] Q. Xu, T. Mak, J. Ko, and R. Sengupta: "Vehicle-to-Vehicle Safety Messaging in DSRC," First ACM International Workshop on Vehicular Ad Hoc Networks (VANET), 2004.

[YEYR04] J. Yin, T. ElBatt, G. Yeung, B. Ryu, S. Habermas, H. Krishnan, and T. Talty: "Performance Evaluation of Safety Applications over DSRC Vehicular Ad Hoc Networks," First ACM International Workshop on Vehicular Ad Hoc Networks (VANET), 2004.

9

WAVE UPPER LAYERS

9.1 INTRODUCTION

In the United States, the 5.9 GHz dedicated short-range communications (DSRC) spectrum is organized into seven 10-MHz channels. This makes it possible for a single wireless access in vehicular environments (WAVE) radio to engage in concurrent operations over multiple DSRC channels. For instance, a WAVE radio can broadcast safety messages over the control channel (CCH) (#178) while supporting nonsafety applications over a service channel (SCH) #182.

DSRC multichannel operations are specified in the Institute of Electrical and Electronics Engineers (IEEE) 1609.4 standard. The standard defines a time division scheme that allows WAVE radios to switch from one DSRC channel to another. This mechanism allows different applications to communicate simultaneously over separated physical channels. The IEEE 1609.4 time division scheme is independent of the spectrum and is therefore directly applicable to other frequency bands.

Vehicle safety applications interact with the WAVE protocol stack through the WAVE Short Message Protocol (WSMP), specified in the IEEE 1609.3 standard. WSMP offers the source the means to specify frame transmission parameters.

In this chapter, we describe the most relevant features of the IEEE 1609.4 standard and discuss several concerns with its design. We present simulation

Vehicle Safety Communications: Protocols, Security, and Privacy, First Edition. Luca Delgrossi and Tao Zhang.
© 2012 John Wiley & Sons, Inc. Published 2012 by John Wiley & Sons, Inc.

results to illustrate the impact of the IEEE 1609.4 time division scheme on CCH load. Furthermore, we present protocol enhancements suggested in the literature. Finally, we provide a brief description of WSMP.

9.2 DSRC MULTICHANNEL OPERATIONS

The first version of the IEEE 1609.4 standard was issued at the end of 2006 [IEEE06]. Version D1.6 was issued in 2010 [IEEE10a]. With respect to the WAVE protocol stack, IEEE 1609.4 sits directly on top of IEEE 802.11p at the upper media access control (MAC) layer, as shown in Figure 9.1.

The IEEE 1609.4 standard specifies a time division scheme for multichannel operations, and defines synchronization intervals timing, channel switching, and channel routing.

One approach to multichannel operations is to use different radios to communicate over different channels. The IEEE 1609.4 standard, however, assumes a single radio operating over multiple DSRC channels. The protocol does not include mechanisms for multiradio operations or for future coexistence of single-radio and multi-radio solutions.

9.2.1 Time Synchronization

The IEEE 1609.4 channel switching scheme requires all devices to maintain synchronization with boundaries of seconds within a common time reference.

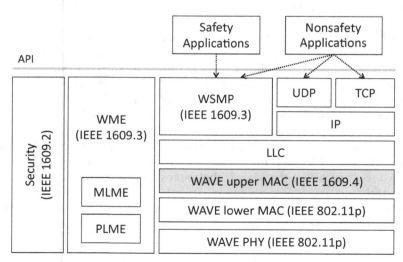

Figure 9.1. IEEE 1609.4 in the WAVE protocol stack. API, application programming interface; WME, WAVE Management Entity; MLME; MAC Layer Management Entity; PLME, Physical Layer Management Entity; UDP, User Datagram Protocol; LLC, logical link control

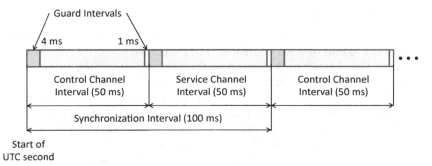

Figure 9.2. Synchronization, CCH, SCH, and guard intervals

For instance, existing implementations exploit the pulse per second (PPS) signal issued by some Global Positioning System (GPS) devices to keep precise timing.

The standard allows radios that do not have direct access to precise timing sources to acquire timing information from other WAVE radios. For instance, this can be achieved by reading the timing information included in WAVE Timing Advertisement frames sent by WAVE upper layer protocols. To become a provider of timing information to other devices, a WAVE radio must meet some minimum Coordinated Universal Time (UTC) synchronization requirements.

9.2.2 Synchronization Intervals

The IEEE 1609.4 time division scheme is based on synchronization intervals. Each second comprises 10 100-ms synchronization intervals (Figure 9.2). Each synchronization interval is divided into a CCH interval followed by an SCH interval. Each CCH interval marks the beginning of a UTC second or multiples of 100 ms thereafter.

In its original form, the IEEE 1609.4 standard assumes WAVE radios use the CCH to broadcast safety messages and advertisements for services available on an SCH. The standard permits the transmission of safety messages over the CCH during SCH intervals. However, vehicles equipped with a single radio tuned to an SCH will not be able to receive these messages.

The standard allows WAVE radios not to engage in channel switching. However, WAVE radios that are permanently tuned to the CCH are expected to keep track of the beginning and end of synchronization intervals. This allows these radios to suspend transmitting messages over the CCH during SCH intervals if necessary. This may happen, for example, when a radio expects that some nearby radios may be using the SCHs to support different applications.

9.2.3 Guard Intervals

IEEE 1609.4 defines front guard intervals at the beginning and rear guard intervals at the end of each control or SCH interval.

A front guard interval is defined as the sum of synchronization tolerance and maximum channel switching time. Synchronization tolerance is the expected precision of a device's internal clock, with respect to its ability to align with the UTC time. Maximum channel switching time is the overhead introduced by the operation of switching from one physical channel to another.

The purpose of rear guard intervals is to allow extra time to complete frame reception before switching to another channel. WAVE radios are not permitted to transmit any frames during rear guard intervals.

With current radio chipset technology, 4-ms front guard intervals and 1-ms rear guard intervals are sufficient to protect the communications.

9.2.4 Channel Switching

At the beginning of a control or SCH guard interval, a WAVE radio suspends MAC layer activities. The radio waits until the end of the guard interval and then starts communications over the channel or resumes activities suspended at the end of the previous cycle.

A radio cannot initiate frame transmission during a guard interval. If frame transmission has not been completed before the start of the next guard interval, a radio may drop the frame and abort the process. However, the IEEE 1609.4 standard suggests that implementations should minimize frame drops through careful transmissions scheduling.

Data frames that are not transmitted during a control or SCH interval are stored in local queues until the next cycle. IEEE 1609.4 defines a mechanism to assign an expiration time to individual data frames. This allows the MAC layer to purge data frames that expired before having a chance of being transmitted.

The standard requires radios to declare channel busy during guard intervals so that transmissions at the beginning of the next channel interval are scheduled through the back-off mechanism.

If a front guard interval begins while a radio is still engaged in frame reception, the process is abandoned with loss of the frame.

9.2.5 Channel Switching State Machine

Figure 9.3 illustrates a state machine for DSRC multichannel operations [Wein06]. A vehicle starts in the No Sync state. Once time synchronization is achieved, the vehicle enters the CCH Guard state and tunes on the CCH. Frames received from the upper layers for transmission while the vehicle is in the CCH Guard state are stored in local queues. When the CCH guard interval

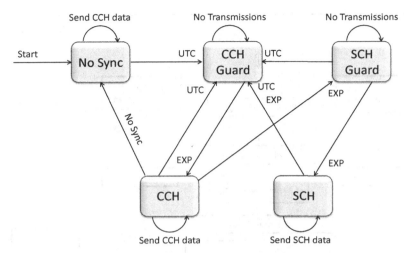

Figure 9.3. State machine for IEEE 1609.4 multichannel operations. EXP, expired timer.

expires, the vehicle enters the CCH state. Here, management or data frames intended for the CCH are transmitted, while data frames intended for any SCH are stored in local queues.

When the CCH interval expires, the vehicle enters the SCH Guard state. No frames are transmitted while the vehicle is in this state. When the SCH guard interval expires, the vehicle transitions into the SCH state and any data frames intended for the SCH associated are transmitted. At the expiration of the SCH interval, the vehicle returns to the CCH Guard state.

At the beginning of a UTC second, the vehicle transitions to the CCH Guard state, regardless of its current state. Should synchronization be lost at any time, the vehicle enters the No Sync state.

9.3 PROTOCOL EVALUATION

The IEEE 1609.4 time division scheme raised several concerns [KeRH09] [MaLS05] [WaHa08]. A primary issue is that it leads to poor CCH utilization impacting the performance of broadcast safety messages. The time division scheme forces vehicles to broadcast safety messages during the CCH interval, even when there are no other vehicles within communication range adopting channel switching. Further concerns are related to the high likelihood of synchronized collisions introduced by the scheme through message queuing during channel busy intervals due to operations over another channel or guard intervals. Simulation studies have provided insight into the effectiveness of the IEEE 1609.4 time division scheme.

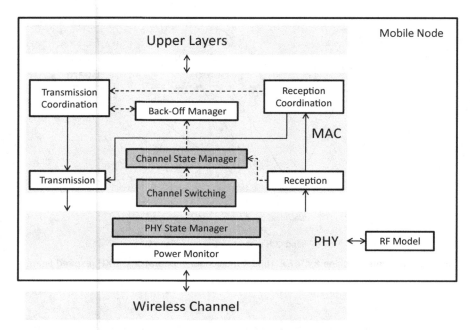

Figure 9.4. Enhanced simulation architecture. RF, radiofrequency

9.3.1 Simulation Study

To model IEEE 1609.4 channel switching, our overhauled network simulator 2 (ns-2) implementation requires a few extensions [ChJD09].

A new module, Channel Switching, is created and inserted between the physical layer (PHY) State Manager and Channel State Manager (Figure 9.4). Modifications to the software in these two modules are necessary. Channel Switching signals to the Channel State Manager the beginning and the end of every CCH interval, and the Channel State Manager declares channel clear or busy accordingly. Channel Switching also informs PHY State Manager of channel switching events. When this occurs, the PHY layer abandons frame reception.

Table 9.1 lists some simulation parameters used in this study. The vehicle density level describes a stressful but plausible scenario. For vehicle safety messages, 75 and 150 bytes can be considered very light and moderate payload sizes. The transmissions would reach 200 and 400 m under ideal channel conditions with no fading or interference. Using the Nakagami fading model, the frame reception rate reduces by 0.25 at about half the transmission range.

Table 9.1. Simulation parameters

Vehicle density	400 cars on 2 km of road
Transmission range	200 and 400 m
Messaging frequency	3, 5, and 10 Hz
Message payload	75 and 150 bytes
Modulation and coding rate	QPSK and 1/2 coding rate (i.e., 6 mbps)
Fading model	Nakagami

QPSK, quadrature phase-shift keying.

9.3.2 Simulation Scenarios

Packet transmission scheduling at the data source affects the network performance. We compare three different scenarios:

- *Off*: Channel switching is off. The radio is always tuned on the CCH. Applications generate safety messages at any time.
- *Naïve Scheduling*: Channel switching is on. The radio alternates between the CCH and an SCH. Applications generate safety messages at any time.
- *Optimized Scheduling*: Channel switching is on. The radio alternates between the CCH and an SCH. Applications generate safety messages only during CCH intervals.

Figure 9.5 illustrates the difference between naïve and optimized scheduling. In the naïve scheduling scenario (a), applications are unaware of the underlying time division scheme. In the optimized scheduling scenario (b), applications generate the same number of messages, but these messages are concentrated during CCH intervals.

9.3.3 Simulation Results

Figure 9.6 shows the message reception rate (*y*-axis) versus distance from sender (*x*-axis) in a simulation scenario characterized by 75 bytes message size, 200 m transmission range, and 5 Hz message rate.

As expected, naïve scheduling leads to significant drops in message reception probability even at short distances. This is due to the high number of synchronized collisions. Thanks to the moderate channel load associated with this scenario, the optimized scheduling curve shows only a marginal difference with respect to the channel switching off curve.

Figure 9.7 illustrates a more stressful scenario: 150 bytes message size, 400 m transmission range, and 5 Hz message rate. In this scenario, as expected, the gap between the optimized scheduling and channel switching off curves is

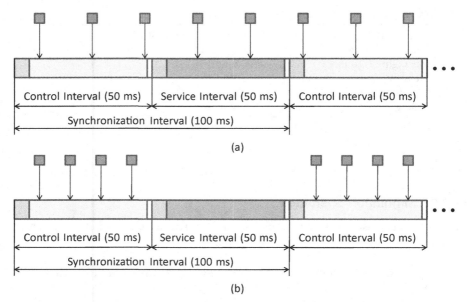

Control Interval (50 ms) | Service Interval (50 ms) | Control Interval (50 ms)

Synchronization Interval (100 ms)

(a)

Control Interval (50 ms) | Service Interval (50 ms) | Control Interval (50 ms)

Synchronization Interval (100 ms)

(b)

Figure 9.5. Optimized safety message scheduling. (a) Naïve scheduling; (b) optimized scheduling

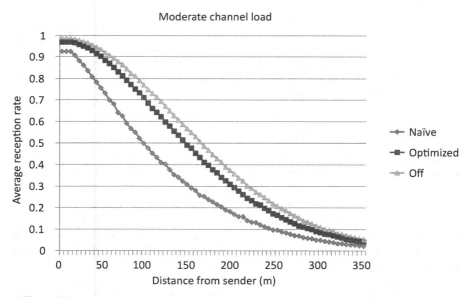

Figure 9.6. Average reception rate versus distance from sender (75 bytes, 200 m, 3 Hz)

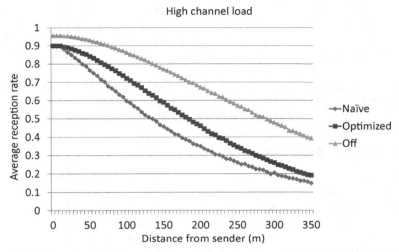

Figure 9.7. Average reception rate versus distance from sender (150 bytes, 400 m, 5 Hz)

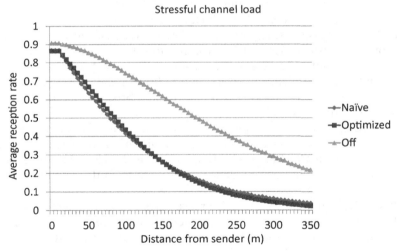

Figure 9.8. Average reception rate versus distance from sender (150 bytes, 400 m, 10 Hz)

significantly larger. The chart shows severe performance degradation introduced by the IEEE 1609.4 time division scheme. At this channel load level, some improvement on the performance can be achieved through careful packet transmission scheduling.

Figure 9.8 shows that, when channel switching is active and channel load is stressful, the network performance is probably not sufficient to support vehicle safety applications, regardless of the scheduling policy adopted [ChJD09].

Similar conclusions are obtained when considering interpacket arrival times. Vehicles need to receive frequent status updates from their neighbors. This calls for low interpacket arrival times. The analysis in [ChJD09] shows that interpacket arrival times for the scenario represented in Figure 9.8 are unlikely to be acceptable for vehicle safety applications.

The results obtained through this simulation study confirm concerns about inefficient CCH utilization. A complete description of this study can be found in [ChJD09]. Consistent results were obtained in [HKRL10].

9.3.4 Protocol Enhancements

A number of protocol enhancements have been proposed to address the poor network performance resulting from the IEEE 1609.4 time division scheme.

The first approach is Dedicated Safety Radio [HKRL10]. The idea is to use DSRC Channel #172 instead of the CCH for safety communications among vehicles. This eliminates the need for the IEEE 1609.4 time division scheme and allows safety applications to broadcast messages at any time. This approach offers the best performance for vehicles equipped with a single radio. Should this scheme be adopted, however, vehicles equipped with a single radio would not be able to access services offered on DSRC SCHs.

The second approach calls for vehicles equipped with single DSRC radios to set a Capability Bit in safety messages they broadcast to inform neighboring radios of their existence [HKRL10]. Multiradio vehicles broadcast safety messages over channel #172 at any time, using full channel capacity. However, when a multi-radio vehicle detects a single-radio neighbor through the Capacity Bit in received safety messages, it additionally broadcasts safety messages over the CCH during CCH intervals.

The third approach also requires a single-radio vehicle to set an Intention Bit in vehicle safety messages to indicate its intention to switch channels at the end of the next CCH interval [HKRL10]. Vehicles can adapt their behaviors based on the additional information received from their neighbors:

- A vehicle equipped with a single radio stays normally tuned to the CCH. It switches to an SCH during SCH intervals upon requests from the application layer. It sets the Intention Bit to indicate its intention to switch channels at the end of the next CCH interval.
- A vehicle equipped with multiple radios tunes one of its radios to the CCH all the time. The vehicle never sets the Intention Bit. When the vehicle encounters another vehicle broadcasting safety messages with the Intention Bit set, it transmits safety messages only during CCH intervals. Otherwise, the vehicle broadcasts safety messages at any time.

The Intention Bit approach does not use channel #172. It allows full utilization of CCH when there are no vehicles within communication range operating in channel switching mode.

Currently, to overcome performance degradations introduced by the IEEE 1609.4 time division scheme and avoid medium access competitions with other types of messages, researchers are planning to broadcast safety messages over DSRC channel #172.

9.4 WAVE SHORT MESSAGE PROTOCOL

The WSMP is positioned between the data link and the application layers (Figure 9.9). This corresponds to the network and transport layers in an Internet Protocol (IP) network. WSMP offers an interface to DSRC safety applications.

Unlike IP networks, DSRC safety communications do not require routing functions. Typically, the communications take place over a single hop. Frames sent over the DSRC link are not fragmented by the network and there is no need for reassembly at the destination. Functions to increase packet reception reliability, such as those offered by the Transmission Control Protocol (TCP), are not required in communication scenarios dominated by pervasive broadcast. Therefore, WSMP is designed to support a few basic services.

WSMP essentially offers the means for the source to specify transmission parameters for individual frames, including channel number, data rate, and transmission power. Also, it allows the sender to indicate which application the message is intended for, facilitating demultiplexing at the destination. This is done through the use of the Provider Service Identifier (PSID) field in the WSMP header. PSIDs are conceptually similar to TCP port numbers. Safety messages share a single PSID, so the PSID mechanism is intended for the coexistence of safety and nonsafety applications.

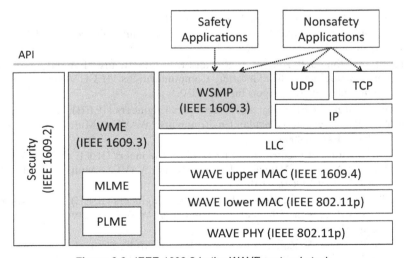

Figure 9.9. IEEE 1609.3 in the WAVE protocol stack

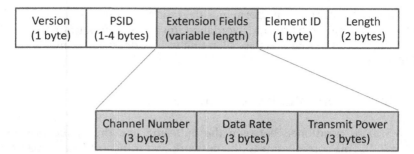

Figure 9.10. WSMP header

WSMP is specified in the IEEE 1609.3 standard [IEEE10b]. The current version of the IEEE 1609.3 standard is version 2. Figure 9.10 illustrates the WAVE Short Message header. The first byte of the header contains information about the protocol version (4 bits are reserved for future use). The second header field describes the PSID associated with the packet. The PSID has a variable length to allow for variable numbers of PSIDs.

The Extension Fields have a variable length to accommodate the description of several elements. Three such elements are specified in the current IEEE 1609.3 standard: channel number, data rate, and transmission power. The Element ID field indicates the end of the Extension Fields. The last 2 bytes of the WSMP header describe the length (expressed in bytes) of the message payload.

REFERENCES

[ChJD09] Q. Chen, D. Jiang, and L. Delgrossi: "IEEE 1609.4 DSRC Multi-Channel Operations and Its Implications on Vehicle Safety Communications," IEEE Vehicular Networking Conference (VNC), Tokyo, Japan, 2009.

[HKRL10] K. Hong, J. B. Kenney, V. Rai, and K. P. Laberteaux: "Evaluation of Multi-Channel Schemes for Vehicular Safety Communications," IEEE Vehicular Technology Conference (VTC), Taipei, Taiwan, 2010.

[IEEE06] Institute of Electrical and Electronics Engineers (IEEE): "Draft Standard for Wireless Access in Vehicular Environments (WAVE)—Multi-channel Operations," IEEE 1609.4, 2006.

[IEEE10a] Institute of Electrical and Electronics Engineers (IEEE): "Draft Standard for Wireless Access in Vehicular Environments (WAVE)—Multi-channel Operations," IEEE 1609.4/D6.0, 2010.

[IEEE10b] Institute of Electrical and Electronics Engineers (IEEE): "Draft Standard for Wireless Access in Vehicular Environments (WAVE)—Networking Services," IEEE 1609.3/D5.0, 2010.

[KeRH09] J. Kenney, V. Rai, and K. Hong: "VSC-A Multi-Channel Operation Investigation: An Update to IEEE 1609," IEEE 1609 Standard Group, 2009.

[MaLS05] T. Mak, K. Laberteaux, and R. Sengupta: "A Multi-Channel VANET Providing Concurrent Safety and Commercial Services," 2nd ACM International Workshop on Vehicular Ad-hoc Networks (VANET), Cologne, Germany, 2005.

[WaHa08] Z. Wang and M. Hassan: "How much of DSRC is available for non-safety use?" 5th ACM International Workshop on Vehicular Ad-hoc Networks (VANET), San Francisco, CA, 2008.

[Wein06] A. Weinfield: "5.9 GHz DSRC Channel Switching Design and Implementation," Vehicle Infrastructure Integration (VII) Radio Working Group, Internal Report, 2006.

10

VEHICLE-TO-INFRASTRUCTURE SAFETY APPLICATIONS

Written in cooperation with Michael Maile and Craig Robinson

10.1 INTERSECTION CRASHES

Crashes occurring within the limits of an intersection and as vehicles approach or exit an intersection constitute a large portion of fatal accidents, accounting for about 1.72 million crashes and 9000 deaths every year in the United States [NHTS06]. In 2004, stop sign and traffic signal violations accounted for 302,000 crashes, resulting in 163,000 functional life years lost and $7.9 billion of economic loss. About 250,000 of those accidents involved vehicles running a red light and colliding with another vehicle crossing the intersection from a lateral direction. These accidents led to $6.6 billion in economic cost in the United States [NaSY07].

Cooperative safety applications, such as those based on 5.9 GHz dedicated short-range communications (DSRC), allow vehicles to interact with each other and with the roadside infrastructure to enable extended traffic awareness and safety measures. This is particularly useful to avoid intersection collisions.

The Cooperative Intersection Collision Avoidance System for Violations (CICAS-V) project (2006–2009) has been a prominent research effort to design a communication-based intersection collision avoidance system.

Vehicle Safety Communications: Protocols, Security, and Privacy, First Edition. Luca Delgrossi and Tao Zhang.
© 2012 John Wiley & Sons, Inc. Published 2012 by John Wiley & Sons, Inc.

CICAS-V was a joint effort by the U.S. Department of Transportation (USDOT) and a consortium of five automotive original equipment manufacturers (OEMs) (Ford, General Motors, Mercedes-Benz, Toyota, and Honda) under the Crash Avoidance Metrics Partnership (CAMP) framework [MABC08] [MaDe09] [NHTS11].

In addition to CICAS-V, the CICAS–Signalized Left Turn Assistance (CICAS-SLTA) and CICAS–Stop Sign Assistance (CICAS-SSA) projects addressed further aspects of intersection collision avoidance.

In Germany, the AKTIV-AS Intersection Assistance project has similar objectives: reducing the number of accidents by supporting the driver while entering, crossing, or turning through an intersection.

This chapter presents the prototype system built for the CICAS-V project as well as the Mercedes-Benz Integrated Safety demonstration that was based on CICAS-V and was publicly showcased at the 15th Intelligent Transportation Systems (ITS) World Congress in New York City in 2008. The Integrated Safety demonstration showed a successful integration of CICAS-V with existing in-vehicle safety features [RoDe10].

10.2 COOPERATIVE INTERSECTION COLLISION AVOIDANCE SYSTEM FOR VIOLATIONS

The goal of CICAS-V is to improve safety at intersections. The system relies on DSRC to exchange safety messages in real time between a roadside unit (RSU) installed at an intersection and an onboard unit (OBU) installed on the vehicle. The OBU monitors vehicle dynamics, position, lane of travel, and distance to the stop line. It assesses the risk of violating intersection signals and warns the driver of imminent danger if a violation is predicted.

A fully functional prototype system was designed, implemented, and tested as part of the CICAS-V project. CICAS-V OBUs were installed in the vehicles of the five participating automotive OEMs. Several intersections in California, Michigan, and Virginia, managed through traffic signal controllers produced by different manufacturers, were instrumented with CICAS-V RSUs and used for testing throughout project execution. Testing included both on-road and test track evaluations. CICAS-V was a groundbreaking project because it demonstrated the first vehicle-to-infrastructure (V2I) prototype safety system based on DSRC ready for a field operation trial (FOT) in the United States [KNMK10].

10.2.1 CICAS-V Design

10.2.1.1 Concept of Operations When the vehicle approaches a signalized intersection, the OBU enters the local RSU's communication range and starts receiving local safety messages broadcast from the RSU. The information included in these messages comprises current signal phase and timing (SPAT) for the local traffic signal controller and a small accurate digital map

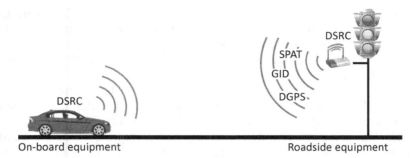

On-board equipment   Roadside equipment

Figure 10.1. Basic CICAS-V concept of operations at signalized intersections

of the local intersection called geometric intersection description (GID). The RSU may optionally broadcast Global Positioning System (GPS) differential corrections and additional GIDs for stop sign-controlled intersections in the vicinity.

The OBU issues a warning to the driver if it predicts that, given the current operating conditions, the vehicle is going to violate the signal to enter the intersection. This warning is intended to raise the driver's attention so that the driver can promptly react and determine the safest course of action, possibly bringing the vehicle to a safe stop before it enters the intersection crash box.

The OBU determines the probability of a violation by continuously assessing the vehicle speed, SPAT, and the vehicle's distance from the stop bar in the lane of travel. If the signal phase is amber, then the OBU determines from the time left in phase whether the vehicle will pass the stop bar before the onset of the red phase. If the vehicle, given its dynamics, is predicted to cross the stop bar after the light has turned red, a warning is issued to the driver (Figure 10.1).

For stop sign-controlled intersections, the vehicle only needs to assess the distance to the stop bar.

10.2.1.2 Positioning Accuracy For CICAS-V, road level positioning accuracy is sufficient at most stop-sign-controlled intersections and at signalized intersections where there are no dedicated turn lanes with movement independent of the signal indication of the through movement. Lane-level accuracy, on the other hand, is necessary at signalized intersections with protected left or right turns where the turn phase differs from the phase for the straight-crossing direction. Assuming that the GPS accuracy can be within 1 m, the error introduced by the local intersection map has to be less than 0.5 m in order to meet the lane level accuracy requirement.

10.2.1.3 Geometric Intersection Description GIDs are maps of an intersection that are detailed enough to support crash avoidance applications. They include the following information:

- accurate road and lane geometry for all approaches to the intersection,
- intersection identification (including whether the intersection is stop sign controlled or signalized),
- stop bar locations and width for all the lanes, and
- information to map individual lanes to the correspondent traffic signal.

The CICAS-V concept of operations does not assume that the vehicles always have an intersection map stored onboard. Also, it does not specify how GIDs for stop-sign-controlled intersections are distributed to vehicles. A possible solution is for RSUs to distribute GIDs for the local intersection and surrounding stop sign intersections.

Broadcasting GIDs over the air imposes constraints on their sizes. To increase reception probability and reduce GID delivery delays, a GID needs to fit within a single DSRC message. The maximum physical (PHY) layer message size specified in the Institute of Electrical and Electronics Engineers (IEEE) 802.11p standard is 1.4 kB. However, it should be assumed that about 400 bytes will be used for the security payload. These constraints call for the design of GIDs of less than 1 kB in size.

To minimize the size of the GID:

- geometry points are represented as Cartesian offsets from an intersection reference point (IRP) specified in latitude, longitude, and altitude coordinates in the World Geodetic System (WGS) 84 system,
- roads and lanes are described as an ordered set of geometry points together with the lane width at each point,
- the lane geometry is described by specifying the centerline of the lane,
- the stop bar location for each lane is the first geometry point for the lane, and
- lane geometries are represented up to a distance of 300 m from the IRP (note that GIDs might overlap).

The basic element of a GID is a point or node. One of the nodes serves as IRP. The set of nodes that describe a lane is called the node list (Figure 10.2).

Two types of lanes are defined: reference lane and computed lane. A reference lane is a lane that is fully specified by a list of points. A computed lane can be derived from a reference lane by a simple parallel shift of the reference lane. This method reduces the GID size by grouping several parallel lanes into a single approach. An approach is defined as all lanes of traffic governed by a single, independent signal phase cycle, moving toward an intersection from one direction. This corresponds to the term "movement" used by traffic engineers.

The GID specifications described above have been adopted by the Society of Automotive Engineers (SAE) J2735 standard [SAE09].

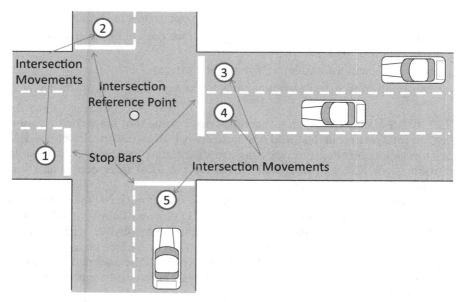

Figure 10.2. Geometric intersection description elements

10.2.1.4 Signal Phase and Timing In CICAS-V, the local RSU broadcasts traffic signal controller SPAT information at 10 Hz. It is up to the vehicle OBU to select the correct signal indication based on its approach to the intersection. The SPAT message contains the signal phase indication of the current phase, the time until the next signal phase change, and information to correlate the signal indication with a particular approach within all the approaches in the GID.

Like the GID, SPAT information is best included in a single DSRC packet. This packet is generated by extracting in real time the data from the traffic signal controller. Depending on the traffic signal controller hardware and the signal controller protocol, SPAT information is either exported directly by the traffic controller to the local RSU or determined through an inference state machine running on the local RSU.

SPAT information is represented in binary Extensible Markup Language (XML), following the proposed SAE standard [SAE09].

10.2.2 CICAS-V Development

10.2.2.1 Hardware Components At signalized intersections, the roadside portion of the system consists of an RSU with a DSRC radio and a local GPS station. When differential GPS (DGPS) corrections are used, the RSU also includes a DGPS base station receiver and its antenna.

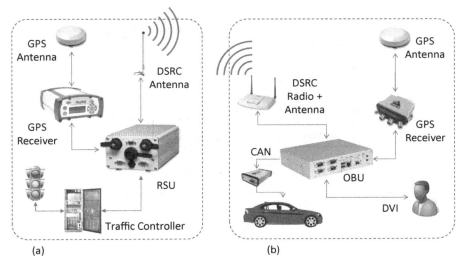

Figure 10.3. CICAS-V hardware components. (a) CICAS-V roadside equipment; (b) CICAS-V onboard equipment

The RSU is connected to the traffic signal controller through a serial or Local Area Network (LAN) connection. Multiple equipment suppliers, regional traffic management authority rules, equipment age, and the policies for roadside installations complicate the deployment of RSUs at traffic intersections.

The RSU can be further connected to a backbone Internet Protocol (IP) network. This allows in-vehicle Internet services to be offered through DSRC service channels. It also allows remote management of the RSU. However, IP connectivity is not required for CICAS-V to function (Figure 10.3).

The vehicle portion of the CICAS-V system includes an OBU computing platform. The OBU is connected to the vehicle controller area network (CAN), a DSRC radio, a GPS receiver, and the driver–vehicle interface (DVI). To facilitate joint system development over proprietary CAN systems, a CAN gateway provides a generic interface with the OBU.

The DENSO Wireless Safety Unit (WSU) was used as the OBU computing platform. It featured a 400 MHz MPC5200B PowerPC processor with 128MB DDR SDRAM running a Linux operating system. The WSU was developed especially for automotive ITS applications and included hardware and software interfaces for DSRC, GPS, and CAN.

10.2.2.2 Software Components The CICAS-V software architecture, of which a simplified version is illustrated in Figure 10.4, comprises message handling and violation detection modules. Message handling modules provide

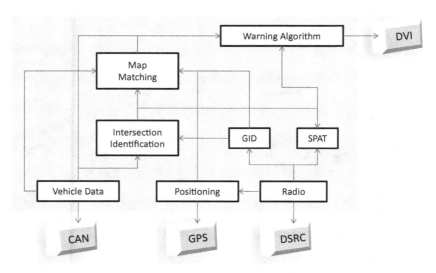

Figure 10.4. CICAS-V OBU software architecture

interfaces with the wireless access in vehicular environments (WAVE) radio, GPS receiver, and CAN.

The Radio Handler processes incoming messages from the RSU, extracts SPAT and GID information, and passes this information to other modules. When the Radio Handler receives DGPS messages from a RSU, it passes this information to the Positioning Handler. This module estimates vehicle position based on input from the GPS receiver and Radio Handler. It makes these estimates available to Intersection Identification and Map Matching. Likewise, the Vehicle Data Handler provides Intersection Identification and Map Matching with up-to-date vehicle data extracted from the CAN.

The other modules in the CICAS-V architecture are violation detection modules. They process the latest vehicle dynamics, GPS, GID, and SPAT data to determine whether an intersection violation is likely to occur. Intersection Identification verifies which intersection is being approached and the correspondent GID. Map Matching positions the vehicle on the GID. Finally, when the CICAS-V warning algorithm determines that the driver should be alerted, it triggers the vehicle DVI.

10.2.2.3 El Camino Real and 5th Avenue Intersection

The first intersection in which DSRC technology was implemented is El Camino Real and 5th Avenue in Atherton, California. The California Department of Transportation (CALTRANS) performed the installation on January 27, 2007.

Some approaches to this intersection are highlighted in Figure 10.5. Approach #2 contains three lanes and approach #5 contains two lanes. Approach #6 consists of three lanes for which the rightmost lane is wider than

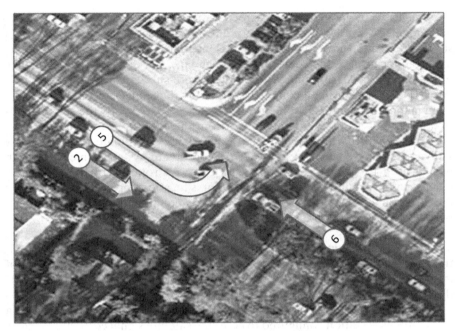

Figure 10.5. El Camino Real and 5th Avenue in Atherton, California

the other two due to parking possibilities. For approach #2, the GID specifies the leftmost through lane as a reference lane and the other two lanes in the approach are represented as computed lanes. The same is true for approach #6, for which the leftmost lane is specified through a node list and the other two lanes can be specified by the offset from the reference lane.

Computed lanes are not a mandatory feature of the GID but only a means to compress its size. All lanes can be specified through node lists when the size of the GID permits, but it is often more practical to specify one lane as a reference lane and compute contiguous lanes from that data. For example, the size of the El Camino Real and 5th Avenue GID is 352 bytes. The GID size for the most complex intersection utilized for the CICAS-V project (Franklin and Peppers Ferry in Christiansburg, VA) is 869 bytes.

10.2.2.4 GID Generation   No existing commercial maps can describe intersection geometries with the required accuracy for CICAS-V. Furthermore, some of the required GID attributes, such as the stop bar location, are not found in existing maps. Therefore, researchers must generate their own GIDs for the intersections under investigation.

To rectify this situation, the CICAS-V project hired HJW GeoSpatial to map intersections using aerial surveying. HJW took high-resolution aerial photographs of the intersections, which were then orthorectified, meaning that the images were geometrically corrected to be on a uniform scale. To verify and enhance accuracy, a number of points on the picture were mapped by a surveyor on site. The company took the lane markings on the image to determine the location of the centerline for each lane and delivered the geometry of the lanes as a set of points. Those points were subsequently converted into the GID message, using a compiler specifically developed for CICAS-V.

In mapping the intersection, work was conducted to specify the North direction accurately as the Geographic North in the WGS 84 coordinate system. Using the "North" direction in the State Plane Coordinate System or the Universal Transverse Mercator (UTM) coordinate system, which are both widely used to specify geography, will lead to a rotation of the GID with respect to the ground truth by an angle that is location dependent. The farther the location of the intersection is from the equator, the larger the angle between UTM north and geographic north.

10.2.2.5 Positioning A major challenge of CICAS-V was to achieve real-time vehicle positioning accurate to less than a meter, in the proximity of instrumented intersections at relatively low cost while using commercial off-the-shelf products available.

The adopted solution for signalized intersections relies on the RSU with a local GPS base station configured to compute satellite signal correction factors for approaching vehicles. These DGPS corrections are sufficient to reconcile positions resulting from estimation algorithms with the base station's known fixed location. This DGPS solution was chosen over the Wide Area Augmentation System (WAAS), another popular correction technique, due to DGPS's higher standards of accuracy. WAAS relies on ground reference stations spaced approximately 500 mi apart and computes corrections on a regional basis. The fine-tuning which is achievable in DGPS is far greater than that provided by WAAS positioning. For example, at the CICAS-V intersection located in Farmington Hills, Michigan, test vehicles consistently achieve absolute real-time vehicle positioning errors of less than 0.5 m.

In a CICAS-V RSU, the GPS receiver is configured in base-station mode, where it computes corrections to GPS satellite signals for vehicle-mounted GPS receivers in its vicinity. The correction information is encoded in Radio Technical Commission for Maritime Services (RTCM) standard format [RTCM06], such as the RTCM Recommended Standards for Differential Global Navigation Satellite Systems (GNSS) service as defined by the Special Committee (SC) 104 on Differential Global Navigation Satellite Systems (DGNSS). For brevity, the corrections data message format is referred to as either RTCM v3.0 or RTCM v2.3, depending on which SC-104 release of the Recommended Standards for DGNSS is used.

The RTCM v3.0 message format used in CICAS-V consists of single-frequency (L1) GPS information. The RTCM 1001 corrections provide per-satellite GPS pseudoranges and carrier phase measurements so the onboard (moving) GPS receivers can estimate positions with much higher accuracy and

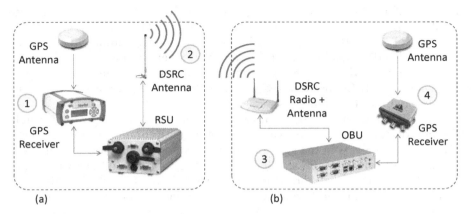

Figure 10.6. CICAS-V aerial positioning correction equipment. (a) RTCM SC-104 data generation and broadcast over DSRC; (b) RTCM SC-104 data reception and use by vehicle's GPS device

reliability. Limiting RTCM 1001v3.0 messages to include L1-only correction information improves accuracy with relatively modest overhead on the communications and GPS receiver workload. For example, only 101 bytes are required for RTCM messages carrying range corrections for 12 satellites and is often smaller, depending on the number of visible GPS satellites in the current constellation [RTCM06].

The amount of correction data that has to be broadcast depends on the RTCM version and on the number of visible satellites. For example, the RTCM v2.3 format requires about 4800 bits per second (bps) to broadcast dual-frequency code and carrier phase observations or observation corrections of 12 satellites. Similar information can be transmitted using 1800 bps in the newer RTCM v3.0 format (i.e., for 12 visible satellites, v2.3 requires 372 bytes to transmit data, whereas RTCM v3.0 requires only 8 + 7.25 × 12 bytes).

Figure 10.6 shows how DGPS corrections are used in CICAS-V. First, the base station GPS receiver generates RTCM SC-104 V3.0 corrections (1). Then, the RSU compiles this data in a DSRC message broadcasted over DSRC at 1 Hz (2). The OBU extracts RTCM-104 V3.0 corrections from the DSRC message (3) and provides them to the onboard GPS unit, which applies the received correction information to its position calculations, yielding much higher accuracy position estimates (4).

RTCM v3.0 is primarily designed to support Real-Time Kinematic (RTK) operations that normally require broadcasting relatively large amounts of information, and generally implies highly sophisticated forms of correction analysis and error removal. However, the L1-only subset of the RTCM v3.0 format can provide good performance improvements for modest system resource requirements, and it works well even with moderately priced receivers. A minimum of two messages from the RSU are required to support local differential L1 solution correction for onboard GPS receivers.

Each RTCM 1001 message contains the satellite observations (in particular the L1-only GPS pseudorange and carrier phase measurements) as derived by the base station GPS receiver by comparing the position estimate determined from current satellite pseudorange observations with the surveyed fixed location of the base station antenna.

The GPS base station works backwards to compute corrections to the satellite pseudoranges that would yield a much more accurate position estimate. Other GPS receivers in the surrounding area will generally face the same set of inaccuracies in the GPS satellite pseudorange observations, so when they apply these pseudorange correction factors to their own observations, they too will be able to significantly reduce errors and obtain a more accurate position estimate.

10.2.2.6 Communications   CICAS-V decided to broadcast SPAT messages over the DSRC control channel (#178) and to broadcast GID and GPS correction messages over a DSRC service channel.

The WAVE Service Advertisement (WSA) messages sent over the control channel included information about the intersection (Intersection ID), the GID version number, and the service channel used for broadcasting the GID and GPS correction messages.

Upon receiving a WSA message, the vehicle OBU may switch its DSRC radio to the indicated service channel to receive the full GID for the intersection as well as GPS corrections. If the OBU already has the latest GID, it will discard the newly received GID but still receive the GPS corrections.

10.2.2.7 Driver–Vehicle Interface   The DVI is an essential component of CICAS-V, which warns the driver of an impending violation. It was not a goal for the CICAS-V project to develop a standard DVI as it is expected that each automotive original equipment manufacturer (OEM) will develop proprietary solutions in the future. The DVI developed for the project included a visual icon, a speech-based warning, and a brake pulse. The test results showed this combination to be highly effective. The CICAS-V DVI is presented in detail in [PeNK09].

10.2.3   CICAS-V Testing

10.2.3.1 System Installation   The components for the CICAS-V roadside equipment installation are shown in Figure 10.7.

Figure 10.7a shows the installation of the RSU, GPS receiver, and data acquisition system (DAS) inside the traffic signal controller cabinet at an intersection in Blacksburg, Virginia. Figure 10.7b shows the installation of the DSRC and GPS antennas at an intersection in Oakland County, Michigan. Figure 10.7c shows the DENSO WSU, Netway CAN Gateway, NovAtel OEMV GPS Receiver, and Virginia Tech Transportation Institute (VTTI)

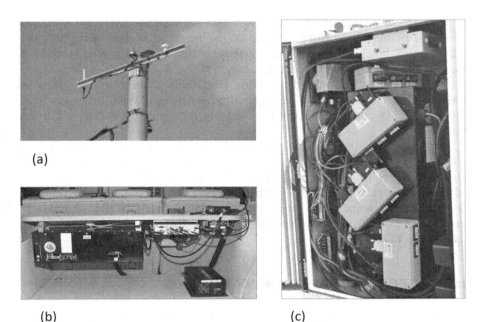

(a)

(b)   (c)

Figure 10.7. CICAS-V RSU and OBE installation. (a) CICAS-V DSRC and GPS antennas; (b) CICAS-V in vehicle equipment; (c) CICAS-V traffic controller

DAS. The DAS recorded messages from the vehicle bus, from the intersections, OBU computation outputs, and camera images from four onboard cameras.

10.2.3.2 Objective Tests A set of objective test procedures (OTPs) was created to validate CICAS-V readiness for a large-scale field operational test (FOT). These OTPs included warning tests (system must issue a warning), nuisance tests (system must not issue a warning), and engineering tests (system limits are tested). The tests covered the typical situations that would be encountered by a vehicle approaching a CICAS-V intersection. They were written in such a way that allows any supplier to use them to test whether their system fulfills the performance specifications.

Tests were conducted on the Virginia Tech Smart Road, a closed test track featuring an intersection provided by VTTI [NeDo09]. To pass a warning test, the system had to alert the driver within a distance equivalent to 200 ms multiplied by vehicle speed with respect to the timing defined by the CICAS-V warning algorithm. All the warning modalities had to be activated within 200 ms from each other.

A series of criteria was defined to determine whether a test was valid. For instance, the speed variance had to be smaller than 2.5 mph with respect to the nominal test speed. Each test had to have at least eight valid runs. Tests for signalized and stop-controlled intersections consisted of approaches at speeds of 25, 35 and 55 mph, each of which needed eight valid runs.

The system passed all the objective tests with almost 100% of the runs passing. The only failed run happened at one intersection approach where the brake pulse failed to activate, even though the other warning modalities warned the driver at the correct distance.

In the SPAT reflection and reception tests, the vehicle followed a tractor-trailer within a distance of 4.5 m to see whether enough DSRC messages from the local RSU could be received to enable the vehicle to issue a correct warning.

In all the tests, the warning was issued such that the vehicle was able to come to a stop before entering the intersection crash box, but in several instances the warning came more than 200 ms late. The complete description of the objective test procedures can be found in [MABC09a], and the complete analysis of the objective testing can be found in [MABC09b]. The system performed well during the objective tests and was judged to be ready for a large-scale FOT.

10.3 INTEGRATED SAFETY DEMONSTRATION

A demonstration based on CICAS-V technology was showcased by Mercedes-Benz at the 15th Intelligent Transportation Systems World Congress in New York City in 2008 [RoDe10]. This demo shows how vehicle safety communications can be combined with existing onboard sensors and in-vehicle safety systems to offer enhanced protection to car occupants.

The prototype was built for demonstration purposes only. It relies on SPAT and GID information broadcast over DSRC by the local RSU. Audible and visual warnings in the vehicle alert the driver of an impending red-light violation. If these warnings are repeatedly ignored and a red-light violation is imminent, the vehicle brakes automatically and comes to a safe stop before entering the intersection crash box. Although the brakes are engaged automatically, the system is not intended to take full control of the vehicle and the driver can override it at any time simply by pressing the brake or gas pedal (Figure 10.8).

The functionality of this prototype is further extended by its integration with existing in-vehicle precrash and brake assistance systems such as Mercedes-Benz PRE-SAFE and BRAKE ASSIST. Precrash safety features are activated as an additional precaution to reduce impact energy and protect vehicle occupants in an unavoidable collision. The prototype provides a basis for a future cooperative safety system that integrates wireless communications with in-vehicle safety systems.

10.3.1 Demonstration Concept

As the vehicle approaches the intersection, it receives SPAT, GID, and DGPS messages from the RSU. The vehicle determines which intersection is being

Figure 10.8. Integrated Safety demonstration vehicle in action at the 15th ITS World Congress

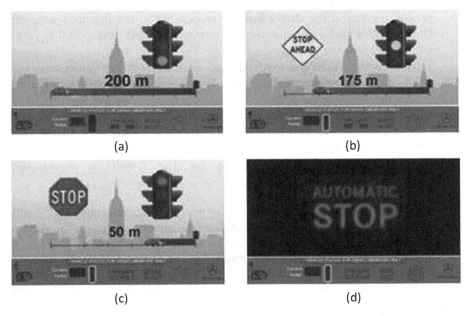

Figure 10.9. Integrated Safety demonstration display. (a) Approach; (b) first alert; (c) final alert; (d) brakes applied

approached, what lane the vehicle is in, and the corresponding traffic signal phase and timing. Figure 10.9a is displayed on the vehicle head unit when no RSUs are detected, whereas Figure 10.9b is displayed when the vehicle is receiving RSU communications. The banner along the bottom of each image provides vehicle status, including brake and accelerator pedal status, activation of active safety systems, and automatic stop.

If the driver is proceeding toward a red light and he or she has not taken appropriate actions, the vehicle warns the driver and attempts to avoid a red-light violation in three stages:

- *Red Light Advisory*: This is an advisory issued to the driver indicating that a red light is being approached and that the driver should have begun to take action already. A "Stop Ahead" icon is displayed, as shown in Figure 10.9b. A subtle acoustic warning is also issued.
- *Red Light Warning*: If the driver continues toward the intersection without taking action, the severity of the advisory is elevated, a "Stop" icon as shown in Figure 10.9c is displayed, and a loud tone is heard. The timing of this warning is such that the driver can still react in time to stop the vehicle before entering the intersection. However, hard braking would be required to stop in time.
- *Automatic Stop*: This is the most critical part of the warning sequence and is only reached if the driver has not responded to the previous warning stages and a red-light violation is unavoidable. At this point, several actions are taken:
  - The driver is issued a loud audible warning and the vehicle display indicates that an automatic stop is being executed, as shown in Figure 10.9d.
  - The vehicle prepares for a potential collision by activating Mercedes-Benz PRE-SAFE. This causes tightening of seat belts, optimizing seat positions, rolling up windows, and raising rear headrests.
  - The vehicle applies the brakes to stop before it enters the intersection.

The three warning stages are activated adaptively, based on vehicle speed and road conditions. They can also be adapted to the characteristics of the driver. However, it is important to note that the onset of autonomous braking is independent of driver style since it is applied at the last possible point that the vehicle can be stopped before entering the intersection and requires no driver input. Thus, driver reaction time can be ignored in the calculation. If the driver begins to brake (or accelerate) at any time, the system is overridden. This allows the driver to maintain final control should the situation warrant it.

10.3.2 Hardware Components

The prototype is heavily based on the CICAS-V system. In the succeeding sections, we focus on additional functionality.

10.3.2.1 *Mercedes-Benz PRE-SAFE* Mercedes-Benz PRE-SAFE was first introduced into the 2003 Mercedes-Benz S-Class. The current system uses a variety of onboard sensors (e.g., radar and accelerometers) to detect

hazardous conditions. When activated, the system tightens seat belts, adjusts seat positions, inflates seat cushions, closes windows and sunroofs, and raises the rear headrests to prepare for a potential crash. When a collision is inevitable, the system can initiate partial braking (1.6 seconds before impact) and full braking (0.6 second before impact) to reduce collision impact energy.

10.3.2.2 CAN Gateways Two separate system functionalities are required for the demo. The first is the activation of PRE-SAFE, the second is autonomous braking. To activate Mercedes-Benz PRE-SAFE, CAN messages from the electronic stability control–electronic control unit (ESC ECU) were intercepted before reaching the vehicle CAN bus. When required, messages were modified before being passed onto the CAN network. A similar gateway was inserted between the radar decision unit (RDU) and the vehicle CAN. To activate autonomous braking, this gateway broadcasts a CAN message to request the ESC unit to apply the desired braking torque.

10.3.2.3 Emergency Shutoff Switch An emergency power shutoff was installed into the dashboard. This switch cuts power to the ESC ECU and the OBU to immediately disable the demo system while still permitting regular vehicle operations.

10.3.2.4 Onboard Software Systems The onboard software can be broken into three separate categories. First, the software which interfaces with the vehicle and sensors was implemented on the CAN interface devices. Second, the warning algorithm was based on the platform developed by the CICAS-V project and used many interfaces provided by WSU hardware and software drivers.

One important design decision was to completely separate (both physically and logically) autonomous braking from the warning generation. In this way, a low-level safety check could be performed before braking was initiated, and a well-defined recovery mechanism was put in place to deal with any higher-level system malfunction. In brief, red-light violations were computed in one program and autonomous braking was implemented in another. The two processes communicated only through a CAN interface and failure of one system did not cause the failure of the other.

10.3.3 Demo Design

10.3.3.1 Positioning The positioning system was an improved version of the one used in the CICAS-V project. Table 10.1 shows the standalone GPS accuracy. We assumed worst-case error ranges and a baseline of 1 km. These values exclude the effect of horizontal dilution of precision (HDOP) that typically increases the expected error. These results show that DGPS corrections could provide sufficient positioning accuracy to support violation warnings.

Table 10.1. One-sigma (1σ) pseudorange GPS error budget with and without DGPS correction

Error Source	L1 GPS	L1–L2 GPS	L1–L2 DGPS
Broadcast clock error	1.1	1.1	0.0
L1 P(Y)–L1 C/A group delay	0.3	N/A	N/A
Broadcast ephemeris	0.8	0.8	0.0006
Ionospheric delay	7.0	0.1	0.04
Tropospheric delay	0.2	0.2	0.04
Receiver noise and resolution	0.1	0.1	0.1
Multipath	0.2	0.2	0.3
Total error (RSS)	**7.1 m**	**1.4 m**	**0.3 m**

RSS, received signal strength.

Figure 10.10. GPS errors due to nearby object interference with satellite ranging

The results in Table 10.1 do not fully reflect the effects of signal multipath and interference on positioning accuracy. Such effects are shown in Figure 10.10. A Kalman filter position estimator can be used to overcome positioning errors caused by multipath and interference. The estimator can combine GPS position information (latitude, longitude, velocity, HDOP, and standard deviation of error in latitude and longitude) with sensor measurements available in the vehicle (wheel rotation speed and direction, yaw rate, and longitudinal acceleration) to produce an accurate vehicle position.

Figure 10.10 shows two "walls," with each panel representing a GPS measurement. The taller wall is the raw GPS measurements and the shorter wall is the output from the estimator. The vehicle was modeled as a free body object, with lateral and longitudinal model error covariance. Model error covariance was chosen to reflect vehicle nonholonomic behavior and to optimize the estimator performance. Using this model and the onboard sensor information, open-loop position prediction produced errors of less than 2 m over 200 m of driving under regular driving conditions (i.e., smooth asphalt roads, moderate speeds, and low wheel slippage).

Position updates are received from the GPS receiver at 10 Hz. Vehicle sensor updates are received at 50 Hz. These updates and measurements are not synchronized. Hence, GPS position update, wheel velocity measurement, and vehicle yaw rate measurement may arrive at the estimator in a random order with different interarrival times. To account for this, when any measurement is received, the Kalman filter time is updated for the time period since the last measurement arrival. A measurement update is then performed using the received measurement.

The estimator resolved two important issues. First, when there is an insufficient number of visible satellites to generate a position estimate or if the position error is too large, the GPS receiver will not provide a position estimate. The estimator filled in these missing observations by generating a position estimate using onboard vehicle sensors and the vehicle model. Different measurement updates were performed depending on which sensor measurements were received. When GPS data are not available, vehicle velocity sensor updates ensure that the vehicle velocity estimation error will remain bounded even though the position error can grow unbounded.

Second, when an individual satellite signal is corrupted (e.g., due to interference), the HDOP and measurement error covariance values do not always reflect this condition immediately. The estimator mitigates the effect of these errors by independently evaluating the reliability of a measurement before using it to generate an updated position.

10.3.3.2 Brake Torque Calculation The Mercedes-Benz S-Class has two mechanisms for applying brake torque. The brake pump is used by the ESC unit to implement features such as Hill Start Assist, Fading Brake Support, and Airgap Reduction. The brake booster is used by the brake assist system (BAS) to augment the driver's applied brake force in emergency braking situations.

For the Integrated Safety demonstration, the brake pump was used to generate braking torque. A brake torque request is sent via the CAN bus to the ESC unit, which activates the brake pump. In our prototype, this incurs a delay of approximately 0.15 second. The pump then begins to build up pressure and the applied brake torque increases linearly. Once the desired torque has been achieved, the pump maintains the required pressure until the torque request is completed.

If the torque is too large and the wheel begins to slip, ESC detects the slip and reduces the torque applied to that particular wheel. If this occurs, predicting a stopping distance and the required braking torque becomes difficult. For this reason, the automatic stopping distances were chosen to avoid excessive wheel slip. Braking distances were chosen for wet and dry conditions and a good quality asphalt surface for the Integrated Safety demonstration.

The system described above reliably stopped the vehicle within 20 cm of the stop bar from any initial speed and distance [RoDe10].

REFERENCES

[KNMK10] S. Kiger, V. Neale, M. Maile, R. Kiefer, F. Ahmed-Zaid, L. Caminiti, J. Lundberg, P. Mudalige, and C. Pall: "Cooperative Intersection Collision Avoidance System Limited to Stop Sign and Traffic Signal Violations (CICAS-V) Task 13 Final Report: Preparation for Field Operational Test," Washington, DC: National Highway Traffic Safety Administration (NHTSA), 2010. (In print.)

[MABC09a] M. Maile, F. Ahmed-Zaid, C. Basnyake, L. Caminiti, S. Kass, M. Losh, J. Lundberg, D. Masselink, E. McGlohon, P. Mudalige, C. Pall, M. Peredo, Z. Popovic, J. Stinnett, and S. VanSickle: "Cooperative Intersection Collision Avoidance System Limited to Stop Sign and Traffic Signal Violations (CICAS-V) Task 7 Final Report: Objective Testing Procedures," National Highway Traffic Safety Administration (NHTSA), Washington, DC, 2009. (In print.)

[MABC09b] M. Maile, F. Ahmed-Zaid, C. Basnyake, L. Caminiti, S. Kass, M. Losh, J. Lundberg, D. Masselink, E. McGlohon, P. Mudalige, C. Pall, M. Peredo, Z. Popovic, J. Stinnett, and S. VanSickle: "Cooperative Intersection Collision Avoidance System Limited to Stop Sign and Traffic Signal Violations (CICAS-V) Task 11 Final Report: Objective Tests," National Highway Traffic Safety Administration (NHTSA), Washington, DC, 2009.

[MABC08] M. Maile, F. Ahmed-Zaid, C. Basnyake, L. Caminiti, S. Kass, M. Losh, J. Lundberg, D. Masselink, E. Mc-Glohon, P. Mudalige, C. Pall, M. Peredo, Z. Popovic, J. Stinnett, and S. VanSickle: "Final Report: Cooperative Intersection Collision Avoidance System for Violations (CICAS-V)," Technical Report, National Highway Traffic Safety Administration (NHTSA), Washington, D.C., 2008.

[MaDe09] M. Maile and L. Delgrossi: "Cooperative Intersection Collision Avoidance System for Violations (CICASV) for Avoidance of Violation Based Intersection Crashes," 21st International Conference on Enhanced Safety of Vehicles (ESV), no. 09-0118, 2009.

[NaSY07] W. Najm, J. Smith, and M. Yanagisawa: "Precrash Scenario Typology for Crash 810 767," National Highway Traffic Safety Administration (NHTSA), Washington, D.C., 2007.

[NHTS06] National Center for Statistics and Analysis (NCSA): "Traffic Safety Facts 2006: A Compilation of Motor Vehicle Crash Data from the Fatality Analysis Reporting System and the General Estimates System," Technical Report, DOT HS 810 818, National Highways Transportation Safety Authority (NHTSA), 2006.

[NeDo09] V. Neale and Z. Doerzaph: "Field Test of a Cooperative Intersection Collision Avoidance System for Violations (CICAS-V)," ESV, Paper Number 09-0478, 2009.

[NHTS11] NHTSA: "Independent Evaluation of the Driver Acceptance of the Cooperative Intersection Collision Avoidance System for Violations (CICAS-V) Pilot Test," DOT HS 811 497, 2011.

[PeNK09] M. Perez, V. Neale, and R. Kiefer: "Test and Evaluation of the Cooperative Intersection Collision Avoidance System for Violations (CICAS-V) Driver Vehicle Interface," ESV, Paper Number 09-0461, 2009.

[RTCM06] Radio Technical Commission for Maritime Services: "Differential Global Navigation Satellite System (GNSS) Services," Technical Report, RTCM Standard 10403.1, 2006.

[RoDe10], C. Robinson and L. Delgrossi: "Integrating In-Vehicle Safety with Dedicated Short Range Communications for Intersection Collision Avoidance," SAE World Congress, Detroit, MI, 2010.

[SAE09] Society of Automotive Engineers: "SAE J2735 Dedicated Short-Range Communications (DSRC) Message Set Dictionary," 2009.

11

VEHICLE-TO-VEHICLE SAFETY APPLICATIONS

Written in cooperation with Michael Maile

11.1 COOPERATION AMONG VEHICLES

This chapter focuses on vehicle-to-vehicle (V2V) safety applications for avoiding imminent collisions. It discusses the basic design philosophy and concept of operations for these safety applications and describes the message formats used to exchange safety information between vehicles. It further describes a system architecture designed to support V2V safety applications, including crucial enabling mechanisms such as target classification, path history, and path prediction. Finally, it provides a high-level description of sample V2V safety application implementations.

The main goal of V2V cooperative safety systems is for vehicles to be aware of other nearby vehicles and continually assess collision risks by exchanging safety messages describing vehicles' current status. Vehicular communications can deliver information beyond onboard sensors' range or field of view. It can also provide otherwise indiscernible, high-quality information such as vehicle weight and size, brake status, and so on.

For low-end vehicles, costs of autonomous sensors can represent a considerable fraction of a vehicle's total cost. Therefore, autonomous sensors are typically introduced first on high-end vehicles today. Radio devices are expected

Vehicle Safety Communications: Protocols, Security, and Privacy, First Edition. Luca Delgrossi and Tao Zhang.
© 2012 John Wiley & Sons, Inc. Published 2012 by John Wiley & Sons, Inc.

to be less costly, thus making communications-based safety solutions more feasible and affordable across the entire vehicle market.

Vehicular communications can provide a foundation for safety systems aimed at alerting drivers of imminent dangers. For safety systems that can activate vehicle control functions, vehicular communications should be combined with onboard autonomous sensors to provide redundancy and to validate the information obtained from wireless channels. Until now, research has mostly focused on communication-based systems aimed at alerting the driver. The prototypes described in this chapter are limited to alerting the driver and are not intended to activate vehicle control functions.

11.2 V2V SAFETY APPLICATIONS

In 2010, National Highway Traffic Safety Administration (NHTSA) published a report estimating the different crash types that could be addressed by various types of Intelligent Transportation Systems [NHTS10]. This NHTSA report focused on safety applications that incorporate V2V communications to increase situational awareness and support driver warnings. The report shows that such V2V systems have the potential to address about 4,409,000 (or 79%) of all police-reported vehicle target crashes, 4,336,000 (or 81%) of all police-reported light-vehicle target crashes, and 267,000 (or 71%) of all police-reported heavy-truck target crashes each year [NHTS10].

In the United States, the first prototype V2V safety applications focused on imminent collision avoidance were designed and implemented as part of the Vehicle Safety Communications–Applications (VSC-A) project [NHTS11] between 2006 and 2009. VSC-A was a collaboration between the U.S. Department of Transportation (USDOT) and the Vehicle Safety Communications 2 (VSC2) Consortium of automakers organized within the Crash Avoidance Metrics Partnership (CAMP) framework. VSC2 Consortium members are Ford, General Motors, Honda, Mercedes-Benz, and Toyota.

Based on a ranking of crash scenarios considering frequency, cost, and functional years lost [NHTS07], the following V2V applications were selected for implementation on the VSC-A project:

- *Emergency Electronic Brake Lights (EEBL)*: This application broadcasts "hard braking" messages to surrounding vehicles. Vehicles use the received hard braking messages to determine when it becomes necessary to warn the driver.
- *Forward Collision Warning (FCW)*: This application warns the driver of the risk of a collision with a vehicle ahead in the same lane and direction of travel.
- *Lane Change Warning (LCW) and Blind Spot Warning (BSW)*: These applications warn the driver when a blind-spot zone is or is about to be occupied by another vehicle traveling in the same direction.

Table 11.1. Vehicle crash scenarios and V2V safety applications

	EEBL	FCW	BSW and LCW	DNPW	IMA	CLW
Lead vehicle stopped		✓				
Control loss without prior vehicle action						✓
Vehicle(s) turning at nonsignalized junctions					✓	
Straight crossing paths at nonsignalized junctions					✓	
Lead vehicle decelerating	✓	✓				
Vehicle(s) not making a maneuver—opposite direction				✓		
Vehicle(s) changing lanes—same direction			✓			
Left turn across path/ opposite direction at nonsignalized junctions					✓	

- *Do Not Pass Warning (DNPW)*: This application warns the driver during a passing maneuver when a vehicle ahead and in the same lane cannot be safely passed.
- *Intersection Movement Assist (IMA)*: This application warns the driver when it is not safe to enter an intersection due to high collision probability with other vehicles.
- *Control Loss Warning (CLW)*: This application broadcasts "control loss" messages to surrounding vehicles. Vehicles used the received control loss messages to determine when it becomes necessary to inform or warn the driver.

Table 11.1 lists a number of selected critical crash scenarios and shows which is addressed by VSC-A V2V applications [NHTS11].

11.3 V2V SAFETY APPLICATIONS DESIGN

Typically, V2V safety applications use the data received over a wireless link without any further interactions with the data source. The data source can often be a vehicle built by another automotive original equipment manufacturer (OEM). V2V safety applications must determine how incoming messages need to be processed, whether the source is reliable, whether the incoming messages are trustworthy, and when and how the driver should be alerted.

11.3.1 Basic Safety Messages

The concept of operations for V2V safety applications requires each vehicle to periodically broadcast safety information in a standard format intelligible by surrounding vehicles.

An analysis conducted as part of the VSC-A project determined that most V2V safety applications require a small set of information such as position, speed, heading, brake status, and vehicle size [NHTS11]. This suggests that a common message set can support multiple V2V safety applications as opposed to separate messages for each safety application. The basic safety message (BSM) was specified as part of the Society of Automotive Engineers (SAE) J2735 dedicated short-range communications (DSRC) Message Set standard [SAE09].

SAE J2735 currently specifies 15 message types, of which the BSM is one of the most prominent. Other DSRC message types include roadside alert, signal phase and timing, map data, probe vehicle data, and traveler information. Although SAE J2735 is intended for use over the DSRC band, the message set definition is independent of the spectrum and can be utilized in different contexts.

A BSM consists of data elements and data frames. A data element is a basic building block and a data frame comprises one or more data elements or other data frames. Vehicles use data elements and data frames to compose BSMs just as words in a dictionary are used to build sentences. For this reason, one may think of SAE J2735 as the data dictionary for V2V communications [Kenn11].

When defining BSM formats, special efforts were made to minimize the message size. Smaller messages can help reduce DSRC channel congestion. To keep BSM sizes small, their content is structured into two parts. Part I (basic vehicle state) is mandatory and contains those data elements and data frames that must always be included in a BSM. Part I has a fixed size of 39 bytes. Part II (vehicle safety extension) includes optional data elements and data frames. BSM data elements and data frames are represented in Table 11.2 and are marked as type "E" and type "F," respectively. Part II data elements and data frames are marked as optional.

To reduce channel load, the vehicle safety extensions is included in a BSM only when necessary. Typically, vehicles periodically broadcast BSM Part I. Only upon the need of transmitting additional data is BSM Part II attached. For example, specific events such as emergency braking and control loss can be described by setting the corresponding event flag in BSM Part II.

The safety information included in BSMs is transmitted in plain text because the information is meant to be seen by all receivers.

11.3.2 Minimum Performance Requirements

The SAE J2735 standard defines the formats of BSM messages, but does not address data accuracy. V2V safety applications can only tolerate a certain level

Table 11.2. BSM data elements and data frames

BSM Data Item	Type	Bytes	Part	BSM Data Item	Type	Bytes	Part
Message ID	E	1	I	Heading	E	2	I
Message count	E	1	I	Steering wheel angle	E	1	I
Temporary ID	E	4	I	Accelerations	F	7	I
Time	E	2	I	Brake system status	F	2	I
Latitude	E	4	I	Vehicle size	F	3	I
Longitude	E	4	I	Event flags (optional)	E	2	II
Elevation	E	2	I	Path history (optional)	F	Var.	II
Positioning accuracy	F	4	I	Path prediction (optional)	F	3	II
Transmission and speed	F	2	I	RTCM package (optional)	F	Var.	II

RTCM, radio technical commission for maritime services.

of error in the data contained in BSMs while still functioning as designed. For example, based on an initial assessment, it is expected that vehicle speed accuracy should be better than 0.35 m/s and heading accuracy should be better than 3° when the speed is greater than 11.5 m/s. However, this depends on the relative positioning accuracy and how specific algorithms are implemented. Further research is required to specify data accuracy requirements.

To address the lack of data accuracy standards, SAE is currently developing SAE J2945 Minimum Performance Requirements to specify BSM data accuracy, broadcast frequency, and transmission power [SAE10]. The current SAE J2945 standard is work in progress. Requirements are being determined through simulation analysis as well as empirical results. It is expected that the experience gained from the USDOT V2V Safety Pilot in 2012 will lead to more conclusive assessments.

Among other BSM usage parameters, SAE J2945 is expected to specify how frequently BSMs should be broadcast. This choice is a trade-off between long interpacket delays experienced by V2V safety applications and heavy wireless channel utilization. It is generally accepted among experts that broadcasting BSMs at 10 Hz is sufficient to meet the requirements of the most demanding V2V safety applications. This establishes an upper limit for BSM transmission frequency. The actual frequency, however, as well as its occasional variations will be determined by current vehicle dynamics and the capabilities of the adopted channel congestion control algorithms.

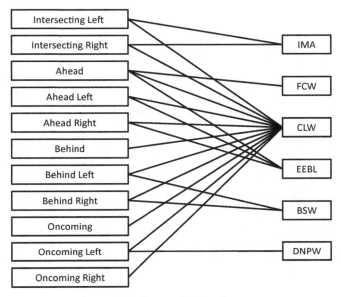

Figure 11.1. Target classification zones

11.3.3 Target Classification

A fundamental building block for V2V safety applications is the algorithm designed to identify, classify, and monitor surrounding vehicles that could introduce collision risks.

In the VSC-A project, this algorithm is denominated Target Classification. We will refer to a vehicle running Target Classification as the host vehicle (HV). We will refer to HV's neighboring vehicles as remote vehicles (RVs). Each vehicle can be an HV, an RV, or both.

Target Classification uses the positioning information included in incoming BSMs to classify RVs into zones of interest for safety applications. When RVs are detected, they are assigned to specific zones based on their positions relative to HV (Figure 11.1). In general, individual applications will be interested only in RVs assigned to specific zones. For instance, blind spot warning is only concerned with targets in the Behind Left or Behind Right zones. One advantage of this scheme is that individual applications do not consume system resources unless there is a target in their zones of interest.

The information included in the BSM basic vehicle state (Part I) describes the vehicle's current position in terms of latitude, longitude, and elevation. This tells a vehicle's neighbors where the vehicle is located but it does not provide any details on where the vehicle has recently been or is expected to be in the immediate future. The Path History and Path Prediction data frames, sent as part of BSM vehicle safety extensions (Part II) provide information for a

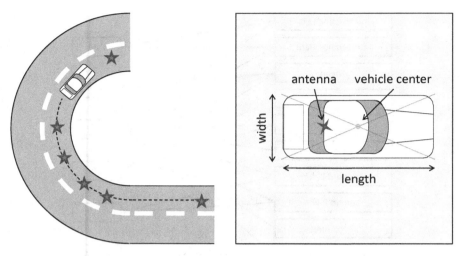

Figure 11.2. Path history bread crumbs (left) and vehicle representation (right)

vehicle to analyze the recent past and immediate future paths of neighboring vehicles.

Path History comprises a series of data elements, referred to as "bread crumbs," each describing a prior position occupied by the vehicle and the correspondent time. For straight paths, a few bread crumbs may suffice for an accurate description. For curved paths, however, a larger number of bread crumbs become necessary and the Path History data element may take up to 100 bytes. The Path Prediction data frame has a fixed length of 3 bytes and describes the path the vehicle expects to traverse. Path History and Path Prediction provide essential information to improve the accuracy of the Target Classification algorithm (Figure 11.2).

11.3.4   Vehicle Representation

Collision detection algorithms need accurate information on the space occupied by each vehicle over time. This requires a model to represent vehicles and the space they occupy as they travel and execute driving maneuvers on the road. In VSC-A, vehicles are represented as rectangles described by their length and width. The BSM position transmitted over the air corresponds to the vehicle center. Each vehicle calculates the vehicle center as an offset from the physical position of the DSRC antenna (typically installed on the roof of the vehicle).

Further research will be necessary to take into account more sophisticated vehicle representations. In particular, describing the space occupied by articulated vehicles (e.g., tractor and trailers) is a challenge. Two approaches are being considered to determine the trailer's position and heading: mathematical functions and separate BSMs. This problem concerns commercial as well as transit and passenger vehicles.

11.3.5 Sample Applications

VSC-A V2V safety applications have been developed by Ford, General Motors, Hyundai-Kia, Honda, Mercedes-Benz, Nissan, Toyota, and Volkswagen-Audi. These implementations have been subsequently refined based on results obtained through the V2V Interoperability (V2V-I) project. In 2011, the original graphical user interface (GUI) was replaced with driver–vehicle interfaces (DVIs) suitable for naïve drivers, and the system was installed in 24 passenger vehicles for use in the USDOT Driver Acceptance Clinics. Further refinements and optimizations are expected to be implemented to support the USDOT V2V Model Deployment activity in 2012–2013.

Here, we briefly describe an implementation of two V2V safety applications: emergency electronic brake lights (EEBL) and intersection movement assist (IMA). These implementations are illustrated at a high level with the only purpose of providing an example of solutions enabled by DSRC. A complete discussion covering VSC-A V2V safety applications can be found in [NHTS11].

11.3.5.1 Emergency Electronic Brake Lights

Upon the occurrence of a hard braking event, EEBL broadcasts a BSM inclusive of the Part II Event Flag data element with the Hard Brake flag set. In VSC-A, hard braking is defined as deceleration of at least $0.4g$. An HV receiving a BSM with Hard Brake flag set uses Target Classification to determine message relevance and severity. If the RV is ahead of the HV and in the same travel lane, EEBL issues the driver an alert (Warning) indicating the hard braking event. If the RV is ahead of the HV but is traveling on the left or right lane with respect to the HV travel lane, EEBL issues a softer alert (Inform).

In a typical EEBL scenario, there can be a third vehicle between the HV and the braking RV that obstructs the driver's line of sight to the braking RV (Figure 11.3). This third vehicle does not need to participate in the communications or even to be DSRC equipped.

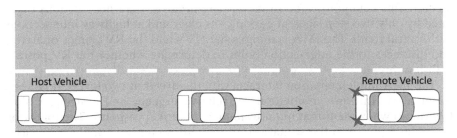

Figure 11.3. Sample EEBL scenario

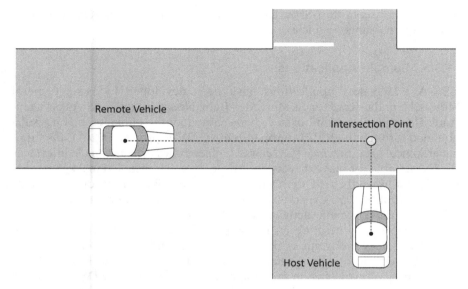

Figure 11.4. Sample IMA scenario

The VSC-A project successfully demonstrated several EEBL scenarios with the HV traveling at constant speed (50 mph) and the RV decelerating while traveling in the same or the left lane, both along straight and curving paths [NHTS11].

11.3.5.2 Intersection Movement Assist The IMA warns the driver when the application determines that it is not safe to enter an intersection due to high collision probability with other vehicles in cross traffic.

The VSC-A IMA implementation checks for targets classified in the Intersecting Left or Intersecting Right zones. For its calculations, IMA uses speed, brake status, and vehicle transmission information from incoming BSMs.

Figure 11.4 illustrates a sample scenario addressed by the IMA application. An HV and an RV approach an intersection. Their paths are roughly perpendicular. This common traffic situation may occur at road intersections regulated by only two stop signs, at parking lots exits, and at highway intersections with lateral roads. The IMA compares the HV's and the RV's times of arrival and distances to the intersection point to determine whether the RV poses a collision threat.

The IMA implementation interacts with the driver at two levels: It issues a softer alert (Inform) in situations of relative urgency and a stronger alert (Warning) when the threat requires immediate driver response. The following examples, relative to the scenario represented in Figure 11.4, illustrate cases in which IMA issues an Inform or Warning to the driver:

- If both the HV and the RV travel at speeds higher than a certain threshold with no brakes applied, the HV (RV) will receive a Warning when it is within the critical stopping distance.
- If the HV is stopped while the RV is traveling on a perpendicular path, posing a collision threat and the HV brakes are engaged, the HV will receive an Inform when the RV is within the critical stopping distance, while the RV will receive neither a Warning nor an Inform.
- If the HV is stopped while the RV is traveling on a perpendicular path, posing a collision threat and the HV creeps forward, the HV will receive a Warning when the RV is within the critical stopping distance, while the RV shall receive an Inform when it is within twice the critical stopping distance.

The IMA critical stopping distance is calculated based on system parameters that can be adjusted to reflect the behaviors of different vehicle makes and models.

The VSC-A project successfully demonstrated several IMA scenarios both with the HV stopped at the intersection (in turn holding or releasing brakes to creep forward) and with the HV approaching the intersection from cross directions [NHTS11].

11.4 SYSTEM IMPLEMENTATION

11.4.1 Onboard Unit Hardware Components

Onboard equipment for the VSC-A project includes a computing platform. This is essentially the same platform adopted for the vehicle portion of the Cooperative Intersection Collision Avoidance System for Violations (CICAS-V) project. The VSC-A onboard unit (OBU) is connected to the vehicle controller area network (CAN), DSRC radio, Global Positioning System (GPS) receiver, and driver–vehicle interface (DVI) (Figure 11.5). To facilitate joint system development over proprietary CAN buses, a CAN gateway provides a generic CAN interface toward the OBU.

The VSC-A project adopted the DENSO Wireless Safety Unit (WSU) as the OBU computing platform. The WSU features a Freescale 400 MHz MPC5200B PowerPC processor with 128MB DDR SDRAM running a Linux operating system. The OBU is connected to an external NovAtel OEMV GPS receiver providing GPS raw data. The GPS receiver further provides Coordinated Universal Time (UTC) pulse per second (PPS) information used to achieve synchronization with other OBUs.

11.4.2 OBU Software Architecture

A simplified version of the VSC-A software architecture is illustrated in Figure 11.6. This diagram is only intended to provide a basic idea of the main

Figure 11.5. VSC-A hardware components

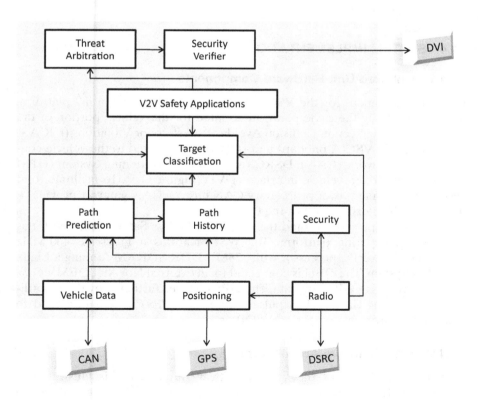

Figure 11.6. Simplified VSC-A software architecture

components and their interactions. The complete VSC-A software architecture is described in detail in [NHTS11].

The lower part of the architecture consists of modules with the task to acquire data from external sources. The Vehicle Data and Positioning handlers extract information from the CAN and GPS receiver, respectively, while the Radio handler processes incoming DSRC messages broadcast by nearby vehicles.

The central portion of the architecture comprises general-purpose modules supporting all the applications. These modules implement key functions such as path history, path prediction, and target classification. Target classification is performed every 100 ms and uses Path History and Path Prediction data frames transmitted by nearby vehicles for improved accuracy.

The upper part of the architecture includes modules for individual V2V safety applications and for activating the DVI. Each application periodically checks for potential threats in its zones of interest.

Multiple applications could simultaneously issue alerts to the driver, thus generating a conflict. The Threat Arbitration module has the task to solve such conflicts by deciding which warning is more relevant to the driver, based on a predefined policy. The Security Verifier forces the system to authenticate any DSRC messages resulting in an alert to the driver.

11.4.3   Driver–Vehicle Interface

In 2010, several VSC-A safety applications were selected by the USDOT for evaluation as part of the Driver Acceptance Clinics activity. This activity was designed to collect feedback from about 650 naïve drivers experiencing V2V safety scenarios in controlled environments at six locations in the United States. For this purpose, VSC3 Consortium OEMs provided the USDOT with 24 vehicles featuring DVIs suitable for naïve drivers. Each DVI solution was developed independently by the automakers and tailored to specific vehicles models. Here, we briefly describe the DVI created by Mercedes-Benz for their C-300 vehicles.

For this prototype system, researchers at Mercedes-Benz Research & Development North America aimed at a DVI able to focus the driver's attention on the direction of the imminent threat. Such a DVI should be simple and intuitive. For instance, it should not contain any cryptic icons or lengthy text messages. At the same time, the DVI should not be a potential cause of distraction for the driver.

The DVI is illustrated in Figure 11.7. It consists of five arrays of light-emitting diodes (LEDs) positioned on the vehicle's dashboard as well as left and right A-pillars. The LEDs can illuminate bright yellow for informing the driver or red for warning the driver. Upon activation of these LEDs, the DVI plays acoustic cues as an additional means to catch the driver's attention. Different acoustic cues are associated with yellow and red LED illuminations.

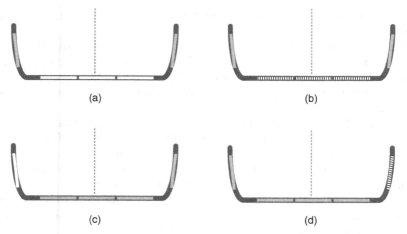

Figure 11.7. Mercedes-Benz C-300 prototype DVI. (a) FCW inform; (b) FCW warning; (c) IMA inform (left); (d) IMA warning (right)

The DVI is designed to cover threats from most directions. Figure 11.7 shows dashboard LEDs activated by FCW to inform (a) and warn the driver (b). The picture also shows A-pillar LEDs activated by IMA to inform (c) and warn the driver (d).

11.5 SYSTEM TESTING

11.5.1 Communications Coverage and Antenna Considerations

Unlike cellular or GPS communications, DSRC is more line-of-sight dependent, due largely to its frequency envelope. At 5.9 GHz, many dense or metallic materials are completely opaque to radiofrequency (RF) communications. As a result, typical V2V communications may be highly influenced by an ever-changing RF landscape surrounding the antenna. Large vehicles, buildings, terrain, metal structures, and dense, wet vegetation can obstruct DSRC transmissions. If the antenna is mounted on the vehicle's roof, transmissions can also suffer from nearby roof racks, luggage, skis, bikes, or anything metallic or dense that blocks the antenna's view of nearby vehicles or infrastructure.

Other considerations for antenna design and placement focus primarily on vehicle structures beneath or near the vehicle's antenna as well as ground reflection phase-shift phenomena. Antenna designs can be selected to either require a ground plane (e.g., metal roof) beneath the antenna or not. Proximity to a sunroof or the roof edge is also a factor, especially if a tilting sunroof can elevate above the roof level, thus partially blocking a roof-mounted antenna. If placed near side mirrors or other locations other than the roof, the metalized glass in the mirror as well as the doors, hood, trunk, and roof pillars can affect antenna coverage. Depending in part on the height of the antenna on the vehicle and the road surface composition, a ground reflection of the DSRC wave can shift in phase. At certain distances and road conditions, this phase-shifted reflection can destructively affect the line-of-sight signal (when the two are 180° out of phase) and cause a loss of communications when the main signal would otherwise be received. This phenomenon is extremely transient due to the movement of the vehicles, yet should still be considered in antenna design and placement.

Once these challenges are addressed, the antenna will need to "see" 360° around the vehicle since potential colliding vehicles can approach from any direction. Vehicles and roadside infrastructure vary in height. Roads can be curved, banked, undulating, or level. Therefore, antennas need to cover potential differences in elevation and orientation. The resulting antenna pattern surrounding the vehicle should optimally be toroidal in shape.

Since driving is mostly in the forward direction and the greatest speed differential between approaching vehicles is seen in head-on situations, it would be prudent to optimize antenna coverage for greatest range in the forward direction.

11.5.2 Positioning

Vehicle collision avoidance systems can tolerate errors in absolute position estimates to a certain degree as long as the relative position estimates meet the accuracy requirements. The purpose of the VSC-A positioning system is to provide relative position estimates with road level (<5 m) and lane level (<1.5 m) accuracies. The system acquires GPS data from the local GPS receiver and from BSMs transmitted by nearby vehicles. This information is used to generate positioning vectors representing the oriented distances between the HV and surrounding RVs. The VSC-A implementation supports two alternative methods to generate these vectors: Single Point (SP) and Real-Time Kinematics (RTK).

The SP method simply calculates the difference between individual vehicle positions, each expressed in terms of latitude, longitude, and altitude. Altitude becomes important when it is necessary to distinguish crossing vehicles on overpasses. This method has the advantage to impose low computational costs and minimum BSM payload overhead. However, empirical results have shown that the resulting accuracy may suffer from cumulative errors introduced by GPS devices produced by different manufacturers, each calculating their position estimates with proprietary algorithms. In other words, when a relative position vector is computed based on SP coordinates obtained from heterogeneous GPS devices, errors introduced by each device may add up to form even larger errors for this vector.

The RTK method addresses this concern by transmitting BSMs including GPS raw data rather than the estimated latitude and longitude values. When a BSM is received, the HV uses the included GPS raw data to calculate the position of the correspondent RV. In this calculation, the HV is likely to introduce roughly the same error as for the calculation of its own position. That is, one error tends to compensate the other, resulting in a reduced error in the relative position of the two vehicles. This improved accuracy comes with higher computational costs and bandwidth consumption, the latter due to the larger size of GPS raw data with respect to the SP representation.

REFERENCES

[Kenn11] J. Kenney: "Dedicated Short-Range Communications (DSRC) Standards in the United States," Proceedings of the IEEE, Vol. 99, No. 7, July 2011.

[NHTS07] W. G. Najm, J. D. Smith, and M. Yanagisawa: "Pre-Crash Scenario Typology for Crash Avoidance Research," National Highway Traffic Safety Administration, DOT HS 810 767, 2007.

[NHTS10] W. G. Najm, J. Koopmann, J. D. Smith, and J. Brewer: "Frequency of Target Crashes for IntelliDrive Safety Systems," National Highway Traffic Safety Administration, DOT HS 811 381, 2010.

[NHTS11] F. Ahmed-Zaid, F. Bai, S. Bai, C. Basnayake, B. Bellur, S. Brovold, G. Brown, L. Caminiti, D. Cunningham, H. Elzein, K. Hong, J. Ivan, D. Jiang, J. Kenney, H. Krishnan, J. Lovell, M. Maile, D. Masselink, E. McGlohon, P. Mudalige, Z. Popovic, V. Rai, J. Stinnett, L. Tellis, K. Tirey, and S. VanSickle: "Vehicle Safety Communications—Applications (VSC-A) Final Report," National Highway Traffic Safety Administration, DOT HS 811 492A, 2011.

[SAE09] Society of Automotive Engineers: "SAE J2735 Dedicated Short-Range Communications (DSRC) Message Set Dictionary," 2009.

[SAE10] Society of Automotive Engineers: "SAE J2945 Dedicated Short Range Communication (DSRC) Minimum Performance Requirements," Work in Progress, 2010.

12

DSRC SCALABILITY

Written in cooperation with Daniel Jiang and Tessa Tielert

12.1 INTRODUCTION

Vehicle safety communications are dominated by pervasive broadcast of safety messages for vehicles to inform each other of their current status. These messages need to be received at a minimum frequency sufficient to satisfy application requirements.

The data traffic over a dedicated short-range communications (DSRC) channel at each location grows with the number of communicating vehicles. For large numbers of vehicles, the radio link can become congested and the network performance can degrade severely, impeding the effectiveness of vehicle safety applications. Early experiments and simulations show that DSRC channel congestions can occur even in relatively simple traffic scenarios [Wein10].

Many congestion control solutions have been proposed for mobile ad hoc networks (MANETs) [AbKu10] [DoLK10] [Grun12] [LoSM07]. These solutions, however, are not well suited for consumer vehicle networks which are highly dynamic with rapid changes in network topology and vehicle density. Vehicles must react quickly to the changing environments by rapidly assessing channel load conditions and adjusting their communications accordingly. Solutions for MANETs have traditionally focused on multihop point-to-point and multicast communications. Intermediate nodes can detect and report channel

Vehicle Safety Communications: Protocols, Security, and Privacy, First Edition. Luca Delgrossi and Tao Zhang.
© 2012 John Wiley & Sons, Inc. Published 2012 by John Wiley & Sons, Inc.

load conditions. Sources can base their routing or message dissemination decisions on feedback from destinations and intermediate nodes. These solutions are not applicable to vehicle safety communications in a consumer vehicle network, where communications are dominated by single-hop broadcasts.

Congestion control solutions for traditional MANETs aim primarily at maximizing data throughput. For vehicle safety communications, however, a main goal is to keep channel usage below a certain threshold to leave a fraction of the bandwidth for the event-driven urgent safety messages.

Over the past few years, a number of congestion control mechanisms for vehicle safety communications have been proposed. This chapter provides an overview of the current state of the art. It illustrates the DSRC scalability problem and the need for effective and fair solutions. It discusses desired properties of DSRC congestion control solutions and describes at a high level several promising approaches. This chapter is only intended as an introduction recognizing that significant research in this area is still underway.

12.2 DSRC DATA TRAFFIC

Congestion control algorithm designs have to be based on assumptions on the data traffic to be expected over the channel. Below, we discuss safety messages expected to be broadcast over a DSRC channel and how individual vehicles can estimate current channel load conditions.

12.2.1 DSRC Safety Messages

Vehicle safety communications are expected to be dominated by broadcasts of safety messages. In the United States, the Society of Automotive Engineers (SAE) J2735 standard specifies basic safety messages (BSMs). Vehicles periodically broadcast BSMs, including the 39-byte basic vehicle state. It is expected that this BSM will contain approximately 400 bytes with the security overheads.

Occasionally, vehicles will broadcast event-driven safety messages. Such messages can be triggered, for example, by the vehicle's sudden sharp decelerations or control losses. The SAE J2735 standard specifies BSM vehicle safety extensions for representing these safety-critical events. These occasional event-driven broadcasts typically account only for a negligible amount of traffic over a DSRC channel. However, these messages are of the highest importance as they represent critical traffic scenarios that often require immediate driver attention. Therefore, one of the main goals of DSRC congestion control is to manage channel load so that critical event-driven messages have reasonably good performance.

Roadside units (RSUs) also contribute to the data traffic over the DSRC link. The SAE J2735 standard specifies several messages RSUs can transmit to vehicles, such as signal phase and timing (SPAT), Radio Technical Commission for Maritime Services (RTCM) Corrections, Traffic Information Message, and Roadside Alert. Furthermore, RSUs can periodically broadcast WAVE Service Advertisements (WSAs) to inform vehicles of services available over a DSRC service channel. Messages generated by the roadside infrastructure typically represent a small fraction of the overall DSRC channel load.

In the rest of this section, we consider only vehicle-to-vehicle (V2V) communications.

12.2.2 Transmission Parameters

A series of transmission parameters influence the data traffic load of DSRC channels [Wein10]. These parameters include message frequency, message size, transmission power or distance, and data rate. The choice of an optimal DSRC data rate has been previously discussed. We assume that all DSRC safety transmissions use quadrature phase-shift keying (QPSK) modulation and 1/2 coding rate, resulting in 6 Mbps data rate.

Message frequency, size, and transmission power are determined by V2V safety applications. Therefore, it is natural to introduce congestion control functions above the wireless access in vehicular environments (WAVE) upper layers. This avoids modifications to lower layer communication protocols or their implementations.

Message frequency or rate is a key parameter. Low message frequencies result in longer time intervals between subsequent BSM messages and less timely support for V2V safety applications. High message rates require high bandwidth and lead to increased risks of channel congestions. Proper message rates depend on vehicle speeds. For example, at 80 mph, a vehicle travels about 36 m/s. Updates sent faster than every 2–3 m will likely not carry significant new information. Currently, message rates between 2 and 10 Hz are being considered. Broadcasting BSMs at 10 Hz should be sufficient to support the most demanding V2V safety applications in a timely fashion.

Transmission power is directly linked to the distance at which BSMs can be received with sufficient received signal strength (RSS). For this reason, transmission power is often described through a range expressed in meters. High-power transmissions can reach faraway nodes, but will also increase channel interference and frame collisions. The WAVE protocol stack allows an application to select the transmission power for each individual BSM message. Existing WAVE radios typically offer transmission powers up to 20 dBm, with 1 dBm step adjustments.

Message size affects network performance because longer messages require longer transmission times and generate longer channel busy intervals. Small messages typically result in overall performance improvements as confirmed by simulations and experiments with real radios [Wein10].

However, adjusting V2V safety application message sizes is impractical. Therefore, congestion control algorithms typically focus on the dynamic adjustment of message rate, transmission power, or both.

12.2.3 Channel Load Assessment

WAVE radios can measure current channel load using the clear channel assessment (CCA) function provided by the Institute of Electrical and Electronics Engineers (IEEE) 802.11 media access control (MAC) layer protocol [IEEE07]. The CCA function, available on all IEEE 802.11 devices, determines whether the medium is busy or idle. The IEEE 802.11 MAC layer considers the medium to be busy when the physical (PHY) layer finds the channel busy or the channel is declared virtually busy, typically due to request to send/clear to send (RTS/CTS) reservations or acknowledgements associated with unicast transmissions. Since vehicle safety communications are predominately broadcasts, we will not consider the case where the channel is declared virtually busy.

The IEEE 802.11 PHY layer declares channel busy when the cumulative power received within a certain time interval exceeds the carrier sensing threshold (CST). For consistency across the network, it is essential that all radios share the same CST. Therefore, WAVE radios equipped with radiofrequency (RF) chipsets of different sensitivity as well as their antennas and cables should be calibrated in such a way that they report channel busy indications in a consistent manner [CJTD11].

WAVE radios periodically invoke the CCA function to calculate the fraction of time when the channel is busy, known as channel busy ratio (CBR). The CBR provides a convenient metric to assess channel load conditions in realistic communication scenarios. Experiments conducted with tens of WAVE radios show that the overall network performance, expressed as broadcast reception rate at receivers, degrades rapidly as the CBR increases [Wein10]. For instance, with 180 vehicles broadcasting 378-byte messages at 10 Hz with 20 dBm transmission power, the average CBR is about 73% and the reception rate degrades to 45% for frames received with an RSS equal to -85 dB [WeKB11]. Such a low broadcast reception rate becomes unacceptable for many vehicle safety applications.

Figure 12.1 describes a typical scenario for congestion control studies. Each vehicle has a carrier sensing range R_{CS} (the distance at which it can sense the signal) and a transmission range R_{TX} (the distance at which it can transmit frames). The transmission range is smaller than the carrier sensing range. Optimal values for R_{CS} are two to three times the R_{TX} [YaVa05]. In the network simulator 2 (ns-2), for instance, the carrier sensing range is by default set to a value of $R_{CS} = 2.2 R_{TX}$ [DeLV04]. In Figure 12.1, vehicle V_1's transmission range allows its broadcast to reach vehicle V_2 but not V_3. However, V_3's transmissions can be sensed by V_1. Interferences from V_2 will be significantly stronger than those created by V_3.

The space dimension needs to be considered when designing congestion control algorithms for DSRC [CJTD11]. For instance, vehicles located at the edge of a congested area might not experience significant performance degradation and therefore not reduce their contribution to the overall channel

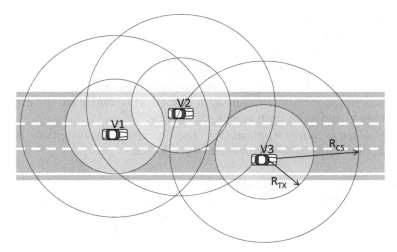

Figure 12.1. Carrier sensing and transmission ranges

load. Some vehicles might suffer from starvation (i.e., no channel access) when caught between other groups of vehicles [WiRo05].

12.3 CONGESTION CONTROL ALGORITHMS

A general framework for designing congestion control solutions has been described in [SBLB10]. Recently, several strategies and algorithms have been proposed [BoSh10] [BSLZ10] [BuKR10] [GSKB07] [HCCC10] [HFSK10] [KeBR11] [KhBR08] [MSKH08] [SGHH11] [SMSH11] [TJCD11] [TMSH09] [ZSCR07].

For vehicle safety communications, the main goal of congestion control is to keep the overall channel load below a certain threshold. When this goal is fulfilled, the channel is not congested and a fraction of the bandwidth is available for urgent event-driven safety messages on top of periodic broadcasts.

Proactive algorithms aim at preventing channel congestions in the first place. These algorithms are based on functions able to predict the overall channel load in the imminent future. Reactive algorithms assume vehicles have the means to assess current channel load. These algorithms typically require vehicles to reduce their contributions to the overall channel load when congestions are detected. In this section, we discuss congestion control schemes properties and describe at a high level a few recently proposed algorithms.

12.3.1 Desired Properties

Congestion control algorithms for vehicle networks need to be distributed, given the ad hoc nature of the network. A key requirement is fairness, which can be expressed in several ways:

- *Participation*: All vehicles contributing to congestion should participate in congestion control.
- *Local Fairness*: Vehicles close to each other should make comparable efforts to control congestion.
- *Global Fairness*: The system should not cause vehicle starvation (no access to the medium) and make a best effort to maximize the minimum channel bandwidth each vehicle can access.
- *Deference to Safety Applications*: The objective of transmitting BSMs is to create a mutual awareness among vehicles.

Congestion control strategies should be designed to meet the requirements of specific safety applications. Different safety applications may have different requirements for transmission power and message rate. Fairness does not necessarily mean that every vehicle should have identical minimum and maximum ranges for transmission power and message rate.

12.3.2 Transmission Power Adjustment

A prominent example of transmission power control algorithm is the distributed fair transmission power adjustment for vehicular ad hoc networks (D-FPAV) scheme proposed in [TMSH09]. D-FPAV controls DSRC channel congestions by dynamically adjusting the transmission power for beacons. The algorithm focuses on keeping transmission power below a predefined threshold called maximum beaconing load (MBL). Rather than reacting to channel congestions, the D-FPAV scheme proactively attempts to prevent congestions. As such, its effectiveness depends on accurate information and suitable models for channel load prediction.

The D-FPAV algorithm works as follows. Each vehicle collects information about neighbors within the carrier sensing range achievable at maximum transmission power. It then computes the maximum common transmission power that does not exceed the MBL threshold at each vehicle while ensuring fairness. Vehicles store transmission power values indicated by neighbors and adjust their own transmission power to the minimum value received.

To reach all vehicles within the carrier sensing range, intermediate vehicles must cooperate to forward neighbors' status. The protocol presents a trade-off between the additional overhead on the communication link and the need for fresh updates to react to network topology changes. Simulations show that piggybacking transmission power information every 10 beacons and using the minimum transmission power value for sending the augmented beacons are likely to offer the best performance [TMSH09]. When transmission power information is distributed to all vehicles within the carrier sensing range, the D-FPAV algorithm is formally proven to achieve fairness among all vehicles.

12.3.3 Message Rate Adjustment

The D-FPAV algorithm first sets a fixed message rate. It then adjusts transmission power based on feedback from vehicles within its carrier sensing range. Algorithms based on message rate adjustment operate on the opposite premise that transmission power is fixed a priori while message rate is dynamically adjusted.

A recent analysis focused on a series of communication scenarios associated with the same 0.6 CBR level generated through different combinations of transmission power and message rate values [TJCD11]. Interestingly, the study shows that the optimal transmission power (fixed a priori by rate control mechanisms) is independent of vehicle density and distribution. However, the same is not true for the optimal messaging rate (fixed a priori by power control mechanisms). This suggests that a natural strategy for DSRC congestion control should be setting the transmission power first, and then adjusting messaging rate based on current channel loads.

This is the strategy adopted by the periodically updated load sensitive adaptive rate control (PULSAR) algorithm [TJCD11].

In PULSAR, the message rate varies between minimum and maximum values defined by V2V safety application. By adjusting message rate, PULSAR achieves different interpacket reception times within the target range. The goal is to obtain the highest possible message rate without violating overall channel load and fairness constraints. It is recognized that V2V safety applications can dynamically adjust transmission power based on current driving contexts. However, it is expected that these changes occur at a slower pace than message rate adjustments.

PULSAR uses CBR as a channel load metric. CBR is considered to be a suitable metric to support mechanisms aimed at maximizing the number of received packets [FHSK11]. CBR values between 0.6 and 0.7 offer the best results and are recommended as target values for algorithms such as PULSAR.

PULSAR adopts a fixed channel monitoring and decision interval (CMDI). The protocol operates as follows. At the end of every CMDI, the CBR value provides an assessment on current channel load. The transmission rate is reduced if current channel load exceeds the target threshold or increased if current channel load is below the target threshold. Rate adjustments follow the additive increase/multiplicative decrease (AIMD) scheme. The AIMD scheme has been shown to converge while ensuring fairness when it applied in wired networks [ChJa89] as well as vehicle networks [KeBR11]. The other combinations, namely additive increase/additive decrease (AIAD), multiplicative increase/additive decrease (MIAD), and multiplicative increase/multiplicative decrease (MIMD), do not offer both convergence and fairness [ChJa89] [WeKB11].

The CMDI fixed size imposes a limit on the algorithm's convergence speed. To achieve faster convergence, PULSAR makes use of a vehicle's target rate, calculated as the average message rate of its neighbors. Increments are doubled or halved when the node is below or above its target rate. This requires piggybacking message rate information on some BSMs. Just as for D-FPAV, multihop message dissemination is necessary to ensure all vehicles within the carrier sensing range are reached. PULSAR proposes a two-hop dissemination scheme, including a mechanism designed to reduce the effects of information dissemination delays. It ensures that all vehicles react at the same time to a given channel load condition to provide fairness across all nodes.

Recently, a new algorithm has been proposed that increases the efficiency offered by systems based on AIMD. This is the linear message rate control (LIMERIC) [KeBR11]. LIMERIC takes advantage of the fact that, with the PULSAR algorithm, vehicles exchange information on the actual channel load. Instead of interpreting channel conditions in a binary manner (congested or not congested), LIMERIC understands actual channel congestion levels and makes rate adjustments accordingly. This results in fast reactions to congestion conditions.

12.3.4 Simulation Study

One difficulty in the study of DSRC congestion control algorithms is the cost and complexity of conducting large-scale field trials. Therefore, simulations are a convenient method to conduct initial assessments and evaluations. Recent measurements conducted with tens of DSRC radios allowed us to calibrate and verify that the ns-2 simulator accurately models the MAC and PHY layer behaviors of IEEE 802.11 hardware implementations [BaKW11].

Experiments conducted with real radios can be simplified by using a small number of radios to emulate a large number of vehicles, for example, as described in [BaKW11].

12.4 CONCLUSIONS

Enhanced congestion control solutions for vehicle networks are expected in the near future. For example, automakers at Crash Avoidance Metrics Partnership (CAMP) are conducting experiments with up to 200 vehicles to evaluate congestion control algorithms in real-world environments. This activity will result in fine-tuned congestion control algorithms that will be used for a 2500 vehicle Model Deployment field operational trial sponsored by the U.S. Department of Transportation (USDOT). Vehicles in this trial will be equipped with fully integrated, retrofit, or aftermarket DSRC devices. Data collected during Model Deployment are expected to provide insight on how to build scalable DSRC systems, including more scalable congestion control solutions.

REFERENCES

[AbKu10] A. M. Abbas and Ø. Kure: "Quality of Service in Mobile Ad Hoc Networks: A Survey," International Journal of Ad Hoc and Ubiquitous Computing, vol. 6, no. 2/2010, pp. 75–98, 2010. DOI: 10.1504/IJAHUC.2010.034322.

[BaKW11] G. Bansal, J. Kenney, A. Weinfield: "Cross-Validation of DSRC Radio Testbed and NS-2 Simulation Platform for Vehicular Safety Communications," Vehicular Technology Conference (VTC Fall), 2011 IEEE, San Francisco, CA, 2011.

[BoSh10] M. Bouassida and M. Shawky: "A Cooperative Congestion Control Approach within VANETs: Formal Verification and Performance Evaluation," EURASIP Journal on Wireless Communications and Networking, vol. 2010, pp. 11:1–11:12, Apr. 2010.

[BSLZ10] R. Baldessari, D. Scanferla, L. Le, W. Zhang, and A. Festag: "Joining Forces for VANETs: A Combined Transmission power and Rate Control Algorithm," in Proc. of the 7th International Workshop on Intelligent Transportation (WIT), 2010.

[BuKR10] H. Busche, C. Khorakhun, and H. Rohling: "Self-Organized Update Rate Control for Inter-Vehicle Networks," in Proc. of the 10th International Workshop on Intelligent Transportation (WIT), 2010.

[ChJa89] D. M. Chiu and R. Jain: "Analysis of the Increase and Decrease Algorithms for Congestion Avoidance in Computer Networks," Journal of Computer Networks and ISDN Systems, vol. 17, pp. 1–14, June 1989.

[CJTD11] Q. Chen, D. Jiang, T. Tielert, L. Delgrossi: "Mathematical Modeling of Channel Load in Vehicle Safety Communications," IEEE International Symposium on Wireless Vehicular Communications (WIVEC) 2011, San Francisco, 2011.

[DeLV04] J. Deng, B. Liang, P. K. Varshney: "Tuning the carrier sensing range of IEEE 802.11 MAC," IEEE Global Telecommunications Conference (GLOBECOM), Dallas, TX, 2004.

[DoLK10] V. T. M. Do, L. Landmark, and Ø. Kure: "A Survey of QoS Multicast in Ad Hoc Networks," Future Internet, vol. 2, pp. 388–416, 2010. DOI: 10.3390/fi2030388.

[FHSK11] Y. P. Fallah, C. Huang, R. Sengupta, and H. Krishnan: "Analysis of Information Dissemination in Vehicular Ad-Hoc Networks With Application to Cooperative Vehicle Safety Systems," IEEE Transactions on Vehicular Technology, vol. 60, no. 1, pp. 233–247, Jan. 2011.

[Grun12] A. M. Grundy: "Congestion Control Framework for Delay-Tolerant Communications," Ph.D. thesis submitted to the University of Nottingham, 2012.

[GSKB07] X. Guan, R. Sengupta, H. Krishnan, and F. Bai: "A Feedback-Based Power Control Algorithm Design for VANET," in Mobile Networking for Vehicular Environments, 2007.

[HCCC10] J. He, H. Chen, T. Chen, and W. Cheng: "Adaptive Congestion Control for DSRC Vehicle Networks," IEEE Communications Letters, vol. 14, no. 2, pp. 127–129, 2010.

[HFSK10] C. Huang, Y. Fallah, R. Sengupta, and H. Krishnan: "Adaptive Inter-vehicle Communication Control for Cooperative Safety Systems," IEEE Network, vol. 24, no. 1, pp. 6–13, 2010.

[IEEE07] Institute of Electrical and Electronics Engineers: "IEEE Standard 802.11-2007, Wireless LAN MAC and PHY Specifications," Section 7.3.2.22.2, 2007.

[KeBR11] J. Kenney, G. Bansal, C. E. Rohrs: "LIMERIC – A Linear Message Rate Control Algorithm for Vehicular DSRC Systems," 8th ACM International Workshop on Vehicular Ad Hoc Networks (VANET), Las Vegas, NV, 2011.

[KhBR08] C. Khorakhun, H. Busche, and H. Rohling: "Congestion Control for VANETs based on Power or Rate Adaptation," in Proc. of the 5th International Workshop on Intelligent Transportation (WIT), 2008.

[LoSM07] C. Lochert, B. Scheuermann, and M. Mauve: "A Survey on Congestion Control for Mobile Ad Hoc Networks," Wireless Communications and Mobile Computing, vol. 7, no. 5, pp. 655–676, 2007.

[MSKH08] J. Mittag, F. Schmidt-Eisenlohr, M. Killat, J. Haerri, and H. Hartenstein: "Analysis and Design of Effective and Low-Overhead Transmission Power Control for VANETs," in Proc. of the 5th ACM international workshop on VehiculAr InterNETworking (VANET), 2008.

[SBLB10] R. R. Schmidt, A. Brakemeier, T. Leinmueller, B. Boeddeker, and G. Schaefer: "Architecture for Decentralized Mitigation of Local Congestion in VANETs," 10th International Conference on ITS Telecommunications (ITST), 2010.

[SGHH11] M. Sepulcre, J. Gozalvez, J. Haerri, and H. Hartenstein: "Contextual Communications Congestion Control for Cooperative Vehicular Networks," IEEE Transactions on Wireless Communications, vol. 10, no. 2, pp. 385–389, 2011.

[SMSH11] M. Sepulcre, J. Mittag, P. Santi, H. Hartenstein, and J. Gozalvez: "Congestion and awareness control in cooperative vehicular systems," Proceedings of the IEEE, vol. 99, no. 7, pp. 1260–1279, Jul. 2011.

[TJCD11] T. Tielert, D. Jiang, Q. Chen. L. Delgrossi, H. Hartenstein: "Design Methodology and Evaluation of Rate Adaptation Based Congestion Control for Vehicle Safety Communications," IEEE Vehicular Networking Conference (VNC), Amsterdam, The Netherlands, 2011.

[TMSH09] M. Torrent-Moreno, J. Mittag, P. Santi, and H. Hartenstein: "Vehicle-to-Vehicle Communication: Fair Transmission power Control for Safety-Critical Information," IEEE Transactions on Vehicular Technology, vol. 58, no. 7, pp. 3684–3703, 2009.

[Wein10] A. Weinfield: "Methods to reduce DSRC Channel Congestion and improve V2V Communications Reliability," Proceedings of the 2010 ITS World Congress, Busan, South Korea, 2010.

[WeKB11] A. Weinfield, J. Kenney, G. Bansal: "An Adaptive DSRC Message Transmission Interval Control Algorithm," 18th ITS World Congress 2011, Orlando, FL, 2011.

[WiRo05] L. Wischhof, H. Rohling: "Congestion Control in Vehicular Ad Hoc Networks," IEEE International Conference on Vehicular Electronics and Safety 2005 (ICVES 2005), pp. 58–63, Xi'an, Shaanxi, China, Oct. 2005.

[YaVa05] X. Yang and N. H. Vaidya: "On the Physical Carrier Sensing in Wireless Ad Hoc Networks," IEEE INFOCOM, Miami, FL, 2005.

[ZSCR07] Y. Zang, L. Stibor, X. Cheng, H. Reumerman, A. Paruzel, and A. Barroso: "Congestion Control in Wireless Networks for Vehicular Safety Applications," in Proc. of the 8th European Wireless Conference, 2007.

13

SECURITY AND PRIVACY THREATS AND REQUIREMENTS

13.1 INTRODUCTION

This chapter discusses security and privacy threats in consumer vehicle networks. We focus on threats that use communication capabilities to attack the vehicle network or breach driver privacy. We identify unique communication-based threats in consumer vehicle networks and discuss what make them different from threats in other types of networks such as enterprise networks or the Internet. Rather than presenting an exhaustive catalog of security and privacy threats and requirements, the goal is to provide a context for discussing privacy-preserving security and certificate management issues and solutions in the following chapters.

The chapter further discusses security and privacy protection capabilities expected to be necessary for consumer vehicle networks and outlines main security and privacy design and performance considerations that are unique to these networks.

13.2 ADVERSARIES

Adversaries are individuals or organizations that would make use of communication capabilities to mount security and privacy attacks to the network. Adversaries can be classified into several categories based on the resources they may possess:

Vehicle Safety Communications: Protocols, Security, and Privacy, First Edition. Luca Delgrossi and Tao Zhang.
© 2012 John Wiley & Sons, Inc. Published 2012 by John Wiley & Sons, Inc.

- Individuals operating mainly on their own with limited monetary resources. Examples of individual adversaries in a consumer vehicle network include computer hackers, automotive enthusiasts, electronics hobbyists, and researchers.
- Loosely coordinated groups with significantly more resources than each individual. For example, one individual who is able to obtain private keys from a vehicle could share the keys with his collaborators, so they could collectively multiply the damages that can be done with these keys.
- Insiders owning sensitive information about the security and privacy protection system for the consumer vehicle network.
- Adversary organizations with abundant monetary resources and sophisticated technologies. Examples include organized crimes.
- Foreign governments that could be interested in mounting security and privacy attacks to a nation's consumer vehicle networks. They can be highly organized and have abundant monetary resources and sophisticated technologies.
- Government agencies permitted to breach driver privacy. This is one of the most serious concerns of the driving public.

13.3 SECURITY THREATS

Potential security threats in consumer vehicle networks can be illustrated using the following example categories:

- send false safety messages using valid security credentials,
- falsely accuse innocent vehicles,
- impersonate vehicles or network entities, and
- denial-of-service attacks unique in a consumer vehicle network.

Discussions on additional security threats can be found in [PCZV07].

A vehicle that misuses vehicle communications capabilities to perform any of these and other security and privacy attacks is referred to as a misbehaving vehicle.

13.3.1 Send False Safety Messages Using Valid Security Credentials

A unique characteristic of consumer vehicle networks is that false safety messages could be sent using a vehicle's valid security credentials. This could be achieved, for example, by altering the input data to a vehicle's onboard unit (OBU), installing malicious software onto the OBU, injecting devices onto the vehicle's controller area network (CAN), or compromising existing OBU software and hardware.

An adversary could cause some vehicles to send false safety messages to other vehicles or the roadside communication infrastructure. These false safety messages could be, for example, counterfeit heartbeat messages carrying bogus information on vehicle positions and speeds, false emergency electronic brake light messages, or fake traffic information messages. In vehicle safety communication systems aimed at raising the driver's attention, these false safety messages could cause vehicles to issue unnecessary warnings to drivers.

An adversary has multiple ways to alter the input data to an OBU. Input data to an OBU often comes from other components and sensors on the vehicle. Examples of such input data elements include the vehicle's position, speed, brake status, activation of the electronic stability control system, tire pressure, and airbag deployment status. A malicious device connected to the CAN could snoop on the packets sent over the vehicle bus, modify their contents, and then forward them to the OBU. A malicious device connected to the CAN bus can also insert extra erroneous packets onto the CAN bus for the OBU to pick up as its input data [KCRP10].

An adversary could also tamper with vehicle onboard sensors to cause them to generate false inputs to the OBU.

When an adversary can inject malicious software onto an OBU, the malicious software can send any false messages designed by the adversary.

Malfunctions of OBU hardware or software could also cause the OBU to send false or erroneous messages.

13.3.2 Falsely Accuse Innocent Vehicles

To help detect misbehaving vehicles, vehicles may be required to report misbehaviors they have detected to a global misbehavior detection system. The global misbehavior detection system can then use the information collected from multiple vehicles to make better judgments—than each individual vehicle would be able to—on which vehicles may be misbehaving vehicles. Misbehaving vehicles, however, could abuse this reporting process. For example, they could send reports to falsely accuse innocent vehicles as misbehaving vehicles. This could negatively impact the misbehavior detection system's ability to detect misbehaving vehicles.

13.3.3 Impersonate Vehicles or Other Network Entities

A malicious vehicle could pose as a different vehicle by using another vehicle's credentials.

A more dangerous form of impersonation attacks in a consumer vehicle network is the Sybil attack [Douc02]. In a Sybil attack, a malicious vehicle sends safety messages to attempt to convince other vehicles that more vehicles are present than there actually are.

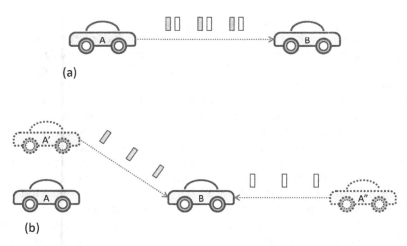

Figure 13.1. Sybil attack in a vehicle network. (a) Malicious vehicle A sends two message streams carrying fake positioning and speed information. (b) Innocent vehicle B thinks there are two different vehicles (A′ and A″) that do not exist

For example, as shown in Figure 13.1, a malicious vehicle A could broadcast two streams of fake safety messages to neighboring vehicles. Each message stream will use a different certificate and will contain a different set of positions and speeds to make them look like messages that come from two fake vehicles: one in front of vehicle B and another by the side of vehicle B. As we can see, the imaginary vehicles created with Sybil attacks could trigger unnecessary warnings to the drivers.

A vehicle needs to have multiple valid identities to mount a Sybil attack. When digital signatures and digital certificates are used to support message authentication, a vehicle may need a separate certificate for each application and may therefore have multiple valid certificates at the same time. A vehicle may also have multiple valid certificates at the same time to protect its privacy. These provide opportunities for Sybil attacks.

13.3.4 Denial-of-Service Attacks Specific to Consumer Vehicle Networks

Denial-of-service (DoS) attacks are to disable, degrade, or disrupt the functionality, capability, and performance of vehicle communication capabilities and cooperative vehicle safety applications. There are several well-known DoS attacks. Adversaries could jam the radio waves used for vehicle communications; flood the consumer vehicle network with wasteful messages to overload the radio communications or the OBUs of innocent vehicles; or compromise roadside units (RSUs) to disable their services. These DoS attacks are not unique to consumer vehicle networks and occur in other types of wireless networks.

There are also new DoS attacks that are unique in large-scale consumer vehicle networks and have not been well addressed. Specifically, attackers could use vehicle communication capabilities to trigger severe overreactions by other vehicles or mission-critical network servers, which could in turn consume excessive system resources or decapacitate mission-critical functions. One major form of such DoS attacks is to use attackers' vehicles to cause the security credential management system, such as a certificate management system, to overreact to overload the network with credential management messages or to cause an excessive number of vehicles to lose their security credentials.

13.3.5 Compromise OBU Software or Firmware

Recent studies have shown that it is often not too difficult to compromise the software or firmware of electronic control units (ECUs) used to control a wide range of functions on a vehicle [KCRP10]. Similar risks would exist for OBUs to be used for vehicle safety communications.

It has been shown that the ability to do software or firmware upgrades to ECUs can be easily used by adversaries to inject malicious code onto the ECUs [BeBF06] [KCRP10]. The challenge–response-based mechanisms used today to protect software and firmware changes to ECUs alone have been shown to provide insufficient protection against malicious firmware updates [KCRP10]. Several ways to insert malicious code onto ECUs on real vehicles have been demonstrated in [KCRP10]. Once malicious code is inserted onto an OBU, it can be used to mount all forms of attacks described previously.

13.4 PRIVACY THREATS

A nationwide consumer vehicle network could offer an opportunity for individuals, organizations, and government agencies to obtain private information about drivers, identify vehicles on the road, and track their movements. Such privacy breaches become an especially serious consumer concern when the government mandates communication capabilities on all vehicles, forcing everyone to drive vehicles connected to this nationwide network.

The ability to provide adequate driver privacy has been recognized as a prerequisite for real-world deployment and broad user adoption of vehicle safety communication systems.

13.4.1 Privacy in a Vehicle Network

Privacy in a vehicle network can be characterized by *vehicle anonymity* and *message unlinkability*.

A widely accepted definition of anonymity is the state of being not identifiable within a set of subjects called the *anonymity set* [DSCP02] [PfHa01]. In other words, anonymity means that an adversary can only identify a subject as one in a set of subjects (the anonymity set). The minimum size of an anonymity set is two because if there is only one subject in the anonymity set it is impossible to protect the subject's identity. In general, the larger the anonymity set, the higher the anonymity level.

In this book, we focus on *sender anonymity* [DSCP02] for vehicles. This is the inability for an adversary to determine which particular vehicle in a set of vehicles is the originator of a message. For ease of expression, we will refer to the sender anonymity for vehicles simply as vehicle anonymity or anonymity.

Message unlinkability is the inability to link multiple messages to the same message originator. The ability to link different messages to the same vehicle allows an adversary to track a vehicle's movements.

Vehicle anonymity and message unlinkability are distinct components of privacy. Vehicle anonymity captures the difficulty of linking an individual message to its originating vehicle. Message unlinkability captures the difficulty of linking multiple messages to the same vehicle without knowing the identity of this vehicle. While a vehicle remains anonymous and the adversary does not know the vehicle's identity, the messages sent from the vehicle can still make the vehicle highly distinguishable and allow the vehicle to be tracked. In other words, an adversary may be able to determine which messages are originated from the same vehicle and hence be able to track this vehicle, although the adversary may not know the identity of the vehicle.

Vehicle anonymity and message unlinkability are related to each other. If an adversary can repeatedly violate vehicle anonymity and be able to identify the same vehicle in different locations, the adversary will also be able to track the vehicle and violate message unlinkability. Similarly, the ability to track vehicles can help the adversary uniquely identify a vehicle because each vehicle usually has its unique movement paths and patterns.

13.4.2 Privacy Threats in Consumer Vehicle Networks

This section discusses three major categories of mechanisms to breach vehicle anonymity and message unlinkability in a consumer vehicle network:

- communication traffic analysis,
- spyware, and
- abuse of private information by network and service operators.

13.4.2.1 Breach Privacy with Communication Traffic Analysis Communication traffic analysis is to capture and analyze messages to and from vehicles to extract information that can be used to identify or track vehicles.

Using existing communication protocols, message headers will carry many information elements that can be used to identify or track a vehicle. These information elements include, for example, media access control (MAC) layer addresses and a vehicle's digital certificates.

Vehicle-to-vehicle (V2V) safety broadcast messages typically contain plaintext information, such as a vehicle's position and speed, which can be used by adversaries to determine which messages are from the same vehicle and to track the vehicle.

An adversary could also perform traffic analysis to identify a vehicle's unique communication patterns and use these patterns to identify or track the vehicle. For example, when vehicles broadcast safety messages periodically, the interarrival times between messages can be used to link multiple messages to the same vehicle. To illustrate this using a simplified example, consider an adversary that has captured two sequences of messages: one sequence of messages is captured at times t_1, $t_1 + \Delta$, $t_1 + 2\Delta$, ..., $t_1 + n\Delta$, whereas the other sequence is captured at times $t_2, t_2 + \Delta, t_2 + 2\Delta, \ldots, t_2 + n\Delta$. Then, there is a high probability that these two sequences come from different vehicles (unless they come from the same vehicle that is carrying out a Sybil attack).

Messages from vehicles can be captured at different places in a consumer vehicle network and by individuals, organizations, and government agencies. For example, V2V messages could be captured by vehicles controlled by adversaries, by RSUs, and by other radio receivers that anyone may deploy along roadsides. V2I messages could be captured by infrastructure network operators and by application providers which the vehicles communicate with.

In general, the more messages captured, the more information about a vehicle and its driver can be inferred from the messages.

13.4.2.2 *Breach Privacy with Spyware*   A spyware on the Internet is a piece of software that tracks the websites visited by an Internet user. An adversary can have several ways to insert spyware or other similar malicious code onto an OBU, which can then record a vehicle's positions, speeds, and movements and report them to the adversaries. For example, spyware could get onto an OBU during OBU software update [BeBF06]. When a vehicle uses Internet Protocol (IP)-based protocols to communicate with application servers, an adversary could also use these communication channels to implant spyware onto the OBU.

13.4.2.3 *Abuse Private Information by Network or Security System Operators*   The systems used to manage vehicles' security credentials maintain private information about drivers and their vehicles, which is required to authenticate vehicles, authorize services to vehicles, and provide security credentials to vehicles. Such information may include drivers' identities and addresses, vehicles' identities, vehicles' makes and models, applications on the vehicles, and the digital keys and certificates assigned to vehicles.

Operators of the security credential management systems could abuse the driver and vehicle private information they have to breach driver privacy. They could also provide the private information to third parties, which could in turn use the information to identify and track vehicles.

13.4.3 How Driver Privacy can be Breached Today

Today, without the help of consumer vehicle networks, organizations and individuals already have means to identify and track vehicles. Besides physically following a vehicle, the following categories of methods exist:

- consumer mobile devices that drivers and passengers elect to use,
- wireless devices and networks that are integrated into the vehicle, and
- traffic monitoring systems.

13.4.3.1 Consumer Mobile Devices   A predominate category of existing vehicle tracking methods uses tracking capabilities that come with the mobile devices and applications consumers elect to use. Today's mobile phones often have positioning capabilities such as Global Positioning System (GPS). Applications on a mobile phone can record the phone's positions and report the information to mobile phone application providers, mobile phone suppliers, or wireless network operators. An increasing number of vehicles are equipped with telematics services that provide traffic information, route guidance, and assistance to drivers. These telematics systems can track vehicle movements, often even when drivers terminated their service subscriptions. An electronic toll collection (ETC) system can track vehicles between toll collection points.

Privacy concerns about consumer mobile devices are sometimes overshadowed by the compelling values these devices bring to the users. People are free to choose not to use these devices and applications that intrude their privacy. People may also temporarily turn off the location-tracking applications on the phones, the GPS receivers on the phones, or the phones themselves. Drivers who do not wish to be tracked by an ETC system can choose not to use it.

The mobile phones and other devices with wireless interfaces in onboard vehicles can also be tracked by radiofrequency (RF) fingerprinting devices [FMTN06] [GDMR06] [UrSe06]. However, deploying a large number of RF fingerprinting devices along roadways will be costly.

13.4.3.2 In-Vehicle Wireless Devices and Networks   Modern vehicles use embedded wireless devices and networks to support an increasing range of functions. For example, tire pressure monitoring systems (TPMSs) are widely deployed in modern vehicles. TPMSs use sensors placed inside tires to monitor tire pressures and use RF transmitters to send measurement data to the tire pressure control unit on the vehicle. A recent study shows that TPMSs emit wireless signals that can be eavesdropped at a distance of roughly 40 m from the passing vehicle [RMMT10]. Since the protocols used to transmit tire pressure measurement data do not use cryptographic mechanisms to protect data integrity today, the data can be spoofed by nearby vehicles or by wireless devices deployed along the roadside.

13.4.3.3 Traffic Monitoring System Today, government agencies often deploy traffic monitoring systems, such as video cameras, at intersections and along roadsides to monitor traffic and road conditions. Monitoring data captured at different locations can be correlated to track a vehicle's movement over a large region.

While an effective vehicle tracking tool, using video cameras to track vehicles continuously will require deploying and operating a large number of cameras in many locations. It will also require image processing and analytics software to correlate the images captured by cameras at different locations in order to track the continuous movement of a vehicle over a large region. These can lead to prohibitive complexity and cost.

13.5 BASIC SECURITY CAPABILITIES

This section introduces security functions that will be required in a consumer vehicle network: authentication, misbehavior detection and revocation, data integrity, and data confidentiality.

13.5.1 Authentication

Vehicles (and their communicating parties) must be able to establish a high level of trust in the messages they use to make safety-related decisions. Therefore, the first and the most fundamental security function for cooperative safety communications is broadcast message authentication: the ability for vehicles to determine whether a broadcast safety message is from another vehicle or RSU authorized to send such messages.

Authenticating a broadcast message does not require uniquely identifying the message-originating vehicle. It is sufficient to determine that the message originator belongs to a group of entities (vehicles or RSUs in this case) entitled to the privilege of sending a given type of messages (safety broadcast messages in this case).

Broadcast message authentication can be accomplished using digital signatures with digital certificates and a public key infrastructure (PKI). A digital signature allows a vehicle to determine that a message originator knows a secret. A certificate attests that this secret is possessed only by a vehicle authorized to broadcast vehicle safety messages. Certificates must be issued by entities trusted by all vehicles to have the authority to issue certificates and the competency to ensure the correctness of the information in the certificates. PKI is a framework for issuing and managing certificates.

13.5.2   Misbehavior Detection and Revocation

The second fundamental security function for a large consumer vehicle network is misbehaving vehicle detection and revocation: the ability to detect misused certificates and misbehaving vehicles and the ability to revoke misbehaving vehicles' privileges to send messages that others will trust. This is required to prevent misbehaving vehicles from harming the transportation system endlessly. Without such capabilities, misbehaving vehicles would accumulate in the network and could eventually be able to cause large-scale harms to innocent vehicles and severe disruptions to the transportation system. The ability to detect and evict misbehaving vehicles is a prerequisite for real-world deployment of a nationwide consumer vehicle network.

13.5.3   Data Integrity

The third major security function for a large vehicle network is data integrity. Data integrity means that data have not been altered in an unauthorized manner since it was created. Data can be altered during transmission over networks or while stored in databases either accidentally (e.g., caused by noisy transmissions or hardware failures) or deliberately (e.g., by adversaries). In a large consumer vehicle network, it is important for a vehicle to be able to determine:

- whether a received message has been altered and
- whether security-related data stored on the vehicle have been altered in an unauthorized manner.

13.5.4   Data Confidentiality

The fourth major security function is data confidentiality, which is the ability to prevent unauthorized parties from knowing the content of the messages. The following aspects of data confidentiality are important in a large consumer vehicle network:

- Every vehicle must protect the confidentiality of security-related data stored on the vehicle. This includes cryptographic keys and certificates.
- Vehicles must be able to protect the confidentiality of the data exchanged with security servers. This includes, for example, data exchanged with a security server to acquire cryptographic keys and certificates.
- Data confidentiality is not required for broadcast safety messages because their contents are meant to be viewed by all vehicles in a broadcast area.

13.6 PRIVACY PROTECTIONS CAPABILITIES

The security functions for a nationwide consumer vehicle network should not provide significant advantages to adversaries for breaching user privacy. In order to build such a privacy-preserving security system, the security credentials and security operations should support vehicle anonymity and message long-term unlinkability.

It is important to realize that adversaries can always use the plaintext contents in V2V vehicle safety broadcast messages to determine which messages are from the same vehicle if the messages are captured close enough in time. That is, message linkability will always be possible by capturing consecutive messages from a vehicle.

The ability to link messages to the same vehicle over a short period of time, or short-term linkability, is usually not a concern and often a requirement for supporting cooperative vehicle safety applications. A vehicle needs to determine the immediate movement trajectory of nearby vehicles to assess collision risks, which requires knowledge of which messages are from the same nearby vehicle over time durations that can last for minutes.

Long-term message unlinkability, on the other hand, is an important privacy requirement. Adversaries can break long-term message unlinkability by capturing consecutive V2V broadcast messages. However, using this approach to track a vehicle is, in general, costly because it requires the adversary to either physically follow the target vehicle or install monitoring capabilities to cover the entire region within which the vehicle might travel.

Therefore, the main issue in protecting long-term message unlinkability becomes: if an adversary captures messages a long time or a long distance apart (say, over a threshold of δ time units or τ meters apart), can the adversary still easily link these messages to the same vehicle?

In this case, the plaintext data, such as vehicle positions and speeds, will no longer provide significant advantages to adversaries for linking the messages to the same vehicle. The security credentials associated with the messages, however, could allow an adversary to identify a vehicle or link different messages to the same vehicle.

Therefore, a privacy-preserving security system should make it computationally or physically hard for adversaries to use the security information in messages to determine whether any two messages were transmitted by the same vehicle if these messages were captured δ time units or τ meters apart for given thresholds δ or τ.

13.7 DESIGN AND PERFORMANCE CONSIDERATIONS

Designing a privacy-preserving security system for a highly dynamic nationwide consumer vehicle network introduces unique challenges. This section discusses some of such design considerations for the overall system and for specific communication modes, including V2V local broadcast and vehicle-to-infrastructure (V2I) bidirectional communications.

13.7.1 Scalability

The privacy-preserving security system for a nationwide vehicle network in the United States must be scalable enough to support over 250 million vehicles. This means that the system must be able to issue, revoke, and replace cryptographic keys and certificates for hundreds of millions of vehicles dynamically. Each vehicle may use multiple private–public key pairs and certificates to support different applications and to protect privacy.

For example, if short-lived certificates are used to protect privacy, each vehicle typically needs to use a different certificate every few minutes. This means that the system needs to support over 2 trillion certificates in operation if each certificate is valid for 5 minutes and each vehicle is loaded each time with enough certificates to use for a month. As vehicles run out of valid certificates, they will need to acquire new certificates. Therefore, 2 trillion new certificates will have to be distributed to vehicles every month on average. If a separate certificate management system is used to serve the vehicles in each state, each statewide certificate management system still needs to support, on average, over 43 billion certificates in operation at any time, which would have to be replaced every month. To put this scalability requirement in perspective, the largest digital certificate service provider for the Internet in the United States—VeriSign—supports only tens of millions of certificates in operations today [Renf02] [RoCh10]. These certificates are mostly for Internet servers and browsers and do not need to change frequently.

13.7.2 Balancing Competing Requirements

Supporting security, protecting privacy, and enabling cooperative vehicle safety applications impose competing requirements. Achieving a proper balance among these competing requirements is a core challenge in the design of privacy-preserving security systems for large consumer vehicle networks. For example, the following competing requirements need to be balanced:

- Security and privacy protection should not jeopardize the ability to effectively support vehicle safety applications. While it is important to support long-term message unlinkability, many safety applications require short-term message linkability.
- Privacy protection should not jeopardize the ability to support crucial security functions such as the ability to detect misbehaving vehicles. When communications are made anonymous and messages unlinkable to protect driver privacy, detecting misbehaving vehicles becomes more difficult.
- Supporting security should not jeopardize driver privacy. Detection of misbehaving vehicles can be simpler if the misbehavior detection system

can identify vehicles or link multiple messages to the same vehicle. However, these same vehicle identification and tracking mechanisms can also be used to breach the privacy of innocent vehicles.

13.7.3   Minimal Side Effects

The security and privacy protection functions should not create excessive side effects. In particular, security and privacy protection should not:

- cause excessive disruptions to the driving public, such as forcing innocent drivers to bring their vehicles to service centers frequently just to reconfigure security parameters on the vehicles, replace security credentials, or help detect misused certificates and misbehaving vehicles;
- create excessive computational burdens and wireless communication overheads on the vehicles; and
- open unmanageable new risks that can result from abusing the privacy-protecting system by adversaries.

13.7.4   Quantifiable Levels of Security and Privacy

The levels of security and privacy protection to be experienced by drivers should be quantifiable and remain measurable over time. It is also desirable to know the quality of security and the quality of privacy. The quality of security (or privacy) refers to the likelihood that the consumer vehicle network can provide a target level of security (or privacy) under various conditions, such as in areas of different vehicle densities and when different numbers of adversaries are present in the network.

13.7.5   Adaptability

It is desirable for the privacy protection mechanisms to be capable of adapting to drivers' varying privacy needs and different privacy demands under different conditions. This is because the acceptable levels of privacy can differ substantially for different people and can change completely under different conditions even for the same person. For example, many people may accept reduced privacy protection in exchange for more effective or cheaper vehicle safety applications while others may view privacy as a higher priority. Faced with imminent collision risks, people typically prefer higher performance of safety applications over privacy protection. When an accident occurs, many people would trade privacy for speedy rescue operations.

13.7.6   Security and Privacy Protection for V2V Broadcast

The most important security requirement for V2V local broadcast is message authentication that can preserve vehicle anonymity and message long-term unlinkability. Digital signatures, privacy-preserving certificates, and a PKI can be used to enable privacy-preserving broadcast message authentication. This will require a vehicle to sign its outbound messages and verify the signatures on all or subsets of the received messages in a timely manner.

Recent studies show that a vehicle may need to receive up to 10 safety broadcast messages per second from each nearby vehicle to support collision avoidance applications [NHTS05]. With dedicated short-range communications (DSRC) radio transmission range up to 300 meters, a vehicle can receive safety broadcast messages from hundreds of vehicles at crowded urban intersections or congested highways. Therefore, a vehicle may need to process thousands of incoming messages per second. This means that each vehicle might have to verify thousands of signatures per second. It also means that a vehicle may have to verify hundreds of times more signatures than it has to sign outgoing messages. Consequently, it is important for the digital signature algorithm to have fast signature verification performance. Furthermore, most V2V broadcast safety messages received by a vehicle will be uneventful in the sense that they will not indicate any dangerous situation. To reduce the number of messages to be verified and the time needed for the verification, methods can be developed for a vehicle to verify only the messages that are detected to be eventful.

13.7.7   Security and Privacy Protection for Communications with Security Servers

A vehicle needs to communicate with security servers such as certificate management servers to acquire new certificates, cryptographic keys, and the latest certificate revocation information. These communications need to be confidential so that adversaries would not know which keys and certificates are assigned to which vehicles. The vehicles and the security servers also need to know whether the messages they receive have been altered in unauthorized manners.

Vehicles need to be able to access security servers frequently enough to ensure proper operation of the security and privacy protection system. For example, before all the certificates on a vehicle expire, the vehicle should be able to access the security servers to acquire new certificates. The vehicle will also need to retrieve the latest certificate revocation information in a timely manner. Exactly how frequently each vehicle should access security servers is a crucial design consideration.

When DSRC RSUs are used to support communications with the security servers, several issues need to be addressed. For example,

- A vehicle may stay inside the radio coverage area of each DSRC RSU for only a few seconds. Therefore, the establishment of a secure connection through the RSU and with the security servers must be completed within the time the vehicle stays inside a RSU radio coverage area [PSCZ08]. The security connection establishment time should be kept to a minimum to leave sufficient time to transport data (such as certificate requests, certificate responses, or certificate revocation lists [CRLs]) between the vehicle and the security servers.

- A large number of vehicles may be inside the radio coverage area of an RSU. This can often happen at busy intersections, for example. Therefore, it is important to minimize the message overhead incurred to establish security connections [PSCZ08].

When long-range radio systems, such as cellular or satellite networks, are used to support vehicle communications with the security servers, important issues to be addressed include determining and reducing the frequency at which a vehicle has to use its long-range radio to communicate with the security servers and the amount of data to be exchanged with the security servers. This is because such long-range radio systems typically have limited bandwidth and can also be costly to the drivers.

REFERENCES

[BeBF06] A. Bellissimo, J. Burgess, and K. Fu: "Secure Software Updates: Disappointments and New Challenges," 1st USENIX Workshop on Hot Topics in Security (HotSec 2006), USENIX, 2006.

[Douc02] J. R. Douceur: "The Sybil Attack," 1st International Workshop on Peer-to-Peer Systems, 2002.

[DSCP02] C. Diaz, S. Seys, J. Claessens, and B. Preneel: "Towards Measuring Anonymity," 2nd International Conference on Privacy Enhancing Technologies (PET'02), 2002.

[FMTN06] J. Franklin, D. McCoy, P. Tabriz, V. Neagoe, J. V. Randwyk, D. Sicker: "Passive Data Link Layer 802.11 Wireless Device Driver Fingerprinting," 15th conference on USENIX Security Symposium, Vancouver, B.C., Canada, 2006.

[GDMR06] R. Gerdes, T. Daniels, M. Mina, and S. Russell: "Device Identification via Analog Signal Fingerprinting: A Matched Filter Approach," Network and Distributed System Security Symposium (NDSS), 2006.

[KCRP10] K. Koscher, A. Czeskis, F. Roesner, S. Patel, T. Kohno, S. Checkoway, D. McCoy, B. Kantor, D. Anderson, H. Shacham, and S. Savage: "Experimental Security Analysis of Modern Automobile," 2010 IEEE Symposium on Security and Privacy, California, 2010.

[NHTS05] National Highway Traffic Safety Administration: "Vehicle Safety Communications Project—Task 3 Final Report—Identify Intelligent Vehicle Safety Applications Enabled by DSRC," DOT HS 809 859, 2005.

[PCZV07] S. Pietrowicz, G. D. Crescenzo, T. Zhang, E. Vandenberg, K. Kavaliauskas: "VII Vehicle Segment Threat and Risk Analysis," VII Consortium Security Work Order Deliverable 1.3, 2007.

[PfHa01] A. Pfitzmann and M. Hansen: "Anonymity, Unobservability and Pseudonymity—A Proposal for Terminology," Designing Privacy Enhancing Technologies, Springer-Verlag Lecture Notes in Computer Science, 2001.

[PSCZ08] S. Pietrowicz, H. Shim, G. Di Crescenzo, and T. Zhang: "VDTLS—Providing Secure Communications in Vehicle Networks," Infocom MOVE (MObile Networking for Vehicular Environments) Workshop, 2008.

[Renf02] S. G. Renfro: "VeriSign CZAG: Privacy Leak in X.509 Certificates," 11th USENIX Security Symposium, California USA, 2002.

[RMMT10] I. Roufa, R. Millerb, H. Mustafaa, T. Taylora, S. Ohb, W. Xua, M. Gruteserb, W. Trappeb, I. Seskarb: "Security and Privacy Vulnerabilities of In-Car Wireless Networks: A Tire Pressure Monitoring System Case Study," 19th USENIX Security Symposium, 2010.

[RoCh10] F. Rosch, A. Chatterjee: "VeriSign Business and User Authentication Overview," 2010.

[UrSe06] O. Ureten and N. Serinken: "Bayesian Detection of Wi-Fi Transmitter RF Fingerprints," Electronic Letters, vol. 41, no. 6, pp. 373–374, 2006.

14

CRYPTOGRAPHIC
MECHANISMS

14.1 INTRODUCTION

This chapter describes fundamental cryptographic mechanisms. The goal is to provide a basis for assessing the suitability of specific cryptographic mechanisms for vehicle communications.

After outlining the main categories of cryptographic mechanisms, we describe digital signature algorithms that form the foundation for most other security functions. We present their main principles, methodologies, and properties that are essential for assessing their suitability for consumer vehicle networks. We then discuss techniques for message authentication and integrity verification. We further describe methods for communicating parties to establish shared secret keys by exchanging plaintext messages over nonprivate and nonsecured networks. Finally, we will describe the public key encryption algorithm used in the Institute of Electrical and Electronics Engineers (IEEE) 1609.2 standard [IEEE11].

14.2 CATEGORIES OF CRYPTOGRAPHIC MECHANISMS

Cryptographic mechanisms or algorithms are methods for generating cryptographic keys, creating and verifying digital signatures, establishing secret keys among communicating entities, and encryption and decryption.

Vehicle Safety Communications: Protocols, Security, and Privacy, First Edition. Luca Delgrossi and Tao Zhang.
© 2012 John Wiley & Sons, Inc. Published 2012 by John Wiley & Sons, Inc.

Existing cryptographic algorithms can be classified into three basic categories based on the number of cryptographic keys they use: hash algorithms, symmetric key algorithms, and public key (asymmetric key) algorithms.

14.2.1 Cryptographic Hash Functions

Hash functions do not require keys. A cryptographically secure hash function, often referred to as cryptographic hash function or hash function for simplicity, converts an arbitrary-length bit string into a small fixed-length bit string called a *hash value* or *message digest* in a way that meets the one-way requirement and the collision resistance requirement.

The one-way requirement means that it is computationally infeasible to find any input that maps to a given hash value. Functions that satisfy this one-way requirement are referred to as one-way functions.

The collision resistance requirement means that it is computationally infeasible to find two distinct inputs to the hash function that map to the same hash value.

Hash functions are, in general, highly computational efficient because they rely on simple bitwise logical operations such as AND, OR, and Exclusive OR (XOR). Furthermore, the hash value of a message is typically much smaller than the message itself. Therefore, hash functions have been widely used as an important component of modern cryptographic algorithms such as digital signature algorithms, key distribution algorithms, encryption and description algorithms, and random number generators. For example, hash functions are commonly used to compress large messages into smaller hash values that can then be used as input to generate digital signatures for the messages.

The Secure Hash Algorithm (SHA) family of cryptographic hash functions is one of the most widely used families of hash functions. SHA is published by the U.S. National Institute of Standards and Technology (NIST) as the U.S. Federal Information Processing Standard (FIPS). Four SHA hash functions are specified by NIST for use by the U.S. federal government and they are SHA-1, SHA-256, SHA-384, and SHA-512 [NIST02]. SHA-1 produces 160-bit long hash values. SHA-256, SHA-384, and SHA-512 produce 256-, 384-, and 512-bit long hash values, respectively.

These hash functions differ primarily in the security strength they provide. The security strength or the level of security of a cryptographic algorithm is typically measured by the amount of work that would be required to violate the security requirements or properties of the cryptographic algorithm. The security strength is typically expressed in bits. k-bit security strength means that it will take on the order of 2^k basic operations, such as additions or multiplications, to break the security properties of a cryptographic algorithm.

The security strength of a hash function is the amount of work that would be required to break at least one of the two fundamental requirements of hash functions: the one-way requirement or the collision resistance requirement.

Table 14.1. Hash value size and security strength of hash functions

	Hash Value Size (bits)	Security Strength (bits)
SHA-1	160	80
SHA-256	256	128
SHA-384	384	192
SHA-512	512	256

The strength of the one-way requirement is determined by the amount of work that would be needed to find an input to the hash function that will generate a given hash value. The estimated strength for the one-way property provided by a well-defined hash function is the length of the hash value produced by the hash function [NIST09b].

The strength of the collision resistance requirement is measured by the amount of work that would be needed to find a collision for a hash function with high probability. The estimated strength for collision resistance provided by a hash function is half the length of the hash value produced by the hash function [NIST09b].

If an application requires more than one property from the cryptographic hash function, then the security strength of the weakest property is the security strength of the cryptographic hash function for the application. For instance, the security strength of a cryptographic hash function when used for digital signatures is defined as its collision resistance strength because digital signatures require both collision resistance and the one-way properties from the cryptographic hash function,

Therefore, the security strength of a hash function is half the length of the hash value it produces. Table 14.1 summarizes the security strengths and the sizes of the hash values of the four SHA hash functions [NIST02]. NIST recommends that the minimum security strength for cryptographic algorithms be 112 bits between the years 2010 and 2030, and 128 bits beyond the year 2030. Therefore, SHA-1 is considered too weak now.

14.2.2 Symmetric Key Algorithms

With symmetric key algorithms, two communicating parties use a single identical secret key for both encryption and decryption. This symmetric key must be kept secret and should only be known to people authorized to access the data protected by the key.

Consequently, symmetric key algorithms have a fundamental limitation: it is difficult for communicating parties to exchange and agree upon secret keys (a process referred to as key establishment) over nonsecure and nonprivate networks without revealing the secrets to others before the communicating parties have the means to secure their communication.

The main advantage of symmetric key algorithms is that they are typically based on simple logical operations, such as AND, OR, and XOR, and therefore are highly computational efficient relative to asymmetric key algorithms which we will discuss next. Therefore, symmetric key algorithms have been commonly used for encryption and decryption, especially for long message flows.

NIST has approved the following two families of symmetric key algorithms for use by the U.S. federal government: the Advanced Encryption Standard (AES) and the Triple Data Encryption Algorithm (TDEA). These algorithms take *blocks* of data as input for encryption and decryption and therefore are commonly referred to as *block ciphers*.

The AES standards are defined in [NIST01] and published in 2001. They consist of three block ciphers: AES-128, AES-192, and AES-256. Each of these AES block ciphers uses the same block size of 128 bits as input but with 128-, 192-, and 256-bit keys, respectively. Triple DEA is defined in [NIST04]. It encrypts and decrypts data in 64-bit blocks, using three 56-bit keys.

Block ciphers have several known weaknesses. The same plaintext block will always encrypt to the same ciphertext block whenever the same key is used. Data patterns in the plaintext, such as repeated blocks, will therefore be apparent in the ciphertext. Furthermore, if multiple blocks in a message are encrypted separately, an adversary can substitute individual blocks and such substitutions are hard to detect.

The security strength of a symmetric key algorithm is measured by the amount of work that would be required to figure out the secret key. When the symmetric keys are k-bit integers, any positive integer of length k may be used as a secret key in a symmetric key algorithm. A brute force attack that tests every possible key in the entire key space would take in the order of 2^k steps to determine the specific key used to encrypt a message. No significantly faster algorithms have been found to figure out a secret symmetric key with only knowledge of the length of the key. Therefore, so far, a symmetric key algorithm using k-bit long keys has been considered to have the security strength of k bits.

Table 14.2 shows the symmetric key algorithms that meet the security strengths recommended by NIST for the year 2011 and after [NIST07].

14.2.3 Public Key (Asymmetric Key) Algorithms

Public key algorithms, or asymmetric key algorithms, represent a fundamental advancement in cryptography. With public key algorithms, each entity has a

Table 14.2. Security strengths and key lengths for symmetric key algorithms

Security Strength (bits)	Symmetric Key Algorithm
112	Three-key TDEA
128	AES-128
192	AES-192
256	AES-256

pair of distinct keys: a private key and a public key. Each entity keeps its private key secret and uses it to generate digital signatures, but makes its public key known to others for them to use to verify the signatures. The private key and the public key in each key pair are mathematically linked in a way that allows the public key to be used to verify signatures generated with the private key while making it infeasible to derive the private key from knowledge of the public key. This enables communicating parties to make their public keys available to each other over nonsecure and nonprivate networks.

Any entity can send digitally signed messages to others without the need for secure exchange of secret keys first. This is important for supporting highly time-sensitive authenticated vehicle-to-vehicle (V2V) broadcast—a crucial capability required by many cooperative vehicle safety applications. Any vehicle can use its private keys to sign the messages it broadcasts to other vehicles without any prior knowledge of, or message exchange with, the other vehicles. Other vehicles can use the broadcasting vehicles' public keys to verify the digital signatures on the received messages to authenticate the messages and verify message integrity.

Today's public key algorithms rely on the availability of *trapdoor functions*. A trapdoor function is a function that is easy to compute in one direction, but computationally infeasible to compute in the opposite direction to find its inverse, unless one has special information called the *trapdoor information* (which corresponds to a private key).

To illustrate the concept of a trapdoor function, consider integers that belong to a very large space S. Let's say that $\oplus$ is an operation that maps two integers P and Q in space S to another integer M that also belongs to space S. Then, function $P \oplus Q = M$ is a trapdoor function if (1) knowing P and Q, it is easy to compute $P \oplus Q = M$; and (2) knowing M and Q, it will be computationally infeasible to find which specific integer among the many possible integers in the very large space S were used as P to generate M. In this case, a public key can be generated from Q and a private key can be generated from P.

Public key signature algorithms are used to create and verify digital signatures. Some public key signature algorithms can also be used for message encryption and decryption. We will describe three leading public key signature algorithms in this chapter and also explain why each of them can or cannot be used for encryption and decryption.

Public key encryption is, in general, more computationally intensive than symmetric key algorithms for the same level of security. Therefore, key establishment protocols using trapdoor functions have been developed to enable communicating parties to establish secret symmetric keys by exchanging plaintext information over nonsecure and nonprivate networks. The first and still widely used key establishment protocol is the Diffie–Hellman Key Exchange Protocol, named after Whitfield Diffie and Martin Hellman who published the protocol in 1976 [DiHe76]. The Diffie–Hellman Key Exchange Protocol will be discussed later in this chapter. Once the secret symmetric keys have been established between the communicating parties, symmetric key algorithms can be used to encrypt the data to be exchanged.

The security strength of a public key algorithm, for digital signature or for encryption, is measured by the amount of work that would be needed to figure out the private key associated with a public key. Since private keys are mathematically linked with their corresponding public keys, the security strength of a public key algorithm is typically linked to the sizes of both the private key and the public key. In general, to achieve a k-bit security level, the lengths of the private key and the public key must be at least k bits long.

14.3 DIGITAL SIGNATURE ALGORITHMS

A digital signature algorithm describes methods for generating private and public keys, creating signatures, and verifying signatures. NIST approved several digital signature algorithms for the U.S. federal government to use [NIST09]. These algorithms include:

- Rivest–Shamir–Adleman (RSA) Algorithm [ANSI98] [PKCS02];
- Digital Signature Algorithm (DSA) [NIST09]; and
- Elliptic Curve Digital Signature Algorithm (ECDSA) [ANSI05] [Kobl87] [MILL85].

14.3.1 The RSA Algorithm

The RSA algorithm is named after the three people who first publicly described the algorithm: Ronald Linn **R**ivest, Adi **S**hamir, and Leonard Max **A**dleman. The RSA algorithm is specified in the American National Standard X9.31 [ANSI98] and the Public Key Cryptography Standard (PKCS) #1 [PKCS02].

14.3.1.1 The Basic Idea The trapdoor functions used in the RSA algorithm is based on the difficulty of solving the RSA Problem. The RSA Problem can be illustrated as follows: One can easily select two large prime numbers p and q, and compute their product $n = p \cdot q$. The RSA Problem is: given an arbitrary large integer C and n, find the eth root of C modulus n. That is, to find u that satisfies $u^e = C(\bmod\ n)$.

Given u and e, computing C is computationally easy. However, solving the RSA Problem to find u for known values of C and e is hard. No polynomial time algorithm has been found for this problem so far, although there has been no formal proof on how hard the RSA Problem is. The belief that the RSA Problem is hard is called the RSA Assumption.

Therefore, n and e can be used as a public key. A private key can be derived from e and the secret prime factors p and q of n.

So far, the most efficient ways to solve the RSA Problem are to first factor the integer modulus n. Solving the integer factorization problem, however, is known to be extremely computational intensive. Although there has been no formal proof of how hard the integer factorization problem is, no polynomial time algorithm has been found for it so far. However, if one can find all the prime factors of n, one can find the particular prime factors p and q used to compute n and, consequently, one can find the private key. Therefore, the RSA digital algorithm is also often said to be based on the difficulty of integer factorization.

14.3.1.2 Generate Private and Public Keys   To generate a private–public key pair, the RSA algorithm first needs to set the length of the public key in bits. In general, the longer the public keys, the stronger the security strength will be for the private–public key pair.

The public key will be a pair of integers n and e, where n is referred to as the public modulus and e is referred to as the public verification exponent.

The private key will be a pair of integers n and d, where d is referred to as the private signature exponent.

The RSA key generation algorithm works as follows:

- *Step 1*: Select two private prime numbers p and q and compute the public modulus n. The prime numbers should be randomly selected large prime numbers of approximately equal size. The public modulus n will be the product of p and q. That is, $n = p \cdot q$. The values of p and q should be chosen so that n will have the required length of the public key. For example, n should be a 1024-bit integer if the required public key length is 1024 bits long.
- *Step 2*: Compute the public verification exponent e. First, compute the totient of n. The totient of an integer k, denoted by $\varphi(k)$, is the number of integers less than k that are relatively prime with k. When k is a prime, its totient $\varphi(k)$ is $k - 1$. Since $n = p \cdot q$, where both p and q are primes, we have $\varphi(n) = (p - 1)(q - 1)$. The public verification exponent e should be a positive integer such that $1 < e < \varphi(n)$ and $gcd(e, \varphi(n)) = 1$, where $gcd(x, y)$ is the greatest common denominator of x and y. That is, e and $\varphi(n)$ are relative primes or coprimes. The public verification exponent can be selected as a fixed value or generated as a random value. If e is randomly generated, it shall be an odd number [ANSI98]. Commonly used fixed values for e include 2, 3, 17, and $2^{16} + 1 = 65,537$ [ANSI98].
- *Step 3*: Compute the private signature exponent d. Find integer d such as $d \cdot e = 1(\bmod \ \varphi(n))$. That is, d is the multiplicative inverse of e modulus $\varphi(n)$.
- *Output*: The public key is the integer pair (n, e) and the private key is the integer pair (n, d).

The size of an RSA key pair is commonly considered to be the length of the modulus n in bits.

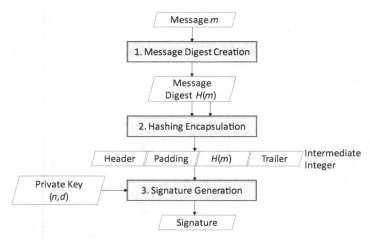

Figure 14.1. RSA signature creation

14.3.1.3 Create a Digital Signature Figure 14.1 illustrates the procedure to create a digital signature for a message. The signature creation algorithm takes as input the message to be signed and the private key for signing the message. The algorithm consists of three steps: message digest creation (message hashing), hash encapsulation, and signature generation.

To create a signature for a message m, the message signer first creates a message digest $H(m)$, where $H()$ is a cryptographic hash function. The message digest, which typically will be significantly smaller than the original message, will be used as an input to generate the signature.

The message signer converts the message digest $H(m)$ into a positive integer h such that h has the same length as n and satisfies $0 < h < n$. This integer h is referred to as the Intermediate Integer. This conversion is achieved using a padding algorithm that appends a carefully designed pattern of bits to the message m.

The padding algorithm needs to be agreed upon by the signature creator (message originator) and the signature verifiers (message receivers). The padding algorithm must be reversible so that a message receiver can remove the padding to recover the message digest $H(m)$.

In the RSA signature generation algorithm defined in [ANSI98], the intermediate integer includes a header and a trailer that contain coded information for message verifiers to determine which hashing function was used to create the message digest.

The message signer uses the intermediate integer h and its private key (n, d) to generate a signature s as follows:

$$s = h^d \pmod{n}. \tag{14.1}$$

The size of the RSA signature will be the size of the modulus n, which is also the size of the RSA keys as illustrated before.

14.3.1.4 *Verify a Signature* RSA signature verification takes as input the received message, the signature of the message, and the message originator's public key. Signature verification is carried out using the following steps: signature opening, encapsulated hash verification, hash recovery, and message hashing and comparison [ANSI98]. This process is shown in Figure 14.2.

To verify a signature s' on a received message m', the signature verifier uses the message originator's public key (n, e) and signature s' as input to open the received signature by computing the following integer v:

$$v = (s')^e (\bmod\, n). \qquad (14.2)$$

This recovers the encrypted intermediate integer received from the message signer. That is, it decrypts the encrypted intermediate integer. A formal proof that Equation 14.2 will recover the intermediate integer h generated using Equation 14.1 can be found in [LiWB06].

If the signature is valid, v should be exactly n bits long. Therefore, if v is not n bits long, the signature is not valid and the algorithm stops. Furthermore, v should be identical to the intermediate integer h used by the message signer to create the signature if the signature is valid and has not been altered.

After verifying the length, header, and trailer of the recovered intermediate integer v, the signature verifier extracts the message digest $H(m)'$ from v by stripping off the padding, the header, and the trailer. This step is called Hash Recovery.

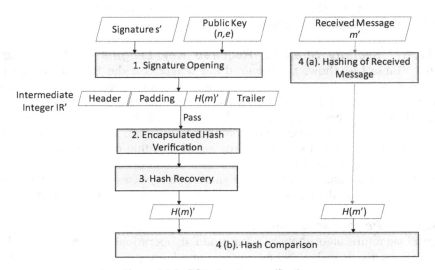

Figure 14.2. RSA signature verification.

In the last step, the signature verifier computes the message digest $H(m')$ directly using the received message m'. It uses the information in the trailer of the recovered intermediate integer v to determine the hashing function used by the message originator to generate the message digest.

If these two message digests $H(m)'$ and $H(m')$ are identical, the signature verification algorithm concludes that the signature is valid and the received message m' is the original message m signed by the message signer (i.e., the message has not been altered since the signature was generated).

14.3.1.5   Encrypt a Message   The RSA signature algorithm signs a message by encrypting the message digest, and verifies the signature by decrypting the encrypted message digest to recover the original message digest. Therefore, the same RSA signature algorithm can also be used for message encryption and decryption.

To encrypt a message m using the recipient's public key (n, e), the message originator first converts the message m into a positive integer h such that $0 < h < n$. This can be done using a padding algorithm that adds a carefully designed pattern of bits to the beginning or end of the message.

The ciphertext c is computed as follows:

$$c = h^e \pmod{n}. \tag{14.3}$$

14.3.1.6   Decrypt a Message   The message receiver can use its private key d to decrypt the ciphertext c to reconstruct h as the following:

$$h = c^d \pmod{n}. \tag{14.4}$$

Given h, the message receiver can recover the original plaintext message m by reversing the padding scheme.

14.3.1.7   Security Strength, Required Key Lengths, and Signature Size   Table 14.3 shows the required key lengths for the RSA algorithm to achieve various levels of security strengths recommended by NIST for use after 2010 [NIST07]. The required key size is significantly larger than the target security strength because only a small subset of all the possible values of the key size can be used as keys. For comparison purpose, Table 14.3 also shows the hash functions and symmetric key algorithms that can achieve the same levels of security strengths.

Table 14.4 summarizes the RSA signature sizes, measured in bit length, for the security strengths NIST recommended for the year 2010 and after.

14.3.1.8   Performance   The RSA algorithm and virtually all other existing digital signature algorithms rely on modular operations. The basic modular operations are described below, where a and b are positive integers and n is an integer modulus:

Table 14.3. RSA key lengths required for different security strengths

Security Strength (bits)	RSA Public Key Length (bits)	Comparable Hash Function	Comparable Symmetric Key Algorithm
112	2,048	SHA-224	Three-key TDEA
128	3,072	SHA-256	AES-128
192	7,680	SHA-384	AES-192
256	15,360	SHA-512	AES-256

Table 14.4. RSA signature sizes for different security strengths

Security Strength (bits)	RSA Signature Size (bits)
112	2,048
128	3,072
192	7,680
256	15,260

Table 14.5. Bit complexity of modular operations

Operations	Bit Complexity
Modular addition: $(a + b) \bmod n$	$O(\log_2 n)$
Modular subtraction: $(a - b) \bmod n$	$O(\log_2 n)$
Modular multiplication: $(a \cdot b) \bmod n$	$O((\log_2 n)^2)$
Modular inversion: $a^{-1} \bmod n$	$O((\log_2 n)^2)$
Modular exponentiation: $a^k \bmod n, k < n$	$O((\log_2 n)^3)$

- Modular addition: $a + b(\bmod\ n)$
- Modular subtraction: $a - b(\bmod\ n)$
- Modular multiplication: $a \cdot b(\bmod\ n)$
- Modular inversion: $a^{-1}(\bmod\ n)$
- Modular exponentiation: $a^k(\bmod\ n)$

Therefore, to understand the computational complexity of public key algorithms, it is important to understand the complexity of the modular operations. The computational complexities of the above modular operations, expressed in terms of bit complexity or the number of bit-wise operations, are summarized in Table 14.5 [MeOV97]. The complexity of a modular operation depends on the size of the modulus. The complexity of a modular exponentiation also depends on the size of the exponent. The complexity of modular exponentiation shown in Table 14.5 assumes that the size of the exponent is not larger than the size of the modulus and that modular exponentiation is performed using the repeated square-and-multiply algorithm in [MeOV97].

The two most time-consuming operations in RSA signature creation or verification are modular exponentiation and hashing.

Table 14.6. Performance of RSA signature generation and verification

	Sign a Message	Verify a Signature
Modular exponentiations	1	1
Hashing operations	1	1

Table 14.6 summarizes the RSA signature generation and verification performance. To sign a message, RSA performs one modular exponentiation and one hashing operation. To verify a message, RSA performs one modular exponentiation and one hashing operation. All modular operations in RSA signature generation and verification use the same RSA modulus. The sizes of the RSA modulus required to meet different security strengths are given in Table 14.3.

14.3.2   The DSA Algorithm

The DSA algorithm was proposed by NIST in August 1991 as a U.S. Digital Signature Standard (DSS) [NIST09].

14.3.2.1   The Basic Idea   The DSA algorithm relies on the difficulty of solving *discrete logarithm problems* to derive the trapdoor function needed for creating private–public key pairs. The principle can be illustrated as follows: Given two integers x and g, raising g to the xth power to find $g^x = y$ is an easy computation. Knowing integers y and g, trying to find integer x such that $g^x = y(\bmod p)$, where p is an integer modulus, is a discrete logarithm problem because $x = \log_g y(\bmod p)$. Although the discrete logarithm problem has not been formally proved to be NP-complete, no polynomial time algorithm has been found for the problem so far.

Therefore, a private key could be a random large integer x and its associated public key could be $y = q^x$. As long as the discrete logarithm problem remains hard to solve, it will be computationally infeasible to derive the private key x from only the knowledge of y and q.

14.3.2.2   Domain Parameters   The DSA algorithm requires that the private–public key pair generation, signature generation, and verification be performed with respect to a set of domain parameters that are known to both signature generators and verifiers. These domain parameters are:

- p: A prime modulus, which satisfies $2L - 1 < p < 2L$, where L is the bit length of p and will be the length of the public key.
- q: A prime divisor of $(p - 1)$, which satisfies $2N - 1 < q < 2N$, where N is the bit length of q and will be the length of the private key.
- g: An integer whose multiplicative order modulo p is q. That is, $g \cdot q = 1$ (mod p).

Table 14.7. DSA key lengths required for different security strengths

Security Strength (bits)	DSA Key Length (bits)	Comparable Hash Function	Comparable Symmetric-Key Algorithm
112	$L = 2{,}048, N = 224$	SHA-224	Three-key TDEA
128	$L = 3{,}072, N = 256$	SHA-256	AES-128
192	$L = 7{,}680, N = 384$	SHA-384	AES-192
256	$L = 15{,}360, N = 512$	SHA-512	AES-256

These domain parameters are used by message originators to create signatures and must be made available to signature verifiers before they can verify the signatures.

All communication entities may use the same set of domain parameters. In this case, the domain parameters can be preselected and established on all the communication entities.

Alternatively, each individual entity can generate its own domain parameter values and send them as part of the public key to signature verifiers. This will eliminate the need to preconfigure and maintain system-wide parameters on all entities. It will also enable each individual entity to adjust the security strength of its keys and signatures independently by selecting its domain parameter values to meet its target security strength. However, as the required security strength increases, the sizes of the DSA domain parameters will increase sharply, as shown in Table 14.7. Consequently, repeatedly sending domain parameters may become resource prohibitive over wireless networks.

14.3.2.3 Generate Private and Public Keys The DSA algorithm provides two basic ways to generate private–public key pairs. Here, we describe one approach that is called key pair generation by testing candidates.

This algorithm works as follows:

- *Step 1*: Take as input, or select, a set of domain parameters that meet the security strength requirement. That is, the length of p should be the required public key length L and the length of q should be the required private key length N.
- *Step 2*: Compute the private key x: x is an integer randomly selected from the interval $[1, q - 1]$.
- *Step 3*: Compute the public key y: $y = g^x (\bmod\ p)$.
- *Output*: Private key $= x$. Public key $= (p, q, g, y)$ if the domain parameters are self-generated. Public key $= y$ if a common set of domain parameters are used by all users.

The length of the DSA public key y will be L bits and the length of the DSA private key x will be N bits.

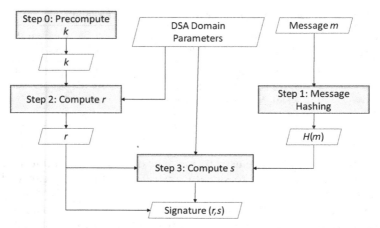

Figure 14.3. DSA signature creation

14.3.2.4 *Create a Digital Signature* The DSA signature generation algorithm takes as input the DSA domain parameters, a message to be signed, and a DSA private key.

The DSA signature is a pair of integers (r, s). The DSA signature creation procedure is illustrated in Figure 14.3.

The DSA signature generation algorithm requires that a new secret random number k be generated prior to the generation of each digital signature for use during the signature generation process. The value of k must be in the interval $(1, q)$. Furthermore, the multiplicative inverse of k, denoted by k^{-1}, with respect to multiplication modulo q, must be derived, which will also be used in the signature generation. The multiplicative inverse of k, which can be precomputed for faster processing, must satisfy the following requirements: $0 < k^{-1} < q$ and $1 = (k^{-1} \cdot k)(\bmod\ q)$.

Then, the DSA signature generation algorithm performs the following three main steps:

- *Step 1*: Hash message m to create a message digest $H(m)$ and convert $H(m)$ into an integer z using a padding algorithm to append a carefully designed pattern of bits to the beginning or the end of the hash value.
- *Step 2*: Compute $r = (g^k \bmod p)(\bmod\ q)$.
- *Step 3*: Compute $s = (k^{-1}(z + x \cdot r))\ (\bmod\ q)$, where x is the private key.

The computation of r does not require the knowledge of the message m. Therefore, r can be computed whenever $k, p, q,$ and g are available. Since the domain parameters $p, q,$ and g are known, and k can be precomputed, r can also be precomputed.

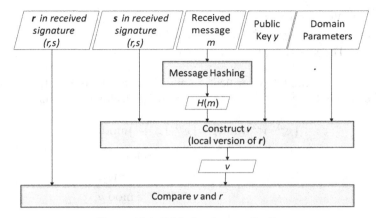

Figure 14.4. DSA signature verification

It is clear from the above algorithm that the length of r and the length of s are both equal to the length of q. The length of q is the length of the DSA private key, which needs to be twice the target security strength measured in bits for the DSA algorithm. Therefore, the DSA algorithm produces digital signatures that are approximately twice the size of the DSA private key, and therefore will be approximately four times the security strength of the DSA algorithm in bits. That is, the sizes of the DSA signatures will be 448, 512, 768, and 1024 bits for 112-, 128-, 192-, and 256-bit security strengths, respectively.

14.3.2.5 Verify a Signature To verify a signature (r, s) for a received message m, the DSA algorithm constructs a local version of the r based on the s in the received signature, the received message m, the message originator's public key, and the DSA domain parameters. If the local version of r, denoted by v, is identical to the r in the received signature, the signature verification algorithm concludes that the received signature is valid.

The DSA signature verification algorithm is illustrated in Figure 14.4.

The main steps of the DSA signature verification algorithm are as follows:

- *Step 1*: Reject the signature if either $0 < r < q$ or $0 < s < q$ is not satisfied.
- *Step 2*: Message Hashing:
  - hash the received message m to create a message digest $H(m)$, and
  - convert $H(m)$ to an integer z using the same padding algorithm used by the signature generator
- *Step 3*: Construct v:
  - Compute $w = s^{-1} \pmod{q}$;
  - Compute $u1 = (z \cdot w) \pmod{q}$;

  ○ Compute $u2 = (r \cdot w) \pmod{q}$;
  ○ Compute $v = ((g^{u1} \cdot y^{u2}) \bmod p)(\bmod q)$
• *Step 4*: The signature is valid if $v = r$.

The correctness of the DSA verification algorithm can be illustrated as follows. Since the public key $y = g^x(\bmod p)$, we know the following is true:

$$\left(g^{u1} \cdot y^{u2}\right)\bmod p = \left(g^{u1} \cdot \left(g^x\right)^{u2}\right)\bmod p$$
$$= \left(g^{u1 + x \cdot u2}\right)\bmod p$$
$$= g^{(z \cdot w + r \cdot w \cdot x)\bmod q} \bmod p$$
$$= g^{w(z + r \cdot x)\bmod q} \bmod p.$$

If the received message is identical to the message used to generate the signature (r, s), the signature (r, s) has not been altered, the same hashing function is used by the signature creator and verifier to create message digests, and the same padding algorithm is used by the signature creator and verifier to convert the hash value to integers, then the following must be true:

$$k^{-1}(z + x \cdot r)\bmod q = s.$$

That is, we should have:

$$(z + x \cdot r)\bmod q = (k \cdot s)\bmod q.$$

Therefore,

$$g^{w(z + r \cdot x)\bmod q} \bmod p = g^{w \cdot (k \cdot s)\bmod q} \bmod p = g^{s^{-1} \cdot (k \cdot s)\bmod q} \bmod p = g^{k \bmod q} \bmod p.$$

Since k is in the interval $(1, q - 1)$, we have $k(\bmod q) = k$. Therefore:

$$g^{k \bmod q} \bmod p = g^k (\bmod p).$$

That is, the following must be true:

$$v = \left(g^k \bmod p\right)(\bmod q) = r.$$

As one can see from the above DSA signature verification algorithm, the DSA signature verification process does not recover the original message used to create the signature (r, s). It verifies the signature by recreating its own version v of the integer r in the signature and verifies that v is identical to the r in the received signature (r, s). As a result, the DSA signature algorithm cannot be used to encrypt and decrypt messages.

14.3.2.6 *Security Strength, Required Key Length, and Signature Size* So far, studies have shown that the integer factoring problems and the

Table 14.8. DSA signature sizes for different security strengths

Security Strength (bits)	DSA Signature Size (bits)
112	448
128	512
192	768
256	1024

Table 14.9. Performance of DSA signature creation and verification

	Sign a Message	Verify a Signature
Modular exponentiation	1	2
Modular inversions	1	1
Modular multiplications	2	2
Hashing operations	1	1

discrete logarithm problems have comparable levels of difficulties. As a result, the DSA algorithm requires the same key length as the RSA algorithm to achieve the same security strength.

Table 14.7 shows the required key lengths for the DSA algorithm to achieve the security strengths recommended by NIST for use after 2010 [NIST07]. For comparison purpose, the table also shows the hash functions and symmetric key algorithms that can achieve the same security strengths.

Table 14.8 summarizes the DSA signature sizes, measured in bit length, for the security strengths NIST recommended for the year 2010 and after.

14.3.2.7 Performance The most time-consuming operations in DSA are:

- Modular exponentiation
- Modular inversion
- Modular multiplication

Modular exponentiation is the most time-consuming operation in DSA signature creation or verification. Table 14.9 summarizes the DSA signature generation and verification performance. To sign a message, DSA performs one modular exponentiation, one modular inversion, two modular multiplications, and one hashing operation. The exponent for the modular exponentiation has the same size as the private key. The modulus for the exponentiation has the same size as the public key. The modulus for all the other modular operations in the signature creation has the same size as the private key.

To verify a message, DSA performs two modular exponentiations, one modular inversion, two modular multiplications, and one hashing operation. The exponent of the modular exponentiation has the same size as the private key. The modulus for the exponentiation has the same size of the public key. The modulus for all the other modular operations in signature verification has the same size as the private key.

14.3.3 The ECDSA Algorithm

Elliptic curve cryptography (ECC), which is developed after RSA and DSA, represents a profound achievement in cryptography. It was first proposed by Victor Miller and Neal Koblitz independently [Kobl87] [Mill85]. Compared with RSA and DSA, elliptic curve algorithms require significantly shorter public keys to achieve the same security strength. This can translate into lower message overheads for transporting public keys over networks and lower computational complexity (processing burden) on mobile devices for achieving the same security strength.

ECC is the digital signature algorithm of choice in the IEEE 1609.2 standard that specifies security services and procedures designed for vehicle communications.

The ECDSA is specified in American National Standard (ANS) X9.62 [ANSI05]. NIST approved the use of ECDSA and specified additional requirements in the FIPS Publication 186-3 [NIST09].

14.3.3.1 The Basic Idea ECDSA relies on the difficulty of solving discrete logarithm problems over points on elliptic curves to derive the trapdoor functions needed for generating private and public keys [BLSS99] [Kobl87] [Mill85]. ECDSA is similar to DSA in that they both rely on the difficulty of solving discrete logarithm problems. The difference is that with ECDSA, the discrete logarithm problems are defined over points on elliptic curves.

Elliptic curves are cubic curves on a two-dimensional plane. A popular elliptic curve is the cubic curve defined in Equation 14.5, where parameters a and b are constants:

$$y^2 = x^3 + ax + b. \qquad (14.5)$$

ECDSA defines a point addition operation, denoted by +, for adding points on an elliptic curve. If P and Q are two points on an elliptic curve, then $M = P + Q$ is defined to be another point M on the curve or at infinity. We use the elliptic curve $y^2 = x^3 - x + 1$ shown in Figure 14.5 to illustrate how the addition of two points on an elliptic curve works.

To find the point on an elliptic curve that represents $P + Q$, first draw a straight line to connect P and Q and let R be the intersection point between this straight line and the elliptic curve. $P + Q$ is the mirror image of R on the elliptic curve with respect to the x-coordinate. Since an elliptic curve is always symmetric around the x-axis, any point on an elliptic curve will always have a mirror image with respect to the x-coordinate.

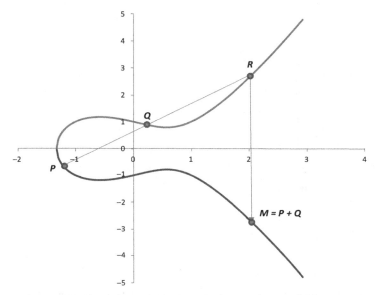

Figure 14.5. Sample elliptic curve and addition of points on the curve

Adding a point P to itself (i.e., computing $P + P$) is illustrated in Figure 14.6. First, draw a tangent line to the elliptic curve at point P and let the intersection of this line and the elliptic curve be R. Then, $P + P$ is the mirror image of R with respect to the x-coordinate.

Two special cases of the point addition operation are illustrated in Figure 14.7. The first special case is to add a point L to the negative of the point L. The negative of L, denoted by $-L$, is the mirror image of L with respect to the x-coordinate. The line connecting L and $-L$ will not intersect with the elliptic curve at any additional point. Therefore, the point at infinity is defined as an extra point in addition to all the points on the elliptic curve and is denoted by O. This point O at infinity is the *additive identity* of the group of points on the elliptic curve. That is, we have $L + (-L) = O$.

The second special case is to add a point P to itself when the tangent line at point P does not intersect with the elliptic curve at any other point on the curve. This case occurs only when the y-coordinate of the point P is zero. The tangent line at a point P will always intersect with the elliptic curve at another point as long as the y-coordinate of the point P is not zero. When the tangent line at point P does not intersect with the elliptic curve at other points besides point P, then $P + P$ is considered to be the point O at infinity, and in this case we have $P + P = 2P = O$.

Now let us consider a discrete logarithm problem over the points on an elliptic curve. Suppose G is a base point on an elliptic curve and k is a random

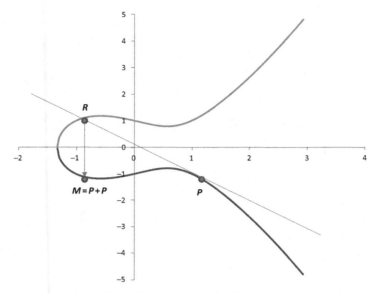

Figure 14.6. Adding a point on an elliptic curve to itself

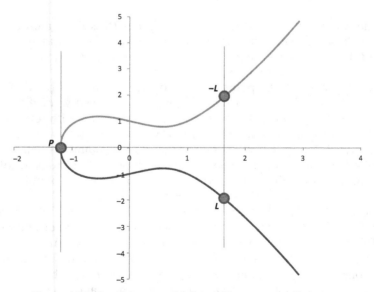

Figure 14.7. Special cases of point addition over an elliptic curve

integer. We add the point G to itself k times to generate a new point C on the elliptic curve. C is denoted by $C = k \cdot G$, which stands for $G + G + G + \ldots + G$, with G appearing k times. Consider an adversary who knows all the parameters that define the elliptic curve, the value of C, and even the base point G, and he wants to determine k. This will be a discrete logarithm problem if all the coordinates of the points on the elliptic curve used in the computation are integers. The adversary could try to determine k by calculating $2 \cdot G$, $3 \cdot G$, $4 \cdot G, \ldots$, and $x \cdot G$ until he finds one of the results that matches C. However, if the set of possible values for k is sufficiently large and if the base point G on the curve is chosen carefully, finding k would be computationally infeasible.

Therefore, a public key can be a point on an elliptic curve and its associated private key can be a random integer that is used to multiply a base point on the elliptic curve to generate the public key.

Elliptic curve algorithms, such as ECDSA, operate only on points with *integer* coordinates on an elliptic curve. This is because operations over real numbers are slow and inaccurate due to rounding errors, and cryptographic operations need to be fast and accurate.

Specifically, the integer coordinates need to form a mathematical structure called a *field* [Alle91]. A field is a set of integers with addition and multiplication operations that combine two elements in the field to a third element in the field. The addition and multiplication operations are commonly denoted by "+" and "·," respectively. The operations must satisfy the following conditions:

- *Closure*: If a and b are in the field, then $a + b$ and $a \cdot b$ are also in the field.
- *Communtativity*: Changing the order of the operands will not change the result of the addition or the multiplication operation.
- *Associativity*: For all a, b, and c in the field, the following must hold: $a + (b + c) = (a + b) + c$ and $a \cdot (b \cdot c) = (a \cdot b) \cdot c$.
- *Distributivity*: For all a, b, and c in the field, the following must hold: $a \cdot (b + c) = (a \cdot b) + (a \cdot c)$.
- *Identities*: The field must have an additive identity O such that for every element a in the field, $a + 0 = a$. The field must also have a multiplicative identity 1 such that for every element a in the field, $a \cdot 1 = a$.
- *Invertibility*: For every element a in the field, there exists an element $-a$ in the field such that $a + (-a) = 0$. Similarly, there exists an element $a - 1$ in the field such that $a \cdot a - 1 = 1$. That is, subtraction and division operations exist for the elements in the field.

An elliptic curve algorithm operates on a finite field that contains a large but finite number of integers that can be used as the coordinates of points on an elliptic curve.

Two categories of finite fields have been used for elliptic curve algorithms: the *prime field* denoted by F_p and the *binary field* denoted by F_2^m.

A prime field F_p contains the integer set $\{0, 1, 2, 3, \ldots, p - 1\}$, where p is a prime number. An elliptic curve over the prime field F_p, denoted by $E(F_p)$, contains all pairs of integers (x, y) that satisfy an elliptic curve when computations are performed modulus p, as illustrated in Equation 14.6:

$$y^2 (\bmod p) = x^3 + ax^2 + b (\bmod p). \tag{14.6}$$

The number of points on the elliptic curve $E(F_p)$ plus the point at the infinity is called the order of the elliptic curve and is denoted by $\#E\ (F_p)$.

A binary field F_2^m contains integers that are m bits long. Each integer in F_2^m is typically represented as a binary polynomial as illustrated in Equation 14.7, where a_i is 0 or 1. The addition and multiplication operations are done modulus an irreducible polynomial:

$$a_{m-1}x^{m-1} + a_{m-2}x^{m-2} + \ldots + a_2x^2 + a_1x^1 + a_0. \tag{14.7}$$

The number of elements in a finite field is called the *field order*. The order of the prime field F_p is p. The order of the binary field F_2^m is 2^m. The security strength of an elliptic curve algorithm, which is the difficulty of deriving the private key with only knowledge of the public key, depends strongly on the order of the finite field over which the elliptic curve is constructed.

The elliptic curves and fields recommended by NIST are described in [NIST99] and [NIST09].

14.3.3.2 Domain Parameters Similar to the DSA signature algorithm, the ECDSA algorithm requires the signature creators and the signature verifiers to operate over a common set of parameters called the *domain parameters*. Here we describe the ECDSA domain parameters for the elliptic curve over the prime field F_p. These domain parameters are (p, a, b, G, n, h):

- p is an odd prime number that defines the prime field F_p.
- a and b are constants that define the elliptic curve $y^2 = x^3 + ax + b$.
- G is the base point on the elliptic curve that everyone uses to compute other points on the curve.
- n is the prime order of the base point G. The order of a point G on an elliptic curve is the smallest positive integer n such that $n \cdot P = O$, where O is the point at the infinity.
- h is a cofactor that equals to $\#E(F_p)/n$, where $\#E(F_p)$ is the number of points on the elliptic curve $E(F_p)$ over the finite prime field F_p.

To select an elliptic curve over the prime field F_p to achieve a given security strength, the prime field is selected so that the length of the order of the elliptic curve is twice the key length of a symmetric key block cipher that can be used to achieve the same security strength. This is because a brute force attack that searches through the entire k-bit block cipher key space will take roughly the

Table 14.10. Prime field sizes recommended by NIST for ECDSA

Security Strength (bits)	Comparable Symmetric Key Length (bits)	Length of p in Prime Field F_p
112	112	224
128	128	256
192	192	384
256	256	512

same time as using the Pollard's rho algorithm to solve an instance of the elliptic curve discrete logarithm problem over a finite field whose order has length $2k$. The Pollard's rho algorithm is currently known as the most effective attack against elliptic curve algorithms [GuPP06] [KoMV00].

Table 14.10 summarizes the lengths of p in prime field F_p required for ECDSA to meet the security strengths recommended by NIST for use after year 2010. For comparison, the table also shows the key lengths of symmetric key algorithms that can achieve the same security strengths.

14.3.3.3 Generate Private and Public Keys
We use elliptic curves over the finite prime field F_p as an example to illustrate the ECDSA key generation algorithm defined in [ANSI05] [NIST09]. The algorithm takes the domain parameters (q, a, b, G, n, h) as input and produces a public–private key pair (Q, d), where Q is the public key and d is the private key. The public key Q is a point on the elliptic curve and the private key d is an integer in the interval $[1, n-1]$.

The ECDSA key generation algorithm has the following main steps:

- *Step 1*: Let N be the length of n in bits.
- *Step 2*: Generate a statistically unique and unpredictable integer d in the interval $[1, n-1]$ to be used as the private key. This can be done with the following steps:
  (a) Generate a string of N random bits.
  (b) Convert the string obtained in Step (a) to a non-negative integer c.
  (c) If c is out of range, that is, if $c > n - 2$, then go back to Step (a) to generate a new c.
  (d) $d = c + 1$.
- *Step 3*: Compute a point Q on the elliptic curve as $Q = d \cdot G$ and Q will be the public key.

The length of the private key d is N, which is the length of n. This also means that the security strength of the ECDSA algorithm is N bits. A range of N values have been specified by NIST in [NIST09] to meet different security strengths. These values are shown in Table 14.11.

Table 14.11. ECDSA private key sizes (length of *n*) recommended by NIST

Security Strength (bits)	N (Length of n)
112	224–255
128	256–383
192	384–511
256	≥512

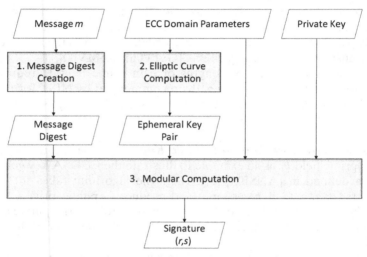

Figure 14.8. ECDSA signature creation

It is not surprising to see that the values of *N* correspond to the sizes of *p* shown in Table 14.10.

The size of an ECDSA public key is twice the size of each coordinate of a point on the elliptic curve. The size of an *x*-coordinate or a *y*-coordinate on the elliptic curve is the order (size) of the finite field over which the elliptic curve is defined.

14.3.3.4 *Create a Digital Signature* The ECDSA signature creation algorithm takes the following information as input: (1) the message to be signed, *m*, which can be of an arbitrary length and is represented by a bit string, (2) a valid set of elliptic curve domain parameters, and (3) a private key *d* associated with the elliptic curve domain parameters.

Like DSA, the ECDSA signature is a pair of integers (*r*, *s*). The ECDSA signature generation algorithm is illustrated in Figure 14.8. It consists of three parts: (1) message digest creation, which produces a message digest used to generate the signature, (2) elliptic curve computations, which produces an ephemeral elliptic curve key pair to be used in generating this particular signature, and (3) modular computations, which produces the signature.

The ECDSA signature generation algorithm works as follows [ANSI05] [NIST09]:

- *Step 1*: Message digest creation:
  ◦ Compute the message digest $H(m)$.
  ◦ Convert $H(m)$ into an integer e.
- *Step 2*: Elliptic curve computation:
  ◦ Select a statistically unique and unpredictable integer k in the interval $[1, n - 1]$.
  ◦ Compute a point $R = (x_R, y_R)$ on the elliptic curve over prime field F_p as $R = k \cdot G$.
- *Step 3*: Modular computations:
  ◦ Convert x_R to an integer $\overline{x_R}$.
  ◦ Set $r = \overline{x_R}(\bmod n)$. If $r = 0$, return to Step 2 to select another pair of k and R.
  ◦ Compute: $s = k^{-1}(e + r \cdot d)(\bmod n)$. If $s = 0$, return to Step 2.
  ◦ Output signature $S = (r, s)$.

This algorithm produces signatures that are twice the length of the private key size. This is because the length of r and the length of s are both N, where N is also the length of the private key. Since the ECDSA private key size is twice the required security strength in bits, the signature size is about four times the security level. For example, to achieve 112-bit security, the signature will be about 448 bits long. In other words, this algorithm generates signatures that have the same size as the DSA signatures; they are both about four times the target security strength in bits.

14.3.3.5 *Verify a Signature* To verify a signature $S = (r, s)$ on a message m, the ECDSA signature verification algorithm takes as input the signature S, the message m, the message signer's public key, and the ECDSA domain parameters. Like DSA, ECDSA verifies the signature by constructing a local version of the r in the received signature (r, s), based on the message originator's public key Q, the s in the received signature (r, s), the received message, and the ECDSA domain parameters. The signature verification algorithm concludes that signature (r, s) is valid if v is identical to r in the received signature (r, s).

The ECDSA signature verification algorithm, illustrated in Figure 14.9, consists of the following main steps: (1) message digest creation, (2) modular computations, (3) elliptic curve computations, and (4) signature checking [ANSI05] [NIST09]:

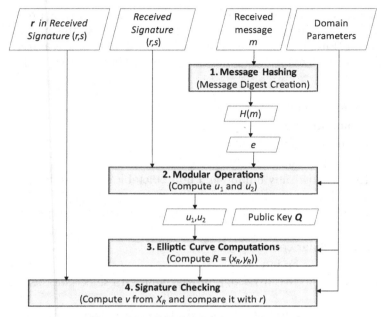

Figure 14.9. ECDSA signature verification

- *Step 1*: Message digest creation:
  - ○ Compute the message digest $H(m)$ of the received message m. If the hash function output is invalid, the signature is invalid and stop.
  - ○ Convert $H(m)$ into an integer e.
- *Step 2*: Modular computations:
  - ○ If r and s are not both integers in the interval $[1, n-1]$, the signature is invalid and stop.
  - ○ Compute: $u_1 = e \cdot s^{-1} (\bmod\ n)$, and $u_2 = r \cdot s^{-1} (\bmod\ n)$.
- *Step 3*: Elliptic curve computations:
  - ○ Compute: $R = (x_R, y_R) = u_1 G + u_2 Q$.
  - ○ If $R = O$, which infinity defined for the addition operation over an elliptic curve, signature is invalid and stop.
- *Step 4*: Signature checking:
  - ○ Convert x_R to an integer $\overline{x_R}$.
  - ○ Set $v = \overline{x_R} (\bmod n)$.
  - ○ Compare v and r: If $v = r$, the signature is valid; if $v \neq r$, signature is invalid.

Similar to the DSA signature algorithm, ECDSA is designed only to create and verify signatures, and cannot be used to encrypt and decrypt messages.

Table 14.12. ECDSA key lengths required for different security strengths

Security Strength (bits)	ECDSA Key Length Using Prime Field F_p (bits)	Comparable Hash Function	Comparable Symmetric Key Algorithm
112	224	SHA-224	Three-key TDEA
128	256	SHA-256	AES-128
192	384	SHA-384	AES-192
256	512	SHA-512	AES-256

Table 14.13. ECDSA signature sizes for different security strengths

Security Strength (bits)	ECDSA Signature Size (bits)
112	448
128	512
192	768
256	1024

This is because the ECDSA signature verification process will not recover the input data used by the ECDSA signature creation algorithm to generate the signature.

Encryption algorithms based on ECC have been developed. An example is the Elliptic Curve Integrated Encryption Scheme (ECIES) [Cert09], which we will describe later in this chapter.

14.3.3.6 Security Strength, Required Key Length, and Signature Size
Compared with RSA and DSA, the ECDSA algorithm requires significantly smaller key sizes to achieve the same security strength.

Table 14.12 shows the required key lengths for the ECDSA algorithm to achieve various levels of security strengths [NIST07]. For comparison purposes, the table also shows the hash functions and symmetric key algorithms that can achieve the same security strengths.

Table 14.13 shows the sizes of the ECDSA signatures required to meet different security strengths.

14.3.3.7 Performance
The most time-consuming operations in ECDSA are:

- Scalar multiplication to compute $k \cdot G$, where k is a random integer and G is a point on an elliptic curve.
- Modular multiplication
- Modular squaring to multiply an integer by itself modulus n.
- Modular inversion.
- Hashing operations.

Table 14.14. Performance of ECDSA signature creation and verification

	Sign a Message	Verify a Signature
Scalar multiplications	1	2
Modular inversions	1	1
Modular multiplications	2	2
Hashing operations	1	1

Table 14.15. Key lengths recommended by NIST for DSA, RSA, and ECDSA

Security Strength (bits)	DSA (bits)	RSA (bits)	ECDSA over F_p (bits)	ECDSA over $F(2^m)$ (bits)
112	2,048	2,048	224	233
128	3,072	3,072	256	283
192	7,680	7,680	384	409
256	15,360	15,360	512	571

Scalar multiplication is by far the most time-consuming operation used in ECDSA. Let T_{kP} be the number of central processing unit (CPU) clock ticks to compute a scalar multiplication $k \cdot P$ (where k is an integer and P is a point on an elliptic curve), T_{MUL} be the number of CPU clock ticks to compute a modular multiplication operation, and T_{SQR} be the number of CPU clock ticks to compute a modular squaring operation. The analysis in [Peti09] shows that:

$$T_{kP} = n\left(5T_{SQR} + 6T_{MUL}\right) \approx 11nT_{MUL}. \qquad (14.8)$$

For 112-bit security strength, n will have to be at least 224 bits long, which means that a scalar multiplication operation could take approximately 2464 times more steps than a modular multiplication operation.

Table 14.14 summarizes the ECDSA signature generation and verification performance. To sign a message, ECDSA performs one scalar multiplication, one modular inversion, two modular multiplications, and one hashing operation. To verify a message, ECDSA performs two scalar multiplications, one modular inversion, two modular multiplications, and one hashing operation. The size of the modulus for all modular operations is the same as the size of n.

14.3.4 ECDSA for Vehicle Safety Communications

14.3.4.1 Advantage of ECDSA The main advantage of ECDSA over other existing public key algorithms, such as DSA and RSA, is that ECDSA uses significantly smaller public keys to achieve the same level of security strength. In other words, ECDSA offers a considerably higher level of security for the same public key size. Table 14.15 shows the minimum public key sizes recommended by NIST for different public key algorithms to meet various levels of security strength [NIST07] [NIST09].

Table 14.16. Signature sizes of DSA, RSA, and ECDSA

Security Strength (bits)	RSA (bits)	DSA (bits)	ECDSA (bits)
112	2,048	448	448
128	3,072	512	512
192	7,680	768	768
256	15,360	1,024	1,024

As computing technology evolves, the strength of cryptographic algorithms must increase accordingly to counter adversaries' increasing computing powers. NIST recommends that the minimum security strength be 112 bits between the years 2010 and 2030, and 128 bits beyond the year 2030. Table 14.15 shows that as the required security strength of a digital signature algorithm increases, the required key lengths for DSA and RSA increase at a much faster rate than the required key lengths for ECDSA. This makes ECDSA more scalable to meet future demands for higher levels of security strength.

ECDSA also produces smaller signatures than RSA. Table 14.16 shows the approximate signature sizes produced by RSA, DSA, and ECDSA for different security strengths.

The smaller key sizes and signature sizes of ECDSA mean lower message overheads when transporting ECDSA public keys over wireless networks compared with transporting RSA or DSA public keys. This is important in a large vehicle network where vehicles may often have to exchange their public keys over bandwidth-limited wireless channels.

The smaller ECDSA key lengths can also translate into savings on computing power, storage and memory space, and energy required to achieve the same security strength [WGEG05]. This makes ECDSA attractive for resource-constrained mobile devices, such as vehicle onboard communication units.

Studies have shown that ECDSA can be tens to hundreds of times faster than RSA in private–public key generation as the required level of security strength increases from 112 to 256 bits [JaAr04] [RaGa05] [TiGr04] [WMPW98] [XuDC09]. ECDSA can have comparable to tens of times higher speeds in signature generations compared with RSA as the required security strength grows, depending on the methods used to perform elliptic curve operations [JaAr04] [RaGa05] [TiGr04] [WMPW98] [XuDC09].

14.3.4.2 *Potential Limitations of ECDSA* A potential limitation of ECDSA for vehicle safety communications is its slower signature verification speeds relative to other digital signature schemes such as RSA and DSA. Studies have shown that for the same security strength, ECDSA signature verification can be several times to tens of times slower than RSA signature verification, depending on the mechanisms used for elliptic curve operations [JaAr04] [RaGa05] [TiGr04] [WMPW98] [XuDC09].

Recent studies show that each ECDSA signature verification operation could take tens of milliseconds for security levels of 112–128 bits when performed on a personal computer (PC) [JaAr04] [WMPW98] and hundreds or even thousands of milliseconds on mobile phones [TiGr04]. An ECDSA signature verification operation for security strengths higher than 128 bits can take over one second even when performed on a PC [JaAr04] [WMPW98].

Fast signature verification is crucial for supporting highly delay-sensitive cooperative vehicle safety communications that rely on vehicles to broadcast timely safety messages to each other. As discussed previously, vehicle safety applications may need to receive and process up to 10 signed broadcast safety messages from each neighboring vehicle every second. A vehicle can often receive such broadcast safety messages from hundreds of neighboring vehicles simultaneously. Using ECDSA to verify such heavy volume of received signatures, or even just a subset of these messages, could cause excessive delays on OBUs that typically have constrained computing power.

Another potential limitation is that ECDSA is for creating and verifying signatures only and cannot be used for encryption and decryption. A separate public key encryption mechanism, such as the ECIES, will be needed to support public key encryption and decryption. Public key encryption will be required for vehicles to establish secure communication channels with the security credential management system servers. For example, the IEEE 1609.2 standard uses ECDSA for signature creation and verification while using a combination of ECIES and symmetric key algorithms for encryption and decryption.

Using a separate public key encryption algorithm such as the ECIES requires the message senders and receivers to be preconfigured with extra parameters in addition to the ECC domain parameters required to support ECDSA. For example, ECIES requires a message sender and a message receiver to agree on a message authentication code (MAC) algorithm, a symmetric key encryption algorithm, and a public key derivation algorithm [Cert09]. The more security parameters to be preconfigured on vehicles, the more difficult it will be to make changes later in the evolution of the security capabilities over time.

14.4 MESSAGE AUTHENTICATION AND MESSAGE INTEGRITY VERIFICATION

Message authentication is to establish and verify the origin of a message to determine whether the message is originated from an authorized entity. The cryptographic mechanisms described earlier in this chapter can be used to achieve a major part of this task: to verify that a message is sent from someone who knows a secret key. Here, we will discuss how this can accomplished using hash functions or digital signatures.

However, the cryptographic mechanisms described earlier in this chapter alone are not sufficient to allow one to determine whether a message is from an authorized entity. To accomplish this part of the message authentication task, extra technologies will be required. This will be the topic of the next chapter.

Message integrity verification is to verify whether a message has been altered. The cryptographic mechanisms described earlier in this chapter are effective tools for message integrity verification. Here, we discuss how message integrity verification can be accomplished using hash functions or digital signatures.

14.4.1 Authentication and Integrity Verification Using Hash Functions

To support message authentication, hash functions are extended to incorporate a secret symmetric key to generate MACs. As illustrated in Figure 14.10,

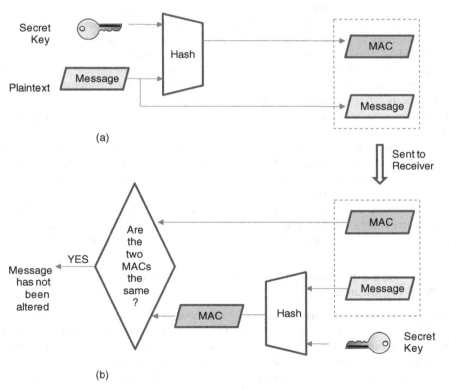

Figure 14.10. Hash function for message authentication and integrity checking. (a) MAC generation by message sender; (b) MAC verification by message receiver

a message sender inputs both the message m and a secret key K into a hash function $H(m, K)$ to generate a hash value called a MAC. The secret key is used to ensure that the MAC value is unique not only for the input message but also for this secret key. That is, only someone who knows the same secret key will be able to generate this particular MAC.

The keyed hash function $H(m, K)$ used to generate MACs must meet the following requirements:

- The MAC can be easily calculated with a message and a key as the input.
- It is computationally infeasible to derive the key from the message and its MAC.
- It is computationally infeasible to find another message that will generate the same MAC that is identical to the MAC of a given message.

If a message receiver also knows the same secret key K, it will be able to verify whether the MAC is generated by someone who knows this same secret key. This secret key is not an encryption and decryption key because it cannot be used to recover the original message from the hash value.

The MAC will be sent together with the message m to message receivers. To authenticate the message, the message receiver computes its own version of the MAC for the message directly from the received message using its own secret key K. If the MAC from its own computation matches the MAC received from the message sender, then the message receiver can conclude that the message is from someone who knows the same secret key K. This is because only someone with the same secret key K could have created the received MAC. The message receiver can further conclude that the message has not been altered after the received MAC was generated. This is because if the received message is different from the message used to generate the received MAC, then the MAC computed by the message receiver from the received message will be different from the received MAC.

Supporting message authentication and integrity using hash functions and secret keys alone suffer from the same limitation as symmetric key encryption—the communicating parties must establish and share secret keys.

14.4.2   Authentication and Integrity Verification Using Digital Signatures

Digital signature provides an effective way to support message authentication and integrity verification in vehicle safety communications. It allows vehicles to send signed messages instantly without having to establish shared secrets and having to exchange any security information a priori.

The digital signature mechanisms described previously in this chapter allow a message receiver to use a public key to verify whether a signed message is from someone who knows the private key that is associated with this public key. The same signature verification process can also be used to verify message

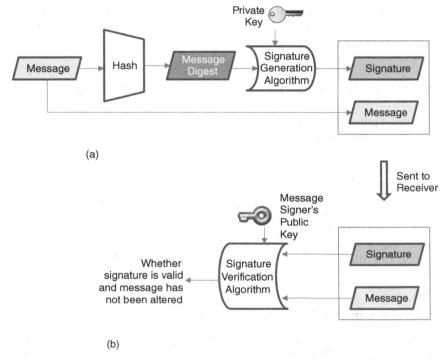

Figure 14.11. Digital signature for message authentication and integrity checking. (a) Signature generation by message sender; (b) signature verification by message receiver

integrity. Authentication and message integrity verification with public keys eliminate the burden to distribute secret keys over unsecure and nonprivate networks among a vast population of vehicles.

Figure 14.11 illustrates how digital signatures can be used to support message authentication and message integrity verification. A message originator first uses a cryptographic hash function to generate a digest of the message to be signed. This allows the signature to be generated over a small digest of the message rather than the potentially large message itself, which helps reduce signature creation time.

The message signer uses a digital signature algorithm and the message signer's private key to sign the message digest to produce a digital signature that will be sent together with the message to the receivers. The message signer may also send its public key together with the message so message receivers do not have to be preconfigured with the public keys of the potentially very large number of possible message originators.

Each message receiver uses the message signer's public key and the corresponding digital signature algorithm to verify the received signature. Successful signature verification means that the signature was created by the private key that is associated with the public key and that the message has not been altered since it was signed. This achieves the dual purpose of authenticating the message and verifying message integrity.

14.5   DIFFIE–HELLMAN KEY ESTABLISHMENT PROTOCOL

The Diffie–Hellman Key Establishment (Exchange) Protocol, first published by Whitfield Diffie and Martin Hellman in 1976 [DiHe76], is a way for communicating parties with no prior knowledge of each other to establish a shared secret key by exchanging information in the open over an insecure communication network. Knowing the information exchanged among the communicating parties will not allow an adversary to derive the shared secret. This shared secret key can then be used to encrypt subsequent communications using a symmetric key algorithm.

Here, we first describe the original Diffie–Hellman Key Establishment Protocol and then the Elliptic Curve Diffie–Hellman Key Establishment Protocol.

14.5.1   The Original Diffie–Hellman Key Establishment Protocol

The Diffie–Hellman Key Exchange Protocol operates over a mathematical structure called a cyclic multiplicative group. A multiplicative group is a group where the multiplication of any two elements in the group is still an element in the group. A group is a cyclic group if there exists an element g in the group such that g^n is also an element in the group for any integer n within a given range. The element g is called the generator of the group because all other members in the group can be generated by raising g to an integer power. The simplest implementation of the Diffie–Hellman Key Exchange Protocol operates over a cyclic multiplicative group modulus p, where p is a prime number. That is, all operations on the elements in the group are performed modulus p.

Using the Diffie–Hellman Key Exchange Protocol to establish a shared secret key, communicating parties Alice and Bob first agree upon a cyclic group G with generator g and modulus p, and then perform the following steps:

- *Step 1*: Alice picks a random integer α, computes $g^\alpha (\bmod p)$, and sends the result to Bob.
- *Step 2*: Bob picks a random integer β, computes $g^\beta (\bmod p)$, and sends the results to Alice.
- *Step 3*: Alice receives g^β from Bob and computes $(g^\beta)^\alpha = g^{\alpha\beta} (\bmod p)$
- *Step 4*: Bob receives g^α from Alice and computes $(g^\alpha)^\beta = g^{\alpha\beta} (\bmod p)$
- *Step 5*: Alice and Bob both know the same value $g^{\alpha\beta} (\bmod p)$ now and will use it as their shared secret key.

Any observer can know g, g^α, and g^β. Alice keeps α secret to herself and Bob keeps β secret to himself. Deriving $g^{\alpha\beta}$ from the knowledge of g, g^α, and g^β is known as the decisional *Diffie–Hellman Problem*. Solving the Diffie–Hellman Problem is known to be computationally hard. Today, the fastest known algorithm is to solve the discrete logarithm problem of determining α knowing g and g^α. Then, solve a similar discrete logarithm problem to determine β.

Although the *Diffie–Hellman Problem* has not been proven to be NP-complete, no polynomial time algorithm has been found for the problem so far. Therefore, it has been assumed to be a hard problem. This assumption has been used as the basis for many modern-day digital signature schemes, such as several state-of-the-art group signature schemes that we will discuss in the succeeding chapters.

14.5.2 Elliptic Curve Diffie–Hellman Key Establishment Protocol

The Diffie–Hellman Key Exchange Protocol can be implemented over elliptic curves. Elliptic Curve Diffie–Hellman Key Exchange protocols have been used to implement elliptic curve encryption algorithms such as the ECIES used in the IEEE 1609.2 standard. One Elliptic Curve Diffie–Hellman Key Exchange protocol is defined in [Cert09]. We illustrate this protocol here using elliptic curves over prime field F_p as an example.

The communicating parties must first establish a set of elliptic curve domain parameters $T = \{p, a, b, G, n, h\}$ as in the ECDSA algorithm. The Elliptic Curve Diffie–Hellman Key Exchange protocol between any two communicating parties, say Alice and Bob, takes the following steps:

- *Step 1*: Alice selects a private key d_A, generates a public key Q_A by computing $Q_A = d_A \cdot G$, where G is the base point in the domain parameters, and sends her public key Q_A to Bob.
- *Step 2*: Bob selects a private key d_B, generates a public key Q_B by computing $Q_B = d_B \cdot G$, where G is the base point in the domain parameters, and sends his public key Q_B to Alice.
- *Step 3*: Alice receives Q_B from Bob and computes the elliptic curve point $P = (x_P, y_P) = d_A \cdot Q_B = d_A \cdot (d_B \cdot G) = (d_A \cdot d_B) \cdot G$.
- *Step 4*: Bob receives Q_A from Alice and computes the elliptic curve point $P = (x_P, y_P) = d_B \cdot Q_A = d_B \cdot (d_A \cdot G) = (d_A \cdot d_B) \cdot G$.
- *Step 5*: Alice and Bob both know the same value elliptic curve point $P = (x_P, y_P)$ now and will use the x-coordinate x_P of this elliptic curve point as their shared secret key.

If Alice already knows Bob's public key, she does not have to wait for Bob to send her another public key and can skip Step 2 and jump to Step 3 directly.

14.6 ELLIPTIC CURVE INTEGRATED ENCRYPTION SCHEME (ECIES)

ECIES is a public key encryption algorithm developed by Certicom [Cert09]. It is used in IEEE 1609.2 for public key encryption and decryption. Here, we illustrate ECIES using elliptic curves defined over prime fields F_p.

14.6.1 The Basic Idea

The basic idea of ECIES can be summarized as follows: First, a message sender and a message receiver use elliptic curve operations and an Elliptic Curve Diffie–Hellman Key Establishment Protocol to establish a shared secret. Then, the message sender uses this shared secret to derive symmetric keys for symmetric key encryption and for generating MAC. Finally, the message sender uses a symmetric key algorithm to encrypt the message and uses a MAC algorithm to create a MAC to support authentication and integrity protection. The message receiver uses the Elliptic Curve Diffie–Hellman Key Establishment Protocol to compute the same shared secret and uses the shared secret to generate the same symmetric keys that it will use to decrypt the message and to verify the MAC.

14.6.2 Scheme Setup

ECIES uses the same set of elliptic curve domain parameters as ECDSA. In addition, a message sender and a message receiver must agree upon the following extra domain parameters:

- a MAC algorithm;
- a symmetric key encryption algorithm;
- an Elliptic Curve Diffie–Hellman Key Establishment Protocol that will be used to establish shared secrets between communicating parties; and
- a key derivation function that will be used to derive symmetric keys.

14.6.3 Encrypt a Message

To encrypt a message M with the recipient's public encryption key Q, the ECIES encryption procedure takes the following steps as illustrated in Figure 14.12:

Step 1: Generate an ephemeral elliptic curve key pair (k, R), with $R = (x_R, y_R)$. Convert R to an octet string $\bar{R}$.
  - *Step 2*: Use the Elliptic Curve Diffie–Hellman Key Exchange Protocol established during the scheme setup procedure to derive a shared secret element $z \in F_p$ from the ephemeral secret key k and the recipi-

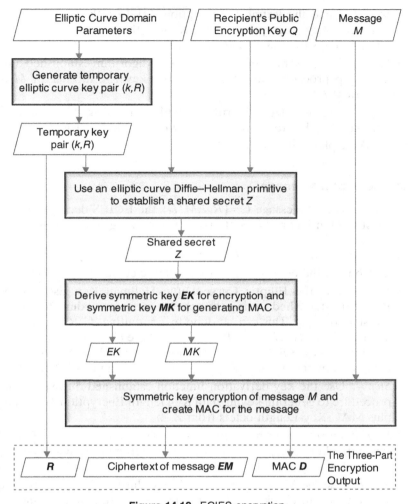

Figure 14.12. ECIES encryption

ent's public encryption key Q. This can be done by computing $P = (x_P, y_P) = k \cdot Q$ and then set $z = x_P$. Convert z to an octet string Z.

- *Step 3*: Use the key derivation function established during the scheme setup procedure to generate keying data K from Z. The length of K should be Encryption-Key-Length plus MAC-Key-Length, where Encryption-Key-Length will be the length for the symmetric encryption key and MAC-Key-Length will be the length of the MAC key, both in octets.

- *Step 4*: Parse the leftmost Encryption-Key-Length octets of K as a symmetric-key encryption key EK and the rightmost MAC-Key-Length octets of K as a MAC key MK.
- *Step 5*: Use the symmetric key encryption algorithm established during the setup procedure and the symmetric key EK to encrypt M into ciphertext EM.
- *Step 6*: Use the MAC algorithm established during the setup procedure and key EK to compute a tag D (i.e., a MAC) of EM.

Step 7: The ciphertext $C = (\bar{R}, EM, D)$.

14.6.4 Decrypt a Message

To decrypt a received message $C = (\bar{R}, EM, D)$, the ECIES decryption procedure is illustrated in Figure 14.13. It takes the following steps [Cert09]:

Step 1: Parse C to recover $\bar{R}$, EM, and D.

Step 2: Convert the octet string $\bar{R}$ to an elliptic curve point $R = (x_R, y_R)$.

- *Step 3*: Use the Elliptic Curve Diffie–Hellman Key Establishment Protocol established during the scheme setup procedure to compute the shared key z. This can be done by computing $d \cdot R$, which should equal to $d \cdot k \cdot G = k \cdot Q = P = (x_P, y_P)$, where Q is the message receiver's public key. Set $z = x_P$, which should be identical to the secret z computed by the message sender. Convert z to an octet string Z.
- *Step 4*: Use the key derivation function established during the setup procedure to generate keying data K of length Encryption-Key-Length plus MAC-Key-Length octets from Z.
- *Step 5*: Parse the leftmost Encryption-Key-Length octets of K as a symmetric key encryption key EK and the rightmost MAC-Key-Length octets of K as a MAC key MK.
- *Step 6*: Use the MAC scheme established during the setup procedure and key MK to check that D is the tag on the received message EM to verify the integrity of the message and that the message is from a sender who shares the same secret key MK.
- *Step 7*: Use the symmetric key decryption algorithm established during the setup procedure and symmetric key EK to decrypt EM into M.
- *Step 8*: Output M.

14.6.5 Performance

With ECIES, two communicating parties use an Elliptic Curve Diffie–Hellman Key Establishment Protocol to establish shared secret symmetric keys and then use a symmetric key algorithm for encryption and decryption. The computational complexities of ECIES encryption and decryption are dominated

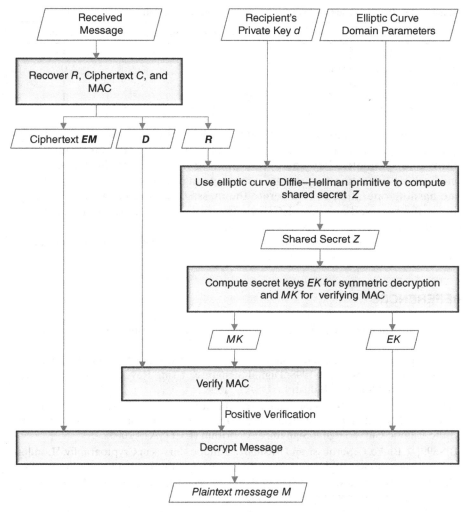

Figure 14.13. ECIES decryption

by the elliptic curve operations performed by the Elliptic Curve Diffie–Hellman Key Establishment Protocol. The most time-consuming operation is elliptic curve scalar multiplication.

ECIES encryption takes two elliptic curve scalar multiplications: one scalar multiplication to generate an ephemeral elliptic curve private–public key pair and another scalar multiplication to generate the shared secret to be used to create symmetric keys.

ECIES decryption takes only one elliptic curve scalar multiplication to compute the shared secret used to derive symmetric keys.

Table 14.17. Performance of ECIES encryption and decryption

	ECIES Encryption	ECIES Decryption
Scalar multiplications	2	1
Symmetric key encryption	1	0
Symmetric key decryption	0	1
Hashing operations	1	1

The other time-consuming operations in ECIES encryption are symmetric key encryption operations and hashing operations. ECIES encryption performs one symmetric key encryption operation to encrypt the message and one hashing operation to generate a message tag. ECIES decryption performs one hashing operation to regenerate the message tag used to verify the integrity of the received message. ECIES performs one symmetric key decryption operation to decrypt the received message.

Table 14.17 summarizes the encryption and decryption performance of ECIES.

REFERENCES

[Alle91] R. Allenby: Rings, Fields and Groups, 2nd edition, Butterworth-Heinemann, 1991.

[ANSI98] American National Standards Institute: "American National Standard for Financial Services X9.31-1998: Digital Signatures Using Reversible Algorithms for the Financial Services Industry (rDSA)," 1998.

[ANSI05] American National Standards Institute (ANSI): "American National Standard X9.62-2005 Public Key Cryptography for the Financial Services Industry, the Elliptic Curve Digital Signature Algorithm (ECDSA)," 2005.

[BlSS99] I. Blake, G. Seroussi, and N. Smart: "Elliptic Curves in Cryptography," London Mathematical Society Lecture Note Series 265, Cambridge University Press, 1999.

[Cert09] Certicom Research: "Standards for Efficient Cryptography, SEC 1: Elliptic Curve Cryptography," Version 2.0, 2009.

[DiHe76] W. Diffie and M. E. Hellman: "New Directions in Cryptography," IEEE Transactions on Information Theory, vol. IT-22, pp. 644–654, 1976.

[GuPP06] T. E. Guneysu, C. Paar, and J. Pelzl: "On the Security of Elliptic Curve Cryptosystems against Attacks with Special-Purpose Hardware," The International Workshop on Special-Purpose Hardware for Attacking Cryptographic Systems (SHARCS'06), Cologne, Germany, 2006.

[IEEE11] IEEE 1609.2/D8: "Draft Standard for Wireless Access in Vehicular Environments—Security Services for Applications and Management Messages," 2011.

[JaAr04] N. Jansma and B. Arredondo: "Performance Comparison of Elliptic Curve and RSA Digital Signatures," Technical Report, University of Michigan College of Engineering, 2004.

[Kobl87] N. Koblitz: "Elliptic Curve Cryptosystems," Mathematics of Computation, vol. 48, pp. 203–209, 1987.

[KoMV00] N. Koblitz, A. Menezes, and S. Vanstone: "The State of Elliptic Curve Cryptography," Designs, Codes and Cryptography, vol. 19, no. 2–3, pp. 173–193, 2000.

[LiWB06] C. Lindenberg, K. Wirt, and J. Buchmann: "Formal Proof for the Correctness of RSA-PSS," Cryptology ePrint Archive, Report 2006/011, 2006.

[MeOV97] A. J. Menezes, P. C. van Oorchot, and S. A. Vanstone: Handbook of Applied Cryptography, CRC Press, 1997.

[Mill85] V. Miller: "Uses of Elliptic Curves in Cryptography," Lecture Notes in Computer Science 218: Advances in Cryptology—CRYPTO '85, Springer-Verlag, Berlin, 1985, pp. 417–426.

[NIST99] United States Department of Commerce, National Institute of Standards and Technology (NIST): "Recommended Elliptic Curves for Federal Government Use," 1999.

[NIST01] United States Department of Commerce, National Institute of Standards and Technology (NIST): "Federal Information Processing Standards (FIPS) Publication PUB 197, Advanced Encryption Standard (AES)," 2001.

[NIST02] United States Department of Commerce, National Institute of Standards and Technology (NIST): "Federal Information Processing Standards (FIPS) Publication PUB 180-2," 2002.

[NIST04] United States Department of Commerce, National Institute of Standards and Technology (NIST): "Special Publication 800-67 Recommendation for Triple Data Encryption Algorithm Block Cipher," 2004.

[NIST07] United States Department of Commerce, National Institute of Standards and Technology (NIST): "Special Publication 800-57 Recommendation for Key Management—Part 1: General (Revised)," 2007.

[NIST09] United States Department of Commerce, National Institute of Standards and Technology (NIST): "Federal Information Processing Standards (FIPS) Publication PUB 186-3 Digital Signature Standard," 2009.

[NIST09b] United States Department of Commerce, National Institute of Standards and Technology (NIST): "Special Publication 800-107 Recommendation for Applications Using Approved Hash Algorithms," 2009.

[Peti09] J. Petit: "Analysis of ECDSA Authentication Processing in VANETs," 3rd International Conference on New Technologies, Mobility and Security (NTMS), 2009.

[PKCS02] RSA Laboratories: "PKCS #1 v2.1: RSA Cryptography Standard," June 14, 2002.

[RaGa05] W. Rao and Q. Gan: "The Performance Analysis of Two Digital Signature Schemes Based on Secure Charging Protocol," the 2005 International Conference on Wireless Communications, Networking and Mobile Computing, 2005.

[TiGr04] S. Tillich and J. Groszschaedl: "A Survey of Public-Key Cryptography on J2ME-Enabled Mobile Devices," Computer and Information Sciences—ISCIS 2004, Lecture Notes in Computer Science, Volume 3280/2004, 935–944, DOI: 10.1007/978-3-540-30182-0_94, 2004.

[WGEG05] A. S. Wander, N. Gura, H. Eberle, V. Gupta, and S. C. Shantz: "Energy Analysis of Public-Key Cryptography for Wireless Sensor Networks," Third IEEE International Conference on Pervasive Computing and Communications, 2005.

[WMPW98] E. D. Win, S. Mister, B. Preneel, and M. Wiener: "On the Performance of Signature Schemes based on Elliptic Curves," Third International Symposium on Algorithmic Number Theory, Springer-Verlag London, UK, 1998.

[XuDC09] Z. Xuan, Z. Du, and R. Chen: "Comparison Research on Digital Signature Algorithms in Mobile Web Services," International Conference on Management and Service Science (MASS '09), 2009.

15

PUBLIC KEY INFRASTRUCTURE FOR VEHICLE NETWORKS

15.1 INTRODUCTION

Digital signatures can solve one part of the message authentication problem: they allow a vehicle to use a public key to verify whether a signature creator possesses the associated private key. However, anyone can create private and public keys and use them to sign messages. Therefore, signatures alone are insufficient to determine the identity or type of the owner of a private–public key pair. Furthermore, signatures alone cannot show what a private–public key pair is entitled to be used for.

To establish a level of trust in who owns a public key, public key certificates and the public key infrastructure (PKI) have been developed. PKI has been widely adopted to support authentication between users and Web servers over the Internet. It has also been used to provide certificates to electronic passports, identification cards, and smart cards for electronic payments.

This chapter discusses PKIs for supporting consumer vehicle networks. We discuss functionality, architectures, protocols, and technical challenges for such vehicular PKIs. We outline technical challenges in designing vehicular PKIs that can preserve driver privacy and how to quantitatively measure privacy.

Vehicle Safety Communications: Protocols, Security, and Privacy, First Edition. Luca Delgrossi and Tao Zhang.
© 2012 John Wiley & Sons, Inc. Published 2012 by John Wiley & Sons, Inc.

15.2 PUBLIC KEY CERTIFICATES

A public key certificate (a digital certificate or a certificate for simplicity) is a digital object that binds a public key with the identity or type of the public key's owner. The owner of a public key identified on a certificate is also referred to as the subject or holder of the certificate. A certificate can also specify what a public key and its associated private key are entitled to be used for.

A conventional certificate consists of the following main information: a public key, identity of the owner of the public key, information identifying the certificate issuer, and the certificate issuer's signature to attest to the correctness of the information on the certificate.

Other information elements on a conventional certificate include the version number of the certificate format, a serial number to identify each certificate, the validity time period for the certificate, and other data specific to each particular certificate such as what functions the keys can be used for.

Figure 15.1 shows the basic information elements typically contained in a conventional public key certificate.

Certificates must be issued by entities that the certificate users trust to issue certificates and to be competent to ensure the correctness of the information

Figure 15.1. Sample information elements on a conventional public key certificate

on the certificates. The PKI is designed to meet this need. A PKI is a set of computer systems, policies, procedures, and people that create, manage, distribute, maintain, and revoke certificates. An entity that performs these functions is referred to as a *certificate authority* (CA).

To allow anyone to verify that a certificate is issued by a trusted CA and to verify the integrity of the certificate content, the CA signs each certificate with its private key and publishes its public key for anyone to verify its signatures.

Certificates usually have finite lifetimes. This ensures that the certificates for retired or obsolete devices will not be valid endlessly. Periodic expirations of certificates trigger certificate owners to update the information on their new certificates periodically. Finite certificate lifetimes also help constrain the size of the certificate revocation list (CRL) that a PKI system uses to inform users of revoked certificates.

The International Telecommunications Union, Telecommunication Standardization Sector (ITU-T) X.509 standard and the Internet Engineering Task Force (IETF) Request for Comments (RFC) 5280 define the certificates currently used over the Internet [CSFB08]. These certificates are commonly referred to as X.509 certificates.

The Institute of Electrical and Electronics Engineers (IEEE) 1609.2 standard defines certificates based on elliptic curve cryptography [BlSS99] for vehicle communication environments.

15.3   MESSAGE AUTHENTICATION WITH CERTIFICATES

Figure 15.2 illustrates how a message receiver can use a certificate to authenticate a digitally signed message. In this example, the message originator sends its certificate with its signed message to the receiver as shown in Figure 15.2a.

The message receiver first verifies the certificate issuer's signature on the certificate using the public key of the CA that is trusted to issue this certificate. A successful verification indicates that the public key on the certificate belongs to the subject of the certificate. The message receiver proceeds to use this public key to verify the signature on the received message. A successful verification tells the message receiver that the message was signed by the subject of the certificate and that the message content has not been altered since it was signed.

Each sender must be sure that all the receivers have its certificate before they need to verify its signatures. In a large consumer vehicle network, it will be impractical to demand each vehicle to store every other vehicle's certificates at all times. A vehicle can include its certificate in each message it sends out. This, however, can lead to heavy message overheads. Therefore, it is important to find ways to reduce the message overheads associated with using certificates for message authentication in a large consumer vehicle network.

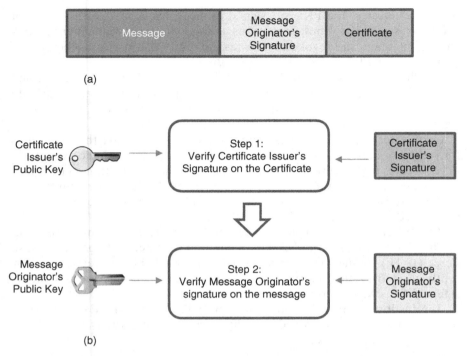

Figure 15.2. Message authentication with digital signature and certificate. (a) Message origina-tor sends a signed message with its certificate; (b) message receiver uses the certificate to authenticate the message

15.4 CERTIFICATE REVOCATION LIST

Misused certificates, which will not expire for a long time, should be revoked. All the certificates assigned to a misbehaving vehicle and the misbehaving vehicle's privilege to receive new certificates should be revoked. A certificate may also be revoked for other reasons such as when the private key is sus-pected to be compromised, when the information on the certificate has changed, or when the CA or the certificate owner wishes to replace the old certificate with a new one for any reasons before the old certificate's natural expiration date.

Vehicles and other certificate users need to know which certificates have been revoked. The PKI uses a certificate revocation list (CRL) to accomplish this goal. A CRL is a digital object that contains information to identify the revoked certificates.

CRLs should be issued by a trusted authority such as a CA. We refer to the entity that creates and issues CRLs as the CRL Issuer. Each CRL should be signed by its issuer so that anyone can verify whether the CRL is issued by a trusted issuer, and can verify the integrity of the CRL content.

The ITU-T X.509 standard and IETF RFC 5280 define the CRLs used over the Internet [CSFB08]. The IEEE 1609.2 standard defines CRLs for certificates based on elliptic curve cryptography [IEEE06].

As additional certificates are revoked, the CRL must be updated to include newly revoked certificates. The latest CRL should be made available to vehicles to allow innocent vehicles to know which certificates can no longer be trusted.

The CRL for a large network can become large in size. To reduce the overheads for transporting CRLs over a network, delta CRLs can be used. A delta CRL contains the certificates revoked since the last CRL was issued. The IETF RFC 5280 [CSFB08] defines delta CRLs for X.509 CRLs. The IEEE 1609.2 standard allows a user to specify the time interval for the reception of a new CRL.

Once a revoked certificate expires, it will no longer need to be on the CRL.

15.5 A BASELINE REFERENCE VEHICULAR PKI MODEL

A baseline reference vehicular PKI functional model is illustrated in Figure 15.3. It consists of the following main functional components:

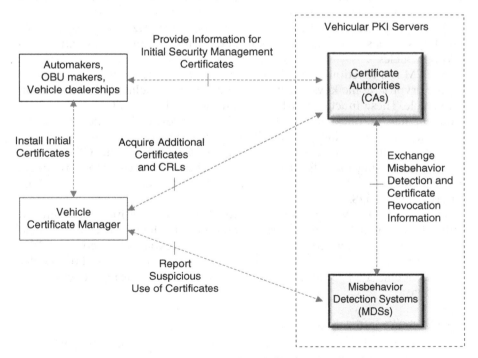

Figure 15.3. A baseline vehicular PKI functional model

- Certificate authorities (CAs)
- Misbehavior detection systems (MDSs)
- Vehicle certificate managers (VCMs)

A CA is a functional entity that is responsible for creating, issuing, maintaining, and revoking certificates. The CA should verify the information provided by certificate requesters to ensure the accuracy of the information to be included on certificates. For example, when a certificate requester provides its identity to be included on a certificate, the CA must verify the evidence provided by the certificate requester to ensure that the certificate requester owns the identity it provides. A certificate requester may create its own private–public key pairs and request certificates from the CA for its public keys. In this case, the certificate requester must prove to the CA that it owns the private keys associated with these public keys; and the CA must verify these proofs to ensure that certificate requester possesses the corresponding private keys.

There can be different types of CA, each responsible for different functions required for certificate management. We will discuss several different types of CAs later in this chapter and in the following chapters. The different types of CAs are often named differently to reflect their specific sets of certificate management functions. We will refer to all types of CAs simply as the CA when there is no need to distinguish the different types of CAs.

An MDS is a functional entity that collects and analyzes information provided by vehicles and other sources to detect misused certificates and misbehaving vehicles.

A VCM is a functional entity on a vehicle, which is responsible for managing the cryptographic keys, certificates, and certificate-related information on the vehicle. These functions include acquiring initial certificates from the CA, acquiring new certificates to replace expired and revoked certificates, storing certificates on the vehicle, determining which certificates to use to sign a message, acquiring the latest CRL from the CA, storing the CRL on the vehicle, processing the CRL to determine whether a certificate is revoked, performing local detection of misused certificates, and sending misbehavior reports to the MDS.

The CAs, MDSs, VCMs, and the protocols for these entities communicate with each other to collectively form an end-to-end vehicular PKI system. The functional entities that are not on end-user devices are referred to as PKI servers. In this reference model, main PKI servers are the CAs and the MDSs.

A vehicular PKI system needs to support the following functions or operations:

- Configure initial system-wide security parameters.
- Assign initial certificates to vehicles and their communicating parties.
- Acquire new certificates.

- Distribute a vehicle's certificates to other vehicles for them to use to verify signed messages.
- Revoke misused certificates and distribute CRLs to vehicles and their communicating parties.
- Detect certificate misuses and misbehaving vehicles.

To support the above vehicular PKI functions, the following protocols will be needed:

- A vehicle-to-CA (V2CA) protocol for vehicles to acquire certificates and CRLs from the CA.
- A vehicle-to-MDS (V2MDS) protocol for vehicles to report misbehaviors to the MDS.
- Protocols to support communications among the vehicular PKI servers.

15.6   CONFIGURE INITIAL SECURITY PARAMETERS AND ASSIGN INITIAL CERTIFICATES

Before a vehicle can participate in secure communications, a set of system-wide security parameters must be configured on every vehicle and on all the other entities that the vehicles will communicate with. The system-wide security parameters for leading public key signature mechanisms (such as Rivest–Shamir–Adleman [RSA], Digital Signature Algorithm [DSA], and Elliptic Curve Digital Signature Algorithm [ECDSA]) and public key encryption mechanisms (such as Elliptic Curve Digital Signature Algorithm [ECIES]) are described in the previous chapter. For example, when ECDSA is used for creating and verifying signatures, a set of ECDSA domain parameters, which define the elliptic curve and specify the values of the public parameters used for signature creation and verification, must be provisioned on every vehicle and their communicating parties before security operations can begin.

Each vehicle needs to be installed with an initial set of *security management certificates* and their corresponding private keys. A vehicle uses these security management certificates to authenticate with the CA to acquire additional cryptographic keys and certificates. Before a vehicle has its initial security management certificates, it will not have the proper credentials to prove its trustworthiness to the CA. Therefore, the vehicles themselves typically will not be able to acquire their initial security management certificates from the CA. Instead, an entity that can provide sufficient evidence to the CA to attest to the binding between a public key and a vehicle has to acquire the initial security management certificates for a vehicle and then install them on the vehicle. This can be accomplished by, for example, a vehicle manufacturer, an onboard unit (OBU) manufacturer, or a vehicle dealership, which can use its own credibility to vouch for the initial keys and certificates to be provided to a vehicle.

Each vehicle may also be installed with an initial set of certificates that it will use to sign vehicle safety messages. We refer to these certificates as *safety application certificates*. After a vehicle is provisioned with its initial security management certificates, it can communicate with the CA directly to acquire its initial safety application certificates. Alternatively, the vehicle manufacturer, the OBU manufacturer, or the vehicle dealership may acquire the initial safety application certificates and install them on a vehicle before the vehicle is sold.

The private–public key pairs can be created using a digital signature algorithm (such as ECDSA).

Either the vehicles or the CA can create the key pairs for the vehicles.

15.6.1 Vehicles Create Their Private and Public Keys

A vehicle can create its own private–public key pairs and then request the CA to issue certificates for its public keys. In this case, the vehicle needs to provide to the CA the public keys to be certified and proofs to show that the vehicle possesses the private keys associated with the public keys and is entitled to receive the requested type of certificates.

A vehicle may prove its possession of a private key by, for example, sending to the CA a token encrypted with the private key. A vehicle can use its security management certificates to prove to the CA that it is entitled to receive new certificates. A vehicle may also be allowed to use its safety application certificates to authenticate to the CA to acquire additional safety application certificates.

The CA authenticates and authorizes the request, generates the requested certificates, and sends the certificates to the vehicle.

The main advantage of vehicles generating their own private–public key pairs is that no one else, not even the CA, will know a vehicle's private keys. Another advantage is that there will be no need to send private keys over the network, making it easier to keep the private keys secret.

If each vehicle uses a large number of short-lived certificates to protect privacy, sending a large number of public keys over wireless networks to the CA for certification can become a challenge. Consider a case in which each short-lived certificate is valid for 5 minutes and a vehicle wishes to acquire sufficient certificates to use for the next 30 days. This means that the vehicle will need to acquire 8640 certificates each time. Using the ECDSA signature algorithm for 112-bit security strength, each public key will be 224 bits or 28 bytes. If a vehicle sends these public keys to the CA in a single certificate request, the request message will have a payload that is approximately 242 kB long in addition to the protocol headers, encryption overhead, vehicle's digital signature, and vehicle's public key certificate, which also need to be added to the request message. It can be highly resource inefficient when a large number of vehicles send such large messages over the wireless network to the CA. Larger messages are more prone to transmission errors. Higher transmission errors will cause packet retransmissions, which waste wireless network resources and result in longer delays. Requesting each certificate individually or requesting certificates in small batches will require a large number of rounds of message exchanges between a vehicle and the CA, which can result in long delays and heavy message processing burdens on both the vehicles and the CA.

15.6.2 Certificate Authority Creates Private and Public Keys for Vehicles

The CA can create private–public key pairs and their certificates for each vehicle. In this case, when requesting keys and certificates from the CA, a vehicle only needs to provide sufficient information to show that it is entitled to receive the requested types of certificates.

A vehicle can use its security management certificates to prove to the CA that it is entitled to receive new certificates. A vehicle may also be allowed to use its safety application certificates to authenticate to the CA to acquire additional safety application certificates.

The CA needs to send the private keys and the certificates for the public keys associated with the private keys to the vehicles.

Compared with the case in which each vehicle creates its own private keys to assign the same number of certificates to each vehicle, each vehicle needs to send a much smaller amount of data to the CA. However, the CA will have to send more data to each vehicle.

15.7 ACQUIRE NEW KEYS AND CERTIFICATES

When a vehicle's certificates are about to expire or have been revoked, the vehicle should acquire new certificates from the CA to replace the expired or revoked certificates so that it can continue to send messages that other vehicles will trust.

If a vehicle requests for a new certificate to replace an expired certificate, the CA needs to verify the request to ensure that it is from a vehicle with an expired certificate.

When a vehicle requests for a new certificate to replace a revoked certificate, the CA will also need to decide whether to grant any new certificate to the vehicle. That is, the CA needs to determine whether the requesting vehicle is a misbehaving vehicle. The fact that a certificate previously assigned to a vehicle has been revoked can be an indication that the vehicle has misused the revoked certificate. Adversaries who misuse certificates should not be allowed to obtain new certificates endlessly.

A main protocol required to support vehicular PKI operations is the protocol for a vehicle to acquire certificates and CRLs from the CA. We will refer to this protocol as the V2CA protocol.

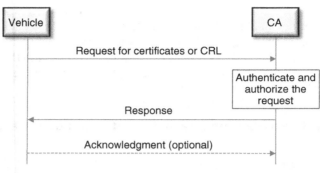

Figure 15.4. Vehicle-to-CA protocol

The V2CA protocol should support one or both of the following ways to assign certificates to a vehicle:

- A vehicle generates its private–public key pairs and requests the CA to issue certificates for its public keys.
- The CA generates private–public key pairs and certificates for vehicles.

Figure 15.4 illustrates a generic model for the V2CA protocol. The request message carries a Type field to indicate whether it is a request for certificates or CRLs. When requesting for certificates, the content of the request message depends on whether the vehicle or the CA generates the private–public key pairs:

- If the vehicle generates its key pairs, the certificate request should include the public keys to be certified by the CA, and proofs that the vehicle has the private keys associated with the public keys without explicitly including the private keys in the certificate request message. The certificate request also specifies the type and the requested lifetimes for the certificates.
- If the CA generates the key pairs, the certificate request will not include public keys but will specify the type, number, and lifetimes of the requested certificates.

The certificate requester signs the request message with its private key so that the CA can authenticate whether the request is from a vehicle. This requires the certificate requester to have a valid certificate that entitles it to receive new certificates.

The request message should be encrypted to the CA in order to protect vehicle privacy if the request carries the public keys to be certified. A privacy-preserving certificate may only identify the owner of a public key as a vehicle without revealing the identity of the vehicle. However, if an adversary knows that a particular public key belongs to a particular vehicle by examining the content of the request message, the public key and its certificate will no longer be privacy preserving.

An encrypted certificate request message can also carry symmetric keys for the CA to use to encrypt its subsequent messages to the certificate requester.

The CA will authenticate each request message and determine whether to provide new certificates to the requester. If the CA decides to provide the requested certificates to the requester, it sends the requested certificates back to the vehicle in a response message. If the CA is responsible for generating the key pairs, the response message will include the key pairs. If the CA denies the certificate request but does not think the requester is a misbehaving vehicle, the response message will carry an error code indicating the reason for the rejection. If the CA deems the requester a misbehaving vehicle, it may discard the certificate request without replying to it.

The CA should sign each response message that carries new certificates or error codes that do not indicate the requesting vehicle as misbehaving. Such a response message should also be encrypted to the requester so that adversaries will not know which certificates have been assigned to the requester. The encrypted response can also carry secret symmetric keys for the requester to encrypt its subsequent messages to the CA later.

When requesting for CRLs, the request message contains information about the last CRL the requester has received. A CRL request message does not need to be encrypted because they should not contain information that can be used to identify or track vehicles. The CRL request message should be signed by the requester. This will allow the CA to authenticate CRL requesters and only send CRLs to authorized requesters. This can reduce denial-of-service attacks adversaries may mount by requesting the CA to send an excessive number of CRLs over the network.

If the CRL requester does not have the latest CRL, the CA sends the latest CRL to the requester in a response message. This response message does not need to be encrypted because the CRL is meant to be seen by everyone. The response message does not need to be signed by the CA either because the CRL itself is signed by the CA.

If the CA denies a CRL request, it may send a response message with an error code indicating the reason for denial. If the CA deems the requesting vehicle as a misbehaving vehicle, the CA may ignore the CRL request without responding to the vehicle.

The vehicle can optionally acknowledge the receipt of a response message from the CA by sending an acknowledgement message back to the CA. This acknowledgement is, however, usually unnecessary especially for CRL requests. If a request or a response message is lost, the requester can resend its request.

The IEEE 1609.2 standard defines messages for requesting certificates and CRLs that are based on elliptic curve cryptography. A protocol for requesting

X.509 certificates over the Internet is defined by RSA Security in its Public Key Cryptography Standards (PKCS) #10 and as an informational RFC by the IETF [NyKa00].

15.8 DISTRIBUTE CERTIFICATES TO VEHICLES FOR SIGNATURE VERIFICATIONS

To verify a signed message, a vehicle must have the public key associated to the private key used to generate the signature and the certificate of the public key. In a large consumer vehicle network, it is impractical to require each vehicle to be preloaded with the certificates for all other vehicles. Storing and processing certificates for all other vehicles become even more challenging if each vehicle uses a large number of short-lived certificates to protect its privacy. This creates an important question: how should vehicles distribute their certificates to each other?

A vehicle can include its certificate in every message it sends. This, however, can lead to a heavy waste of wireless bandwidth. Innovative methods can be developed to allow each vehicle to include its certificates only in a small subset of the messages. Here, we use an example to illustrate one approach. Consider a group of vehicles broadcasting messages to each other periodically. Vehicles tell each other whether they have the certificates to verify all received signatures by setting a Certificate Request Flag in the messages they send out. A vehicle decides whether to include its certificate in its outbound messages based on its knowledge of whether the other vehicles have its certificate. Vehicles cache the certificates they have received from other vehicles for a finite time. A simplified sample procedure is shown in Figure 15.5.

As a vehicle receives messages, it checks whether it has the certificates for verifying these messages. If it is missing any certificate required to verify any message, it will set the Certificate Request Flag in the next message it broadcasts to ask the other vehicles to include their certificates in their next messages. If a vehicle receives any message with the Certificate Request Flag set, it will include its certificate in the next message it sends out. When the Certificate Request Flags in all received messages are null, the vehicle will stop including its certificate in the subsequent messages it sends out. When a new vehicle joins this communication group, it can discover from the received messages whether it has the certificates from all the neighboring vehicles. If not, it will set the Certificate Request Flag in its next outbound message to request other vehicles to include their certificates in their next messages.

The above approach allows vehicles to quickly distribute their certificates to each other and converge to a steady state in which they will no longer need to include their certificates in the messages. Before the steady state, a vehicle may receive messages and does not have the certificates to verify them. However, each vehicle should be able to obtain all required certificates within four message broadcast intervals, as illustrated in Figure 15.6. Consider two

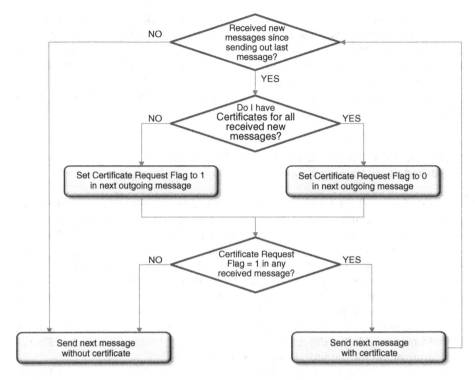

Figure 15.5. Procedure for vehicles to distribute certificates to each other

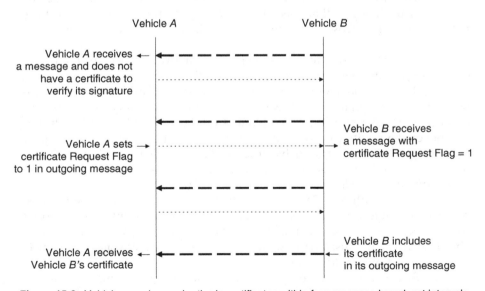

Figure 15.6. Vehicles receive each other's certificates within four message broadcast intervals

vehicles A and B broadcasting messages to each other periodically. When vehicle A receives a message from vehicle B and does not have the certificate to verify its signature, vehicle A sets the Certificate Request Flag in the next message it broadcasts. Upon receiving this message, vehicle B will include its certificate in the next message it broadcasts. Therefore, vehicle A should receive vehicle B's certificate by the time it receives the third message from vehicle B.

15.9 DETECT MISUSED CERTIFICATES AND MISBEHAVING VEHICLES

When PKI is used for Internet applications today, it is typically the certificate owners' responsibility to detect misused certificates. For example, when Web server certificates are assigned to an Internet Service Provider (ISP), it is usually the ISP's responsibility to detect any misuse of its certificates. When end users are issued certificate-enabled smart cards today, the end users are typically responsible for detecting and reporting unauthorized use of their cards. The end users do not have to understand or know what certificates are. They just need to report unauthorized use of their cards.

In a consumer vehicle network, however, drivers as the end users cannot be held responsible for detecting the misuse of certificates by vehicles. Therefore, a misbehavior detection system will be needed.

Methods to detect misused certificates and misbehaving vehicles are dependent on the specific privacy-preserving certificate schemes and on whether the global misbehavior detection system is capable of identifying and tracking vehicles. This suggests that misbehavior detection for a consumer vehicle network should be closely coordinated with the design of the vehicular PKI system.

Misbehavior detection in a consumer vehicle network poses several unique challenges, requirements, and design considerations that need to be addressed.

First, the CA's judgment on whether a vehicle is a misbehaving one must be made with minimal errors. This is because losing the ability to acquire new certificates can cause a vehicle to run out of valid safety application certificates and hence the ability to protect its privacy or even the ability to send safety messages that can be trusted by other vehicles. Losing the ability to send trustworthy safety messages can reduce the level of safety for this vehicle and its surrounding vehicles.

In a country with over 250 million vehicles, a very small error rate in misbehavior detection could cause a large number of innocent vehicles to be wrongly accused as misbehaving. When a large number of vehicles lose their ability to transmit trustworthy safety messages, the effectiveness of vehicle safety applications could suffer.

Second, misbehavior detection should incur negligible interruptions to innocent drivers. For example, innocent vehicles should not be forced to regain

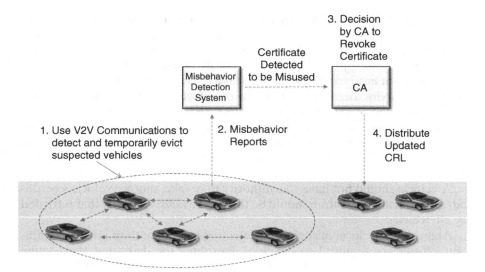

Figure 15.7. Functional architecture for misbehaviors detection. V2V, vehicle to vehicle

their privileges through physical inspections to receive new security keys and certificates.

Third, a fundamental issue is whether the misbehavior detection system should be allowed to have sufficient private information that will enable it to breach vehicle privacy. Having the ability to identify or track vehicles will facilitate the detection of misbehaving vehicles. However, this capability can be abused by misbehavior detection system operators or exploited by adversaries to identify and track innocent vehicles.

Figure 15.7 illustrates a functional architecture for misbehavior detection. It contains two main components:

- Local misbehavior detection
- Global misbehavior detection

15.9.1 Local Misbehavior Detection

Vehicles should have local misbehavior detection capabilities to determine whether messages received from other vehicles can be trusted. A vehicle can discard untrustworthy messages to reduce the negative impact of malicious or erroneous messages.

A vehicle may treat messages as untrustworthy even if these messages have valid signatures and certificates. This may be the case when, for example, the information contained in a message contradicts with the majority of other messages or with the information collected through reliable onboard sensors.

Effective vehicle local misbehavior detection could allow vehicles to continue to operate safely in the presence of misbehaving vehicles. As long as a vehicle's local misbehavior detection system can detect untrustworthy messages to maintain sufficient effectiveness of its vehicle safety applications, the vehicle will not have to retrieve the next CRL. The effectiveness of a vehicle's local misbehavior detection capability therefore directly impacts how frequently this vehicle has to obtain the latest CRL from the CA.

As vehicles use their local misbehavior detection mechanisms to effectively detect untrustworthy messages for longer periods of time without having to retrieve the next CRL, more options will open up for how CRLs can be distributed to vehicles.

A vehicle should not have the authority to revoke another vehicle's certificates. Certificate revocation should be the responsibility of a CA that is trusted by all vehicles.

Algorithms for local misbehavior detection are still in their infancy. Significant efforts are required to develop algorithms for dedicated short-range communications (DSRC) radios to detect, with high confidence and low error rates, whether other DSRC radios are misbehaving. This will include detecting intentional and unintentional misbehaviors. Effective local misbehavior detection solutions will be critical to real-world deployment of consumer vehicle networks.

15.9.2   Global Misbehavior Detection

Each vehicle's local misbehavior detection capabilities can be limited due to the limited local information each vehicle will have. As the number of misbehaving vehicles and the severity and the scope of misbehaviors grow, the effectiveness of individual vehicles' local detection systems could diminish.

Global Misbehavior Detection Systems can collect and analyze information from a large population of vehicles to make more reliable judgments on whether a specific vehicle is misbehaving.

The misbehavior detection results will then be provided to the CA, which in turn decides whether to revoke the certificate, to revoke all the certificates previously assigned to the vehicle, or to revoke the vehicle's privilege to receive future certificates.

Once a vehicle's privilege to receive new certificates is revoked, it needs to undergo physical inspection to regain the privilege.

15.9.3   Misbehavior Reporting

To enable global misbehavior detection, vehicles need to provide information that an MDS can use to detect misused certificates and misbehaving vehicles.

A fundamental question is: *What information should vehicles report to the MDS?* The answers to this question directly impact an MDS's ability to detect misbehaviors and the volume of network traffic generated by the misbehavior reporting.

It is in general difficult to require vehicles to provide trustworthy and verifiable evidence about alleged misbehaviors in a resource-efficient manner over wireless networks. First, the reporting vehicles can be misbehaving vehicles sending erroneous misbehavior reports. Second, sufficient evidence to show that a certificate has been misused or a vehicle has misbehaved will typically contain a large amount of data. The most direct evidence showing a vehicle has misbehaved will be the messages received from the alleged misbehaving vehicle. A single or a small number of received messages taken out of its context will often provide little value to an MDS. Sending a large number of messages received from the alleged misbehaving vehicles to the MDS can create excessively heavy network traffic.

One approach to misbehavior reporting is that each misbehavior report contains the reporting vehicle's vote that another vehicle has misbehaved or a certificate has been misused. The message can contain additional information such as the type of the alleged misbehavior. However, a misbehavior report does not need to contain evidence for the vote. The MDS collects and analyzes votes from multiple vehicles to decide whether a certificate is misused and whether a vehicle is misbehaving. This eliminates the need for vehicles to send a large volume of data to the MDS.

Several voting mechanisms and voting-based misbehavior detection mechanisms are available [BrEM08] [ClMo06] [MRCP08] [RPAJ07] [SoAR08]. New voting-based misbehavior detection mechanisms with lower error rates still need to be developed to support a large consumer vehicle network.

We refer to the protocol for vehicles to send misbehavior reports to an MDS as the V2MDS protocol. A sample V2MDS protocol is illustrated in Figure 15.8.

Since the same misbehaving vehicle or misused certificate will likely be detected and reported by multiple vehicles, the transport of misbehavior

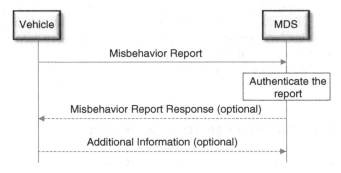

Figure 15.8. Vehicle-to-MDS protocol

reports from vehicles to the MDS needs not be 100% reliable. In addition, a response from the MDS is usually unnecessary.

The MDS may choose to send a Misbehavior Report Response message back to the vehicle to request the vehicle to send further data about itself or about the alleged misbehavior. Upon receiving a Misbehavior Report Response message that requests the vehicle to provide further data, the vehicle should respond to the MDS with the requested extra data.

Initially, vehicles can report misbehaviors anonymously. However, the more reporting performed by a vehicle, the more credibility the vehicle may need to show to the MDS. Otherwise, a misbehaving vehicle could send an arbitrarily large number of false reports to wrongfully accuse innocent vehicles. Therefore, when an MDS receives an excessive number of Misbehavior Reports from the same vehicle, it may request the vehicle to reveal more about itself so that the MDS will have more information about the reporter to judge the credibility of its reports.

If a vehicle fails to respond to a MDS's request for further information, the MDS may discard the previous misbehavior reports from this vehicle.

If the Misbehavior Report Response message does not request the vehicle to send further data regarding an already reported misbehavior, the vehicle does not have to acknowledge the Misbehavior Report Response message from the MDS.

The misbehavior report message should be signed by the reporting vehicle and encrypted to the MDS. The vehicle's signature allows the MDS to verify that the report is from an entity entitled to send such reports. To protect vehicle privacy, a vehicle may sign its misbehavior reports with its privacy-preserving certificates. This will allow the MDS to verify that the reports are from a vehicle without knowing from which particular vehicle.

Not all vehicles detecting the same event need to send misbehavior reports to the MDS. To balance the accuracy of the reports and the volume of network traffic generated by the misbehavior reports, algorithms can be developed to allow each vehicle to autonomously determine whether it should send misbehavior reports each time it detects a suspected misbehaving vehicle or misused certificate.

Misbehavior reporting can be exploited by adversaries to attack the security system and the consumer vehicle network. For example, multiple misbehaving vehicles can collectively claim that an innocent vehicle is misbehaving. When the adversaries form a local majority, their reports may convince an MDS that an innocent vehicle has misbehaved. Therefore, the global misbehavior detection system must be designed to minimize the impact of such abuses.

15.10 WAYS FOR VEHICLES TO ACQUIRE CRLs

Vehicles can proactively retrieve the latest CRL from the CA either periodically or when triggered by special events. The CRL issuer can include an estimated time in each CRL to indicate when the next update to the CRL is likely to be available. The vehicle may communicate with the CA at these estimated times to retrieve the next CRL.

Sample events that can trigger a vehicle to request the latest CRL include:

- The vehicle has been out of contact with the CRL issuer for an excessive period of time that may cause its current CRL to be out of date.
- The vehicle detects an excessive number of misused certificates that are not on the CRL it has.
- The vehicle enters a region where a different CRL issuer is responsible for issuing CRLs.

An entity that is separate from the CRL issuer can be used to store and distribute CRLs. For example, CRL distribution points (CDPs), which can be separate from the CRL issuer geographically, are used today to store CRLs and to provide an interface for applications over the Internet to download CRLs.

A CRL issuer or a CDP can also proactively distribute CRLs to vehicles. For example, the latest CRL can be supplied to roadside units (RSUs) that can broadcast it to the vehicles that pass through their radio coverage zones.

Vehicles that can use different types of networks to acquire CRLs are illustrated in Figure 15.9. For example, vehicles can use DSRC or Wi-Fi RSUs

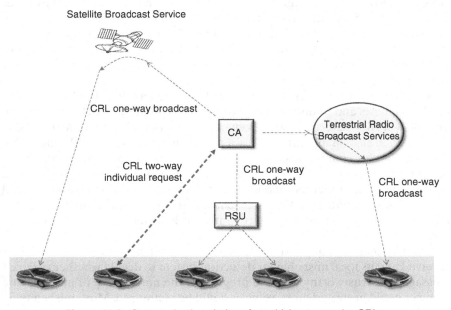

Figure 15.9. Communication choices for vehicles to acquire CRLs

deployed along the roads, at vehicle service centers, at garages, and in parking lots to connect to the CRL issuer or a CDP to retrieve CRLs.

Vehicles can also use cellular networks to retrieve CRLs. This can be achieved with either cellular communication capabilities embedded into the vehicles or cellular devices drivers and passengers brought into the vehicles. Today, a rapidly increasing population of vehicles is equipped with cellular communication capabilities to support telematics applications such as route guidance and emergency assistance. These cellular communication capabilities can also be used to retrieve CRLs.

Satellite and terrestrial radio and TV broadcast services, which are increasingly available on vehicles and have been used to distribute traffic information to vehicles, can also be used to broadcast CRLs to vehicles.

15.11 HOW OFTEN CRLs SHOULD BE DISTRIBUTED TO VEHICLES?

An important question in designing a vehicular PKI is how often the latest CRL or delta CRL should be disseminated to vehicles. Answers to this question have a direct and significant impact on several important aspects of vehicular PKI design.

If PKI design requires each latest CRL to be immediately distributed after it is created, all vehicles need to have frequent network connectivity with the CRL issuer. Frequent distributions of CRLs to a large number of vehicles can consume heavy wireless bandwidth. On the other hand, if PKI design assumes vehicles can rely on their local misbehavior detection capabilities for a long period of time, the distribution of the latest CRL can be less frequent.

How long vehicles can wait for a new CRL depends on the effectiveness of their local misbehavior detection functions and the target level of security for the consumer vehicle network.

To provide a framework for analyzing how often a CRL should be distributed to vehicles and to understand the correlations between CRL distribution and misbehavior detection, we introduce the concept of *sustainable interval* T_s [CrVZ09]. A sustainable interval for a vehicle is a time interval during which the vehicle can communicate with other vehicles safely without having to receive the next CRL. The sustainable interval concept can be used to link together important factors such as how often a CRL needs to be distributed to vehicles, capabilities of local misbehavior detection, capabilities of global misbehavior detection, and how frequently a vehicle has to connect to the PKI servers.

Before receiving the next CRL, a vehicle must rely on the last CRL it received and its local misbehavior detection system to decide whether received messages are trustworthy. As time progresses, the number N_m of misbehaving vehicles could grow and more vehicles could be victimized. More misbehaving vehicles can generate more malicious messages that could impact the effectiveness of safety applications on a larger number of innocent vehicles. More misbehaving vehicles could also cause the certificates of a larger number of innocent vehicles to be revoked by falsely reporting innocent vehicles as misbehaving vehicles. When the number N_v of victimized vehicles exceeds certain thresholds, the effectiveness of cooperative safety applications would reduce. For example, when misbehaving vehicles cause the certificates of too many innocent vehicles to be wrongly revoked, vehicle safety applications that rely on messages from neighboring vehicles may become less effective. Furthermore, when the number of misbehaving vehicles exceeds certain thresholds, an MDS's ability to distinguish misbehaving vehicles from innocent vehicles could decrease because there could be too many false misbehavior reports from the misbehaving vehicles.

Figure 15.10 illustrates the concept of sustainable interval. First, each vehicle must be able to rely on its local misbehavior detection capability to determine whether messages from other vehicles are trustworthy for at least as long as $T_s \geq T_{update} + T_{delivery}$ time units. T_{update} is the interval between the time a vehicle received its last CRL and the time a new CRL is created by the CA; and $T_{delivery}$ is the time it takes a vehicle to receive a CRL after the CRL is created.

T_{update} depends on the effectiveness of the MDS's ability to detect misbehaviors, which in turn depends on the misbehavior reports vehicles can provide to the MDS.

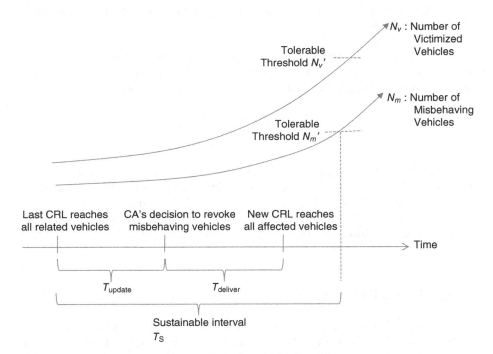

Figure 15.10. Sustainable interval

If a vehicle has an always-on network connectivity to the CRL issuer, T_{delivery} will be dominated by the end-to-end message transport delay. For a vehicle that does not have always-on network connectivity, T_{delivery} will be dominated by the time it takes for a vehicle to obtain network connectivity to the CRL issuer.

Let N'_m be the maximum number of active misbehaving vehicles and N'_v be the maximum number of victimized vehicles that the consumer vehicle network can tolerate. During a sustainable interval T_s, the following conditions need to be satisfied:

$$N_m \leq N'_m \text{ and } N_v \leq N'_v. \tag{15.1}$$

The sustainable interval T_s will last as long as the local and global misbehavior detection mechanisms can keep $N_m \leq N'_m$ and $N_v \leq N'_v$. That is, the length of the sustainable interval depends on the following main factors:

- How quickly the misbehaving vehicle population grows.
- How effectively vehicles can use their local misbehavior detection capabilities to detect untrustworthy messages to keep the value of N_v low.
- How effectively misbehaving vehicles can be detected by the MDS and revoked by the CA. Effective misbehaving vehicle detection keeps the value of N_m low by revoking misbehaving vehicles, which in turn will help keep the value of N_v low.

Let r be the rate at which new misbehaving vehicles enter the consumer vehicle network per time unit (e.g., a day or month). Let q be the rate at which vehicles are victimized. The sustainable interval T_s should satisfy:

$$T_s \leq \min\left\{\frac{N'_m}{r}, \frac{N'_v}{q}\right\}. \tag{15.2}$$

Given sustainable interval T_s, the frequency at which vehicles should receive the latest CRL can be estimated. In particular, each vehicle needs to communicate with a CRL issuer at least once in every T_s time units.

Several methods have been developed to estimate the sustainable interval and to design roadside networks that can achieve a target sustainable interval [CrVZ09].

15.12 PKI HIERARCHY

To support a large consumer vehicle network, a distributed and possibly hierarchical PKI architecture will be required. To prevent each individual CA to have sufficient information to breach vehicle privacy, the CA functions could be split to be implemented on separate CA servers operated by independent organizations. Here, we discuss some basic mechanisms for enabling distributed and hierarchical systems of CAs.

15.12.1 Certificate Chaining to Enable Hierarchical CAs

Certificates can be chained to enable a hierarchical and distributed CA architecture. A Root CA, which is trusted by all users without requiring anyone else to certify its public key, can authorize a subordinate CA to issue certificates. It can do so by issuing a certificate to certify that the subordinate CA owns a public key and that this public key entitles the subordinate CA to issue certificates. That is, the Root CA uses its credibility to vouch for the authority and competence of a subordinate CA to issue certificates.

A subordinate CA can also be authorized by the Root CA to authorize other subordinate CAs to issue certificates.

Certificate chaining described above enables multiple CAs to be organized into a hierarchical architecture while users only have to be preconfigured to trust a single or a very small number of Root CAs.

Different CAs can be used to issue different types of certificates. For example, the IEEE 1609.2 standard distinguishes between CAs that issue certificates used by vehicles to send safety messages and CAs that issue certificates used by vehicles to communicate with the CA. The IEEE 1609.2 standard also separates the CAs that issue CRLs from the CAs that issue certificates. Separate CAs may also be used to serve different geographical regions or vehicles from different automakers.

Figure 15.11 shows an example of certificate chaining. A Root CA issues a certificate to a subordinate CA_1 (operated by an automaker) authorizing CA_1 to issue safety application certificates to the automaker's vehicles. CA_1 uses its own private key to sign the certificates it issues to the vehicles ("vehicle certificates"). To verify a certificate issued by CA_1, a vehicle first verifies the Root CA's certificate for CA_1 by verifying the Root CA's signature on this certificate. A positive verification tells the vehicle that CA_1 owns the public key specified on the Root CA's certificate. The vehicle proceeds to use the public key on the Root CA's certificate to verify CA_1's signature on the vehicle certificate. If this verification is also positive, the vehicle can trust the certificate issued by CA_1.

15.12.2 Hierarchical CA Architecture Example

Figure 15.12 illustrates a sample distributed CA hierarchy. The country is divided into several regions, each served by a set of regional CAs. For each region, separate CAs can be used to issue different types of certificates or for different groups of vehicles. The CAs may be operated by different organizations. For example, an automaker may operate CAs to issue certificates to its own vehicles.

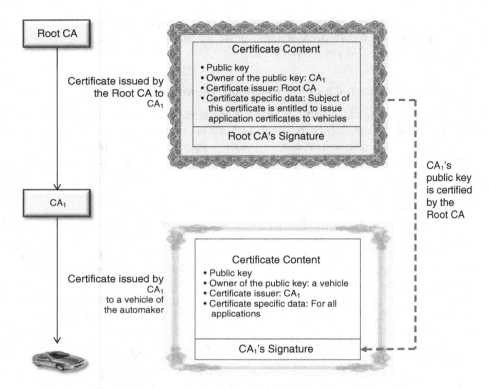

Figure 15.11. Certificate chains for hierarchical and distributed CA architecture

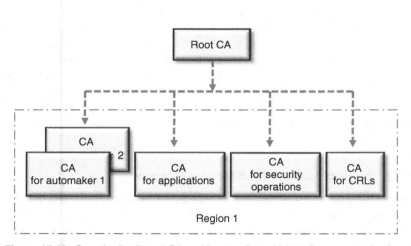

Figure 15.12. Sample distributed CA architecture for vehicle safety communications

A single logical Root CA can be used so that every vehicle, no matter where it was sold and where it travels to subsequently, can be preconfigured only with the public key of a single Root CA.

Alternatively, each region could have its own separate Root CA. This would require each vehicle to be provisioned with the public keys of all the Root CAs so that it can verify certificates issued by CAs in the different regions.

15.13 PRIVACY-PRESERVING VEHICULAR PKI

A conventional certificate contains plaintext information that adversaries can use to identify the subject of the certificate. This breaks the vehicle anonymity requirement. A conventional certificate also contains multiple data elements with unique values, such as certificate serial number, public key, and certificate-specific data, which adversaries can use to determine which messages are from the same vehicle. This breaks the long-term message unlinkability requirement and allows adversaries to track vehicles.

Therefore, the first step in designing a privacy-preserving PKI is to make certificates privacy preserving. That is, to make certificates anonymous and unlinkable. An anonymous certificate does not provide information for adversaries to identify the subject of the certificate. Certificates are unlinkable when it is practically infeasible for adversaries to use the certificates alone to determine which certificates belong to the same vehicle.

Designing a privacy-preserving PKI for a large consumer vehicle network continues to be a challenge. Making certificates anonymous and unlinkable is easy. The challenge is to make certificates anonymous and unlinkable while meeting other important design goals such as achieving high scalability and robustness [CrZh12] [WPVC09]. For example, the highest level of privacy can be easily achieved when all vehicles use a single identical certificate. However, revoking the certificate will require changes on all vehicles, making the system unscalable. Furthermore, detecting misbehaving vehicles will also be extremely difficult, which means the system is not robust.

In the rest of this book, we will discuss three categories of methods to make certificates privacy preserving and discuss their scalability and robustness (statistically shared certificates, short-lived unique certificates, and group signatures).

In a conventional PKI, each CA maintains complete knowledge of which certificate is assigned to which vehicle. CA operators may abuse this information to identify and track vehicles. Therefore, a second major area to be addressed in making a PKI privacy preserving is how to eliminate or substantially reduce the chances that any CA operator will have sufficient information to identify and track vehicles. We will describe one potential solution later in the book.

Before a detailed discussion of privacy-preserving PKIs, it is important to first discuss how to quantitatively measure privacy. The quantitative measurements can then be used to evaluate different privacy-preserving PKI solutions.

15.13.1   Quantitative Measurements of Vehicle Anonymity

When designing privacy-preserving PKI solutions, we consider vehicle privacy as the inability for an adversary to determine which particular vehicle in a set of vehicles is the owner of a certificate.

In a system of V vehicles, the maximum degree of anonymity is achieved when an adversary sees all vehicles as equally probable of being the subject of a certificate. That is, maximum anonymity is achieved when adversaries have no better ways than guessing completely at random which particular vehicle is the subject of a certificate. Therefore, the size of the anonymity set determines the maximum level of anonymity achievable by a privacy-preserving PKI solution.

Privacy-preserving solutions can leak different amounts of private information that can help adversaries do better than randomly guessing which vehicle is the subject of a certificate.

A way to measure anonymity while taking into account privacy leakage from specific privacy-preserving solutions is to use *entropy* [DSCP02] [HuCL04]. Entropy measures the level of uncertainty or unpredictability associated with a random variable.

Consider a privacy-preserving strategy that establishes an anonymity set of size V. An adversary wants to determine which vehicle in this anonymity set is the subject of a certificate. Let X be the discrete random variable that denotes the subject of the certificate. The adversary wants to determine the value of X, which can be $1, 2, 3, \ldots,$ or V. The adversary uses the certificates he captured from the vehicles to decide with probability p_i that the certificate belongs to vehicle i in the anonymity set. That is, $p_i = P_r(X = i)$, where each i corresponds to a vehicle in the anonymity set. Let $H(X)$ be the entropy of variable X after the adversary has analyzed a set of messages from vehicles. $H(X)$ measures how uncertain the adversary's decision is. A larger value of $H(X)$ means higher uncertainty and higher level of privacy.

$H(X)$ can be calculated as [DSCP02]:

$$H(X) = -\sum\nolimits_{i=1}^{V} p_i \log_2(p_i).$$ (15.3)

The maximum anonymity $H_{\max}$ is achieved when $pi = 1/V$ for every i. $H_{\max}$ is given by:

$$H_{\max} = \log_2 V.$$ (15.4)

15.13.2   Quantitative Measurement of Message Unlinkability

Message unlinkability is the inability to link multiple messages to the same vehicle. Message unlinkability can be expressed as the probability that an adversary can use only the messages to successfully determine whether a message $M(t_1)$ captured at time t_1 and a message $M(t_2)$ captured at time $t_2 > t_1$ are originated from the same vehicle.

When designing privacy-preserving PKI solutions, we consider message unlinkability as the probability that an adversary can use only the certificates for the messages to successfully determine whether two messages $M(t_1)$ and $M(t_2)$ are originated from the same vehicle.

We use $M(t_1) \leftrightarrow M(t_2)$ to denote the fact these two messages are originated from the same vehicle. Message unlinkability U can be expressed as:

$$U = 1 - P_r(M(t_1) \leftrightarrow M(t_2)). \tag{15.5}$$

The long-term message unlinkability for time threshold τ can be expressed as:

$$U(\tau) = 1 - P_r(M(t_1) \leftrightarrow M(t_2) \mid t_2 - t_1 \geq \tau). \tag{15.6}$$

REFERENCES

[BlSS99] I. Blake, G. Seroussi, and N. Smart: "Elliptic Curves in Cryptography," London Mathematical Society Lecture Note Series 265, Cambridge University Press, 1999.

[BrEM08] M. Braverman, O. Etesami, E. Mossel: "Mafia: A Theoretical Study of Players and Coalitions in a Partial Information Environment," Annals of Applied Probability, vol. 18, no. 3, pp. 825–846, 2008.

[ClMo06] J. Clulow and T. Moore: "Suicide for the Common Good: A New Strategy for Credential Revocation in Self-Organizing Systems," ACM Special Interest Group on Operating Systems (SIGOPS) Operating Systems Review, Volume 40, Issue 3, 2006.

[CrVZ09] G. D. Crescenzo, E. Vendanberg, and T. Zhang: "Analysis of Infrastructure and Communications Requirements for V2V PKI Security Management," Deliverable to the Vehicle Safety Communications—Applications (VSC-A) Program, 2009.

[CrZh12] G. D. Crescenzo, and T. Zhang: "Privacy-Preserving PKIs with Reduced Server Trust," G. D. Crescenzo, T. Zhang, IEEE ICC 2012, Communication and Information Systems Security Symposium, Ottawa, Canada, 2012.

[CSFB08] D. Cooper, S. Santesson, S. Farrell, S. Boeyen, R. Housley, and W. Polk: "Internet X.509 Public Key Infrastructure Certificate and Certificate Revocation List (CRL) Profile," Internet Engineering Task Force (IETF) Request For Comments (RFC) 5280, 2008.

[DSCP02] C. Diaz, S. Seys, J. Claessens, and B. Preneel: "Towards Measuring Anonymity," 2nd International Conference on Privacy Enhancing Technologies (PET), 2002.

[HuCL04] J.-P. Hubaux, S. Capkun, and J. Luo: "The Security and Privacy of Smart Vehicles," IEEE SECURITY & PRIVACY, 2004.

[IEEE06] IEEE P1609.2: "Trial-Use Standard for Wireless Access in Vehicular Environments—Security Services for Applications and Management Messages," 2006.

[MRCP08] T. Moore, M. Rayay, J. Clulow, P. Papadimitratosy, R. Anderson, and J.-P. Hubaux: "Fast Exclusion of Errant Devices From Vehicular Networks," 5th Annual IEEE Communications Society Conference on Sensor, Mesh and Ad Hoc Communications and Networks (IEEE SECON), San Francisco, CA, 2008.

[NyKa00] M. Nystrom and B. Kaliski: "PKCS #10: Certification Request Syntax Specification Version 1.7," Internet Engineering Task Force (IETF) Request for Comments (RFC) 2986, 2000.

[RPAJ07] M. Raya, P. Papadimitratos, I. Aad, D. Jungels, and J.-P. Hubaux: "Eviction of Misbehaving and Faulty Nodes in Vehicular Networks," IEEE Journal on Selected Areas in Communication (JSAC), vol. 25, no. 8, 2007.

[RSA78] R. Rivest, A. Shamir, and L. Adleman: "A Method for Obtaining Digital Signatures and Public-Key Cryptosystems," Communications of the ACM, vol. 21, no. 2, pp. 120–126, 1978.

[SoAR08] V. Sood, T. Antal, and S. Redner: "Voter Models on Heterogeneous Networks," Physics Review, 77, ID 041121, 2008.

[WPVC09] R. White, S. Pietrowicz, E. Vandenberg, G. Di Crescenzo, D. Mok, R. Ferrer, T. Zhang, and H. Shim: "Privacy and Scalability Analysis of Vehicular Combinatorial Certificate Schemes," IEEE Consumer Communications and Networking Conference (CCNC), Las Vegas, Nevada, USA, 2009.

16

PRIVACY PROTECTION WITH SHARED CERTIFICATES

16.1 SHARED CERTIFICATES

The idea of shared certificates is to assign each certificate to a large enough group of vehicles so that it will be difficult for adversaries to link a certificate to any particular vehicle.

We focus on statistically shared certificates. That is, a certificate is assigned to each vehicle according to a probability distribution so that a random group of vehicles shares each certificate.

Statistically shared certificates were first proposed for vehicle communications around 2006 [ReKW06] and these initial schemes are called combinatorial certificate schemes. Since then, statistically shared certificate schemes have been further analyzed and enhanced [WPBG09] [WVZM07].

We will first discuss the properties and limitations of the original combinatorial certificate scheme. Then, we will discuss potential solutions to improve the scalability and robustness of statistical certificate sharing.

16.2 THE COMBINATORIAL CERTIFICATE SCHEME

The original combinatorial certificate scheme works as follows [ReKW06] [WVZM07] [ZCPV07]:

Vehicle Safety Communications: Protocols, Security, and Privacy, First Edition. Luca Delgrossi and Tao Zhang.
© 2012 John Wiley & Sons, Inc. Published 2012 by John Wiley & Sons, Inc.

- *Certificate Generation*: A certificate authority (CA) creates a Shared Certificate Pool, which contains N pairs of private and public keys and their certificates to be used by all vehicles. These keys and certificates can be valid for long periods of time such as months or even years.
- *Certificate Assignment*: Every vehicle is given $n \leq N$ certificates, with their associated private–public key pairs, which are selected uniformly at random from the Shared Certificate Pool.
- *Certificate Revocation and Replacement*: The CA revokes a certificate by posting its identifier on a certificate revocation list (CRL) and making the CRL available to all vehicles. The CA removes each revoked certificate, and its associated private and public key pair, from the Shared Certificate Pool and replaces it with a new certificate. The CA assigns this new certificate to every vehicle that shared the revoked certificate. This ensures that the probability distribution of each certificate among all the vehicles remain unchanged after a certificate is revoked and replaced.
- *Processing of CRL by Vehicle*: Each vehicle verifies whether a certificate has been revoked by checking whether the certificate's identifier is on the CRL.

The level of privacy provided by any shared certificate scheme depends on the number of vehicles sharing each certificate. When V vehicles use the combinatorial certificate scheme, the probability ρ that an arbitrary vehicle is assigned any particular certificate is given by Equation 16.1, where

$$\binom{x}{y}$$

is the number of ways to choose y distinct numbers from x different numbers:

$$\rho = \frac{\binom{N-1}{n-1}}{\binom{N}{n}} = \frac{\frac{(N-1)!}{(n-1)!(N-n)!}}{\frac{N!}{n!(N-n)!}} = \frac{n}{N}. \tag{16.1}$$

The expected number N_S of vehicles sharing each certificate, which will be the expected size of the anonymity set, is given by Equation 16.2:

$$N_S = \frac{n}{N}V = \rho V. \tag{16.2}$$

An important property of the combinatorial certificate scheme is that the probability ρ that each certificate is assigned to any vehicle is independent of time. That is, probability ρ is stationary. This stationary distribution can be maintained even after certificates are revoked and replaced with new certificates over time, allowing a target level of privacy to be maintained over time.

Furthermore, probability ρ is also independent of the size of the vehicle population. This makes it easy to set the value of ρ for a dynamically changing vehicle population.

Next we use the expected size N_S of the anonymity set to estimate the level of anonymity and the long-term certificate unlinkability. Based on the anonymity definition in the previous chapter, the maximum expected level of anonymity achievable by the basic statistically shared certificate scheme can be estimated in Equation 16.3:

$$H_{max} = log_2(\rho V) = log_2\left(\frac{nV}{N}\right). \tag{16.3}$$

Now let us consider the long-term certificate unlinkability $U(\tau)$ for time threshold τ. An adversary tries to determine whether certificate C_2 for message M_2 belongs to the same vehicle that sent a message M_1 with certificate C_1 τ time units ago. If these two certificates are identical, that is, $C_1 = C_2 = C$, then the adversary knows that the owners of certificates C_1 and C_2 are among the ρV vehicles that are expected to share certificate C. The owner of certificate C_2 can be any one of the ρV vehicles with equal probability. Therefore, the probability that the owner of certificate C_2 is the same as the owner of certificate C_1 will be $1/\rho V$.

If C_1 and C_2 are different, then the adversary knows that the owner of C_1 is one of the ρV vehicles that are expected to share C_1 and the owner of certificate C_2 is one of the ρV vehicles that are expected to share C_2. The set of vehicles that owns either certificate C_1 or certificate C_2 has an expected size of $(2\rho - \rho^2)V$. Therefore, $U(\tau)$ can be estimated as:

$$U(\tau) = \begin{cases} 1 - \dfrac{1}{\rho V} \text{ if } C_2 = C_1 \\ 1 - \dfrac{1}{(2\rho - \rho^2)V} \text{ if } C_2 \neq C_1 \end{cases}. \tag{16.4}$$

One can see that the long-term certificate unlinkability achievable by the original combinatorial certificate scheme is independent of time. That is, the original combinatorial certificate scheme could maintain consistent certificate unlinkability over time.

16.3 CERTIFICATE REVOCATION COLLATERAL DAMAGE

When a certificate assigned to a misbehaving vehicle is revoked, all the other vehicles sharing this certificate will also not be able to use it. We refer to this effect as certificate revocation collateral damage. Collateral damage is a fundamental issue associated with all shared certificate schemes and has profound impact on the level of privacy, system scalability, and misbehavior detection [WPBG09] [WVZM07].

The more vehicles share each certificate, the more collateral damage will occur when a certificate is revoked. Since privacy is achieved by sharing each certificate among multiple vehicles, reducing collateral damage by reducing the number of vehicles sharing each certificate will also reduce privacy.

When an innocent vehicle shares a certificate with a misbehaving vehicle and the certificate has been revoked, we say this certificate on the innocent vehicle is "covered." A group of misbehaving vehicles may collectively share all the certificates assigned to an innocent vehicle. When these certificates are revoked, all the certificates assigned to the innocent vehicle will also be revoked and we say this innocent vehicle is covered. When all the certificates assigned to a misbehaving vehicle have been revoked, we say this vehicle is revoked.

First, let us consider the collateral damage a single misbehaving vehicle can introduce by repeatedly misusing and causing the revocation of a certificate and its subsequent replacement certificates. Consider a group $G(C)$ of vehicles that share certificate C. Assuming $n = 5$ and $N = 10,000$ in Equation 16.2, $G(C)$ is expected to contain over 125,000 vehicles in the United States, where over 250 million vehicles are on the road. Suppose a misbehaving vehicle misused certificate C and caused it to be revoked. All the vehicles in group $G(C)$ will need a new certificate to replace certificate C and they will all receive the same replacement certificate C_1 from the CA using the original combinatorial certificate scheme. That is, the same group $G(C)$ of vehicles will share certificate C_1. If the misbehaving vehicle subsequently misuses certificate C_1 and causes it to be revoked, the misbehaving vehicle and all the other vehicles in $G(C)$ will need a new certificate to replace C_1 and they will all receive the same replacement certificate C_2 from the CA. If the misbehaving vehicle continues this pattern of misbehavior, every time the misbehaving vehicle needs to replace a certificate, all the other vehicles in group $G(C)$ will also need to replace the exact same certificate. This leads to two major consequences:

- A single misbehaving vehicle can cause collateral damages to a large number of innocent vehicles.
- It is difficult to detect which vehicle is misbehaving because all the vehicles in group $G(C)$ will appear to have misused the same exact sequence of certificates as the misbehaving vehicle does.

Now let us consider the collateral damage introduced by multiple misbehaving vehicles. If m misbehaving vehicles have been revoked, the probability that any given certificate on an innocent vehicle is not covered will be:

$$\left(1 - \frac{n}{N}\right)^m = (1-\rho)^m. \tag{16.5}$$

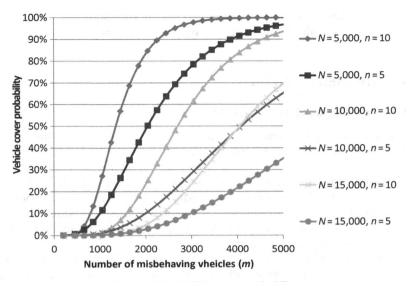

Figure 16.1. Vehicle cover probability

Therefore, the probability that an innocent vehicle is covered, referred to as the vehicle cover probability, can be estimated as [WVZM07]:

$$P_{\text{cover}}(N, n, m) = \left(1 - \left(1 - \frac{n}{N}\right)^{m}\right)^{n} = \left(1 - (1 - \rho)^{m}\right)^{n}. \qquad (16.6)$$

Several important observations can be made based on Equation 16.6. First, the vehicle cover probability is independent of the size V of the vehicle population. It depends only on the size N of the Shared Certificate Pool, the number n of certificates assigned to each vehicle, and the number m of the misbehaving vehicles. Second, the vehicle cover probability increases rapidly with the number of misbehaving vehicles. This rapid increase is illustrated in Figure 16.1.

These observations suggest that a relatively small number of misbehaving vehicles could cause an excessively large number of innocent vehicles to be covered. For example, when $N = 10,000$ and $n = 5$, over 50% of the entire vehicle population would be covered when there are only 4250 misbehaving vehicles. This means that over 125 million vehicles would be covered in the United States, where more than 250 million vehicles are on the road. Physical inspections by a trusted entity to reassign these vehicles their privileges to acquire new certificates would result in unnecessary costs and significant inconvenience to the driving public. This makes the original combinatorial certificate scheme impractical to deploy.

Collateral damage can be reduced by reducing n or increasing N. Unfortunately, this will simultaneously reduce the level of privacy.

16.4   CERTIFIED INTERVALS

16.4.1   The Concept of Certified Interval

We introduce the certified interval of a vehicle as the time period during which the CA deems the vehicle as innocent and will provide new certificates to the vehicle. Certified interval is a crucial indicator of a certificate scheme's robustness. Ideally, each innocent vehicle should have an unlimited certified interval and can therefore always obtain new certificates from the CA. Misbehaving vehicles, on the other hand, should have short certified intervals. This means that misbehaving vehicles will be detected and privileges to acquire new certificates will be revoked quickly.

An innocent vehicle's certified interval ends when the CA believes the vehicle has been misbehaving and decides to revoke its privileges to receive new certificates. In other words, the CA and MDS are no longer able to determine the vehicle's innocence.

A practical privacy-preserving certificate scheme should produce very long certified intervals for innocent vehicles to minimize the likelihood that they need to undergo physical inspections to regain their privileges to receive new certificates. For example, if a vehicle's certified interval is significantly longer than the vehicle's routine maintenance and repair cycles, it can be physically inspected only during routine maintenance or repairs.

16.4.2   Certified Interval Produced by the Original Combinatorial Certificate Scheme

We compute an innocent vehicle's certified interval for the original combinatorial certificate scheme and analyze how the certified interval will be impacted by the number of active misbehaving vehicles. The results characterize the robustness of the original combinatorial certificate scheme and will serve as a baseline for comparison with enhanced statistically shared certificate schemes.

For analysis purpose, we assume that the CA sets an upper bound or certificate quota on the number of certificates each vehicle can acquire to prevent misbehaving vehicles from obtaining new certificates endlessly [ReKW06] [WVZM07]. The CA rejects certificate requests from any vehicle that has reached its certificate quota.

We assume that the same certificate quota q is used for every vehicle where $q \geq n$. Each vehicle can have n initial certificates when it first enters the vehicular communication system and acquire $(q - n)$ more certificates later.

We assume that m new misbehaving vehicles are detected, on average, during each time unit. We refer to m as the misbehaving vehicle rate. A misbehaving vehicle is one that has n of its certificates revoked by the time the vehicle is detected as misbehaving. If a misbehaving vehicle has $y > n$ certificates and only $x < n$ of them have revoked when it is detected as misbehaving, the vehicle's remaining certificates will be revoked immediately once the vehicle is detected as misbehaving, and this vehicle will count as y/n misbehaving vehicles.

Consider an innocent vehicle that was introduced into the consumer vehicle network at time t_0 with n initial certificates. The vehicle's certified interval spans from time t_0 to the time it will have acquired $(q - n)$ new certificates. The probability that the vehicle has all of its q certificates revoked within the first t time units is given by:

$$F(t, N, n, m, q) = \left(1 - \left(1 - \frac{n}{N}\right)^{m \cdot t}\right)^q = \left(1 - (1 - \rho)^{m \cdot t}\right)^q. \qquad (16.7)$$

When ρ is small, using the Poisson approximation of binomial distribution, we have:

$$\left(1 - \frac{n}{N}\right)^{m \cdot t} \approx e^{-\frac{n \cdot m}{N} t}. \qquad (16.8)$$

For $\lambda = (nm/N) = m\rho$, we have:

$$F(t, N, n, m, q) = \left(1 - e^{-\lambda t}\right)^q. \qquad (16.9)$$

The probability density function of a new vehicle being revoked within the first t time units can be estimated as:

$$f(t, \lambda, q) = q\lambda e^{-\lambda t}\left(1 - e^{-\lambda t}\right)^{q-1}. \qquad (16.10)$$

Figure 16.2 shows the probability density function $f(t, \lambda, q)$ for $N = 10,000$, $n = 5, b = 15$, and $m = 300$.

The mean certified interval $E(T_{\text{certified}})$ can be estimated as:

$$E(T_{\text{certified}}) = \int_0^\infty q\lambda t e^{-\lambda t}\left(1 - e^{-\lambda t}\right)^{q-1} dt. \qquad (16.11)$$

Figure 16.3 shows numerical examples of an innocent vehicle's expected certified interval (using month as the time unit) as a function of m for $N = 10,000$ and $n = 5$. We observe that the expected certified interval decreases rapidly as the active misbehaving vehicle rate m increases. For example, when $q = 20$, the monthly misbehaving vehicle rate m has to be lower than 300 to maintain an average certified interval of 20 months for innocent vehicles. When the average monthly misbehaving rate increases to 600, the average certified interval drops below 12 months.

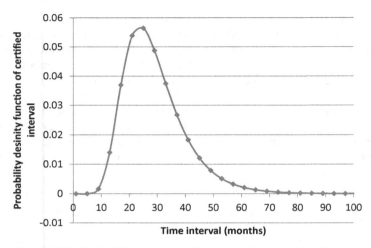

Figure 16.2. Probability density function of a vehicle's certified interval

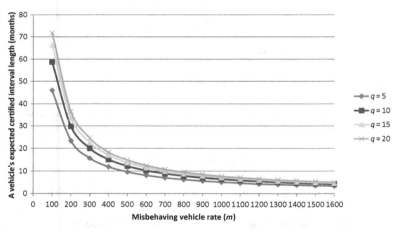

Figure 16.3. Innocent vehicle's expected certified interval

16.5   REDUCE COLLATERAL DAMAGE AND IMPROVE CERTIFIED INTERVAL

Statistically shared certificate schemes have some interesting characteristics that can be used to reduce collateral damages, detect misused certificates, and revoke misbehaving vehicles anonymously. This section describes several such characteristics and ways to use them to reduce collateral damages and improve an innocent vehicle's certified interval.

16.5.1 Reduce Collateral Damage Caused by a Single Misused Certificate

There are several strategies to reduce collateral damages caused by repeatedly misusing a single certificate and its replacement certificates.

The first category of strategies randomizes the replacement certificates for each revoked certificate so that each time a certificate is revoked, its collateral damage will be spread across a different group of vehicles. This makes a misbehaving vehicle increasingly distinguishable each time it requests for a new certificate.

A second category of strategies uses a random back-off process for certificate requests to constrain misbehaving vehicles' abilities to misuse certificates and make misbehaving vehicles distinguishable from innocent vehicles.

16.5.1.1 Randomized Certificate Replacements Let us use an example to illustrate how randomizing the replacement certificates can force a misbehaving vehicle to become increasingly distinguishable as it requests for more certificates. We consider a misbehaving vehicle that misuses certificate C_1 and causes it to be revoked. Let us say that this misbehaving vehicle received a new certificate C_2 to replace C_1. Suppose $G(C_i)$ is the set of vehicles that share certificate C_i. When replacement certificates are randomized, the other vehicles in $G(C_1)$ will receive randomly selected new certificates to replace C_1. Therefore, the set $G(C_2)$ of vehicles that share certificate C_2 will be different from the set $G(C_1)$ of vehicles that shared C_1. If the misbehaving vehicle subsequently causes certificate C_2 to be revoked, this misbehaving vehicle and the other vehicles in $G(C_2)$ will receive randomly selected certificates to replace C_2. Let us say that the misbehaving vehicle receives certificate C_3 to replace C_2. As this pattern continues and the misbehaving vehicle requests replacement certificates for a sequence of revoked certificates C_1, C_2, C_3, and so on, fewer and fewer other vehicles will request replacement certificates for the exact sequence of revoked certificates as the misbehaving vehicle. Specifically, only the vehicles in the intersection of the sets $G(C_1), G(C_2), \ldots, G(C_k)$ will be requesting replacement certificates for the exact sequence of revoked certificates $C_1, C_2, \ldots, C_k$. As the value of k increases, this intersection set will shrink rapidly. This effect can be used to detect the misbehaving vehicle.

Different methods can be used to randomize the replacement certificates. Examples include:

- *Individually Randomized Replacement*: Each time a vehicle requests for a certificate to replace a certificate C, the CA randomly selects a certificate from the Shared Certificate Pool as the replacement certificate.
- *Group Revocation with Group Replacement* [ReKW06]: The CA revokes $g > 1$ certificates at a time and replaces them with g new certificates. A vehicle that shares x of the g revoked certificates will be given x new certificates selected uniformly at random from the g new certificates.

• *Full Set Replacement* [ReKW06]: The CA keeps track of the new certificates it has generated to replace revoked certificates. When there are $g \geq 1$ new certificates in the Shared Certificate Pool, each replacement certificate for a vehicle will be selected with probability p from the g new certificates and with probability $(1 - p)$ from the old certificates in the Share Certificate Pool.

These randomized certificate replacement strategies share a major limitation: they introduce uneven distributions of certificates among the vehicles, which change as certificate revocations and replacements occur. This makes it difficult to know what level of privacy the system will provide over time.

16.5.1.2 *Certificate Requests with Random Back Off*

When a certificate is revoked, instead of requesting for a replacement certificate immediately, a vehicle will wait for a random time before requesting for a replacement certificate. This random wait time, which we refer to as a random back off, can increase with the number of times a vehicle requests for replacement certificates. Here, we show how a random back-off process can reduce a misbehaving vehicle's ability to misuse certificates and help make misbehaving vehicles distinguishable.

A basic form of random back-off strategies works as follows:

• All vehicles sharing a revoked certificate will be given the same new certificate to replace it. This ensures that the probability distribution of each certificate across all vehicles remain the same after certificate revocations and replacements.
• Each innocent vehicle keeps track of its certificate revocation history. If certificate C is revoked, the vehicle waits for a random time with a mean of d time units before requesting for a replacement certificate. Here, d is referred to as the back-off step. If the first replacement certificate C_1 is revoked, the vehicle waits for a random time with a mean of $2d$ time units before requesting a replacement certificate. If this second replacement certificate C_2 is revoked, the vehicle will wait for $2^2 d$ time units before requesting for a new replacement, and the process continues on.

If a misbehaving vehicle follows the same back-off strategy as innocent vehicles, then misusing a certificate and a small number of its replacement certificates will force the vehicle to wait for a long time before it can get new replacement certificates to continue its misbehaviors.

If a misbehaving vehicle chooses to request replacement certificates sooner without following the back-off procedure, it will become distinguishable in the certificate request process as someone who either requests replacement certificates excessively more frequently than innocent vehicles. This will help the MDS to detect misbehaving vehicles.

A drawback of the random back-off strategies is that an innocent vehicle may also have to wait for a long time to get its replacement certificate for a revoked certificate. Therefore, the goal is to force misbehaving vehicles to become distinguishable before each innocent vehicle loses too many of its certificates. The misbehaving vehicles that become distinguishable can be detected and revoked before an innocent vehicle loses all their certificates. Innocent vehicles will eventually be able to receive new certificates to replace their revoked certificates after the misbehaving vehicles are detected and revoked.

Next, we use a simplified example to illustrate how to select the value of the back-off step d and how to determine how many replacements of a single certificate can be tolerated by innocent vehicles.

If a misbehaving vehicle caused certificate C to be revoked and replaced with C_1, then C_1 to be revoked and replaced with $C_2, \dots$, and then C_{r-1} be revoked and replaced with C_r, we say that certificate C has been replaced r times.

Assuming a misbehaving vehicle caused certificate C to be replaced r times since time t_0, then each innocent vehicle will have to wait for d time units on average for its first replacement certificate for C, $2d$ time units on average for its second replacement certificate, and $2^r d$ time units on average for its rth replacement certificate.

Let us assume that an innocent vehicle cannot tolerate waiting for longer than $2^r d$ time units on average for any single replacement certificate. Suppose that in every E time units an innocent vehicle can have its innocence verified by a trusted entity (such as a service center) so that it can receive its next replacement certificate immediately without waiting for a back-off time. This means that an innocent vehicle will only be able to receive up to $(r - 1)$ replacement certificates within E time units. Let D_{r-1} be the time a vehicle has to spend to receive its first r replacement certificates. D_{r-1} can be computed as:

$$D_{r-1} = d + 2d + 2^2 d + \dots + 2^{r-1} d = (2^r - 1)d \leq E. \tag{16.12}$$

Therefore, the tolerable number of replacements for a certificate is:

$$r \leq \log_2 \left(\frac{E}{d} + 1 \right). \tag{16.13}$$

If $E = 180$ days and we set $d = 1$ day, then an innocent vehicle will be able to tolerate, on average, to have a certificate replaced eight times during each 180-day period.

For given E and r, the value of d can be set as:

$$d \leq \frac{E}{(2^r - 1)}. \tag{16.14}$$

16.5.2 Vehicles Become Statistically Distinguishable When Misusing Multiple Certificates

When a vehicle misuses multiple certificates and causes them to be revoked, the collateral damages associated with these revoked certificates will spread across different groups of vehicles. This will make the misbehaving vehicle distinguishable to the CA in the certificate replacement process as it will tend to request for new certificates more frequently than innocent vehicles.

Consider a misbehaving vehicle that has two different certificates C_1 and C_2. Let S_1 and S_2 be the sets of vehicles that share certificate C_1 and C_2, respectively. Using the basic statistically shared certificate scheme, the two sets S_1 and S_2 will be largely different. Therefore, when this misbehaving vehicle has caused both certificates C_1 and C_2 to be revoked, it is likely that fewer than two certificates on each innocent will be revoked as a result.

Next, we show the mathematical underpinning for the above observation. Specifically, we compare the probabilities that an innocent vehicle and a misbehaving vehicle will have the same number α of certificates revoked after m misbehaving vehicles have been revoked. We assume that each revoked misbehaving vehicle has n certificates.

For each innocent vehicle: The probability $P_I(x)$ that x of its certificates will be revoked after m misbehaving vehicles have been revoked is given by Equation 16.15, where $\lambda = (nm/N) = m\rho$:

$$P_I(x) = \left(1 - \left(1 - \frac{n}{N}\right)^m\right)^x \approx \left(1 - e^{-\lambda}\right)^x. \tag{16.15}$$

For each misbehaving vehicle that has exactly k revoked certificates, given the assumption that these k certificates have been revoked, the probability $P_M(x)$ that x of the vehicle's certificates will be revoked after m misbehaving vehicles have been completely revoked equals to the probability that $(x - k)$ additional certificates of this vehicle will be revoked due to collateral damages and is given by:

$$P_M(x) = \left(1 - \left(1 - \frac{n}{N}\right)^m\right)^{x-k} \approx \left(1 - e^{-\lambda}\right)^{x-k}. \tag{16.16}$$

We refer to the ratio $P_M(x)/P_I(x)$ as the distinguishing ratio and denote it by δ. The distinguishing ratio indicates how much more likely a misbehaving vehicle will have the same number of certificates revoked than an innocent vehicle. In other words, it represents how likely a misbehaving vehicle will become distinguishable from innocent vehicles during the process of certificate revocation and replacement. When $\delta = 1$, misbehaving vehicles are indistinguishable from innocent vehicles. The higher the value of δ is above 1, the more distinguishable misbehaving vehicles are. The distinguishing ratio is given by:

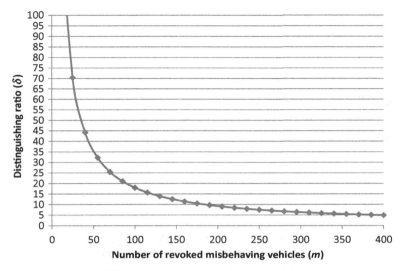

Figure 16.4. Distinguishing ratio for $N = 10,000$, $n = 5$, and $k = 1$

$$\delta = \left(1 - e^{-\lambda}\right)^{-k}. \tag{16.17}$$

Figure 16.4 shows δ for $n = 5$, $N = 10,000$, and $k = 1$ as a function of m. It shows that a misbehaving vehicle will be highly distinguishable from innocent vehicles. As the number of malicious vehicles increase, each misbehaving vehicle will become less distinguishable. However, even when there are hundreds of misbehaving vehicles, each misbehaving vehicle will still be several times more likely to have the same number of certificates revoked than an innocent vehicle.

Each time a misbehaving vehicle requests for a new certificate, its distinguishing ratio will increase significantly. Figure 16.5 illustrates this effect when a misbehaving vehicle causes $k = 2$ and 3 certificates to be revoked and have to request for their replacements.

When the number n of certificates allocated to each vehicle increases, each revoked certificate will provide less information to the CA for distinguishing misbehaving vehicles because each certificate will be shared by more vehicles. However, even when n is increased by several times, for example from 5 to 50, the value of δ remains large enough for misbehaving vehicles to be readily distinguishable. As illustrated in Figure 16.6, the distinguishing ratio is higher than 1.2 when $k = 1$ and there are 100 misbehaving vehicles. When each misbehaving vehicle caused $k = 3$ certificates to be revoked, the distinguishing ratio will become higher than 10 when the number of misbehaving vehicles stays the same.

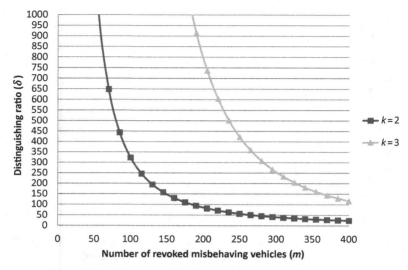

Figure 16.5. Distinguishing value for $N = 10,000$ and $n = 5$

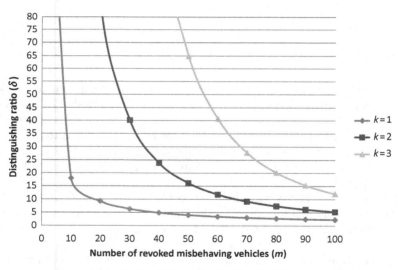

Figure 16.6. Distinguishing value for $N = 10,000$ and $n = 50$

16.5.3 The Dynamic Reward Algorithm

We describe an algorithm that the CA can use to revoke misbehaving vehicles anonymously by revoking misused certificates, to reduce collateral damages, and to increase innocent vehicles' certified intervals. We refer to this algorithm as the Dynamic Reward Algorithm. The basic form of this algorithm was first introduced in [WVZM07]. The idea is to prolong innocent vehicles' certified intervals beyond the certified intervals of misbehaving vehicles so that the misbehaving vehicles will be revoked before innocent vehicles are covered. Additional enhanced statistically shared certificate schemes can also be found in [WVZM07].

The Dynamic Reward Algorithm works as follows:

- Let $t_1, t_2, t_3, \ldots$, be consecutive time intervals such as calendar months.
- The CA maintains a black mark $B(t)$ for each vehicle. This means that this vehicle is allowed to acquire $q - B(t)$ new certificates in and after time period t, where q is the certificate quota for the vehicle. The value of $B(t)$ for each new vehicle is set to zero: $B(0) = 0$.
- The CA increments a vehicle's black mark by an integer value δ_1 whenever the vehicle requests for a new certificate. No certificate requests from a vehicle will be granted when the vehicle's black mark reaches the certificate quota q.
- The CA decrements a vehicle's black mark by an integer value δ_2 if the vehicle did not request for any new certificates during the previous time period.
- A black mark will not be decremented once it reaches the certificate quota.

This Dynamic Reward Algorithm enables the black marks for innocent vehicles to grow slower than the black marks for misbehaving vehicles. This means that a misbehaving vehicle will more likely reach its certificate quota earlier than innocent vehicles. As misbehaving vehicles reach their certificate quotas and are revoked, the black marks for innocent vehicles will be decremented and can reduce to zero eventually once there is no misbehaving vehicle.

Next, we compute a vehicle's certified interval when the Dynamic Reward Algorithm is used, based on the mathematical model in [WVZM07] and [WPBG09]. For the analysis, we assume:

- $\delta_1 = \delta_2 = 1$.
- $B(t) \geq 0$. That is, a black mark will not be decremented to lower than 0.
- The misbehaving vehicle rate is m per month. That is, m new misbehaving vehicles will be detected and n certificates on each of these misbehaving vehicles will be revoked during each month.

Let $S(k, t)$ be the probability that a vehicle V's black mark equals k in time period t. Therefore:

$$S(k, 0) = \begin{cases} 1, k = 0 \\ 0, k > 0 \end{cases}. \tag{16.17}$$

For any subsequent time interval $t \geq 1$, vehicle V's black mark will be 0 if and only if its black mark was either 0 or 1 in the previous time period $(t - 1)$ *and* it does not request any new certificate in the current time period t. Therefore, we can derive $S(0, t)$ as in Equation 16.18, where $P(k)$ is the probability that k of vehicle V's certificates have been revoked within one time period:

$$S(0, t) = S(1, t-1)P(0) + S(0, t-1)P(0); \forall t \geq 1. \qquad (16.18)$$

Probability $P(k)$ is given by the binomial probability that there are exactly k successes in m trials with success probability n/N for each trial. Here, a trial is to check whether a given certificate of vehicle A is shared by any of the m misbehaving vehicles evicted during one time interval. Therefore, we have:

$$P(k) = \binom{m}{k}\left(\frac{n}{N}\right)^{k}\left(1 - \frac{n}{N}\right)^{m-k}, \qquad (16.19)$$

where

$$\binom{m}{k} = \frac{m!}{k!(m-k!)}.$$

For $t \geq 1$ and $1 \leq k \leq b - 2$ we have:

$$S(k, t) = S(k+1, t-1)P(0) + \sum_{i=0}^{k-1} S(i, t-1)P(k-i). \qquad (16.20)$$

The first term on the right-hand side of Equation 16.20 represents the case where the vehicle's black mark was $(k + 1)$ in the previous time period and the vehicle has no certificate request during the current time period, and as a result, the vehicle's black mark is decremented by 1 in the current time period to reach k. The second term on the right-hand side of Equation 16.20 represents the case when the black mark value was any integer i between 0 and $k - 1$ in the previous time period and there are $(k - i)$ new rekey requests from the vehicle in the current time period, causing the vehicle's black mark to be incremented by $(k - i)$ in the current time period.

If a vehicle's black mark was k in the previous time period, it will not become k in the current time period. In this case, the black mark will become higher than k if there is one or more rekey requests from the vehicle in the current time period or lower than k if the vehicle has no rekey request in the current time period.

Since a vehicle's black mark will not be decremented once it reaches the certificate quota q, we have:

$$S(q-1, t) = \sum_{i=0}^{q-2} S(i, t-1)P(q-1-i) \qquad (16.21)$$

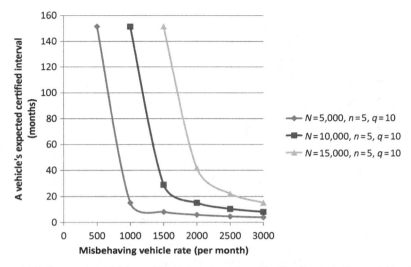

Figure 16.7. Innocent vehicle's certified interval produced by the Dynamic Reward Algorithm

and

$$S(q,t) = 1 - \sum_{i=0}^{q-1} S(i,t). \tag{16.22}$$

The mean value of a vehicle's certified interval is given by:

$$\sum_{t=1}^{\infty} t(S(q,t) - S(q,t-1))^t.$$

Assuming time is measured in months, Figure 16.7 shows sample numerical results for a vehicle's certified intervals as a function of the misbehaving vehicle rate per month. The results in this figure assumed $b = 10$ and $n = 5$. Comparing Figure 16.7 with Figure 16.3, we observe that the Dynamic Reward Algorithm can significantly improve an innocent vehicle's certified time interval. For example, it extends an innocent vehicle's certified interval from 6 months to approximately 151 months for $N = 10,000$ and $q = 10$ when $m = 1000$.

16.6   PRIVACY IN LOW VEHICLE DENSITY AREAS

16.6.1   The Problem

Consider a Shared Certificate Pool that contains N certificates for use by V vehicles. Now consider a specific geographical area, such as a parking lot or a road intersection, where an arbitrary vehicle A and v other vehicles are present where $v < V$. The level of privacy vehicle A experiences in this geographical

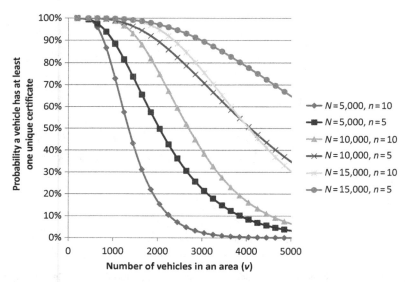

Figure 16.8. Probability that a vehicle has at least one unique certificate among $(v + 1)$ vehicles

area depends on how many of the other v vehicles currently in this area share the certificates assigned to vehicle A.

If vehicle A has a certificate that is not assigned to any of the other v vehicles, this locally unique certificate can be used to track vehicle A inside this geographic area.

The probability that vehicle A has at least one unique certificate that is not assigned to any of the other v vehicles, denoted by $P_{1-\text{unique}}(N, n, v)$, can be estimated as follows [WVZM07]:

$$P_{1-\text{unique}}(N, n, v) = 1 - \left(1 - \left(1 - \frac{n}{N}\right)\right)^n = 1 - \left(1 - (1 - \rho)^v\right)^n. \qquad (16.23)$$

Figure 16.8 shows the numerical samples of probability $P_{1-\text{unique}}(N, n, v)$.

Probability $P_{1-\text{unique}}(N, n, v)$ will decrease (and hence the level of privacy will increase) when:

- v increases, that is, when there are more vehicles in the area, or
- N reduces, that is, when the size of the Shared Certificate Pool is smaller, or
- n increases, that is, when more certificates are assigned to each vehicle.

Unfortunately, reducing N or increasing n will also increase collateral damages when a certificate is revoked because each certificate will be shared by more vehicles on average.

Next, we consider the probability $P_{n-\text{unique}}(N, n, v)$ that all of vehicle A's n certificates are unique among the other v vehicles. The probability that any given certificate of vehicle A is not assigned to any of the other v vehicles is:

$$\left(1 - \frac{n}{N}\right)^{v}.$$

Therefore we have:

$$P_{n-\text{unique}}(N, n, v) = \left(1 - \frac{n}{N}\right)^{v \cdot n} = (1 - \rho)^{v \cdot n}. \qquad (16.24)$$

In fact, $P_{n-\text{unique}}(N, n, v)$ is the binomial probability for 0 successes in $v \cdot n$ trials with a success probability of ρ for each trial. Here, each trial is to check whether any of the $v \cdot n$ certificates assigned to the other v vehicles is also shared by vehicle A.

Figure 16.9 presents numerical samples of $P_{1-\text{unique}}(N, n, v)$ as a function of v. The results suggest the following:

- In a lightly populated area (e.g., with a few tens of vehicles), it is highly likely that all the certificates of each vehicle are unique within this area. For example, when $N = 10,000$ and $n = 5$, the probability that a vehicle's certificates are all unique in an area of $v = 50$ other vehicles is 0.88. This probability increases to over 0.90 in an area with $v = 25$ other vehicles.

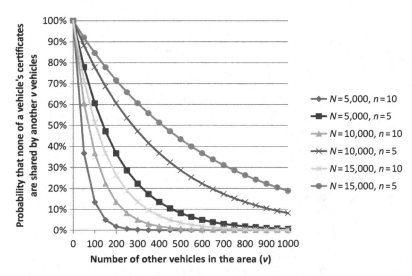

Figure 16.9. Probability that all the certificates of a vehicle are unique among $(v + 1)$ vehicles

This means that it will be fairly easy for an adversary to track a vehicle in such areas.

- Probability $P_{1-\text{unique}}(N, n, v)$ decreases sharply when ρ increases, that is, when either the number n of certificates assigned to each vehicle is increased or the size N of the Shared Certificate Pool is decreased. However, as discussed before, increasing ρ will also increase the collateral damages when certificates are revoked.

16.6.2   The Blend-In Algorithm to Improve Privacy

We show that privacy in low vehicle density areas can be improved significantly if vehicles use smart algorithms to decide which of its n certificates to use.

An algorithm, called the Blend-In Algorithm, is presented in [BeZP09]. It improves privacy in low vehicle density based on the following observation: although there is a high likelihood for a vehicle to have unique certificates in a lightly populated area, it is also highly likely that at least one of the vehicle's certificates will be shared by the other vehicles even when there are not so many other vehicles. More precisely, the probability p_1 that a vehicle has at least one certificate that is shared by at least one of the other v vehicles is given by:

$$p_1 = 1 - \left(1 - \frac{n}{N}\right)^{nv}. \tag{16.25}$$

Numerical samples of p_1 are shown in Figure 16.10.

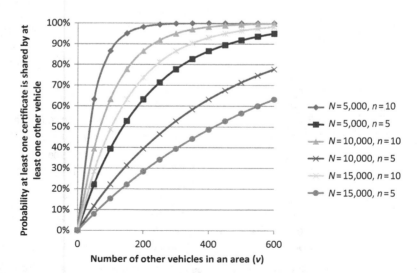

Figure 16.10. Probability that a vehicle shares at least one certificate with other vehicles

Therefore, the idea to improve privacy in lightly populated areas is as follows. A vehicle monitors the certificates being used by neighboring vehicles. If the vehicle discovers that it has a certificate that is also being used by other vehicles, this vehicle will also select this same certificate to sign its own messages. This allows a vehicle to "blend in" to its environment. This method, which we refer to as the Blend-In Algorithm, is first described in [BeZP09].

Each vehicle runs the Blend-In Algorithm autonomously. The basic form of the Blend-In Algorithm works as follows:

- Each vehicle A monitors the certificates used in the messages it has recently received from other vehicles. When vehicle A finds a certificate used in a received message, we say it heard this certificate.
- When vehicle A has a message to send, and none of its n certificates have been heard from other vehicles, it will pick one of its n certificates uniformly at random to sign its message.
- Suppose now that vehicle A has heard one of its certificates from the other vehicles, vehicle A will mark this certificate as SHARED to indicate that it knows that this certificate is shared by at least one other vehicle.
- If vehicle A has further messages to send, it picks a certificate uniformly at random from the set of its certificates that are marked as SHARED.

We use a simplified case to illustrate that if a vehicle A has at least one certificate that are shared by some other vehicles, then vehicle A will be able to quickly converge to using a certificate that is also being used by some other vehicles.

Let us say vehicle A has exactly one certificate C in common with some other vehicles. Let's further assume that vehicle A shares this certificate C with exactly one other vehicle B. Each vehicle has n certificates to use to sign the messages it sends. Let's say that each time a vehicle picks one of its certificates to use to sign outgoing messages, it will use this certificate for T time units before changing to another certificate. Then, within each T consecutive time units, vehicle A will change to a new certificate and vehicle B will also switch to a new certificate. We refer to each T consecutive time units as a certificate-selection round. According to the Blend-In Algorithm, as soon as vehicle A or vehicle B begins to use certificate C, the other vehicle will hear this certificate and will also start to use it. From this point on, the two vehicles will both use certificate C.

Let α be the probability that neither vehicle A nor vehicle B picks certificate C to use during a certificate-selection round. The probability that a vehicle does not pick certificate C in a certificate-selection round is $1 - 1/n$. Therefore, we have,

$$\alpha = \left(1 - \frac{1}{n}\right)^2. \tag{16.26}$$

Let S be the number of certificate-selection rounds it takes for both vehicles A and B to start using certificate C. Then the probability that it will take at least x rounds for the two vehicles to converge to using the shared certificate C (i.e., the probability of $S \geq x$,) is α^x. Therefore, we have:

$$P_r(S < r) = 1 - \alpha^x = 1 - \left(1 - \frac{1}{n}\right)^{2x}. \tag{16.27}$$

Given that vehicle A shares exactly one certificate C with exactly v other vehicles, we refer to the probability that vehicle A and at least one other vehicle will start to use this shared certificate C within x certificate-selection rounds as the convergence probability and denote it by $P_r(S < x)$. $P_r(S < x)$ is given by:

$$P_r(S < x) = 1 - \alpha^x = 1 - \left(\left(1 - \frac{1}{n}\right)^v\right)^x = 1 - \left(1 - \frac{1}{n}\right)^{vx}. \tag{16.28}$$

Figure 16.11 shows numerical samples of the convergence probability for $v = 2$, 4, and 8. We observe from the results that a vehicle will quickly converge to using a certificate that is also being used by at least one other vehicle. Furthermore, if a vehicle shares at least one certificate with just a very small number of other vehicles, the time it takes for this vehicle to converge to using a certificate that is used by at least one other vehicle will decrease sharply.

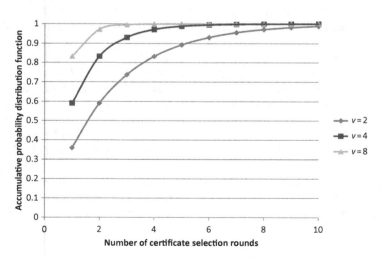

Figure 16.11. Convergence probability of the Blend-In Algorithm

REFERENCES

[BeZP09] E. van den Berg, T. Zhang, and S. Pietrowicz: "Blend-In: A Privacy-Enhancing Certificate-Selection Method for Vehicular Communication," IEEE Transactions on Vehicular Technology, vol. 58, no. 9, 2009.

[ReKW06] E. Rescorla, J. Kelsey, and D. Whiting: "Vehicle Safety Communications Consortium Report to National Highway Traffic Safety Administration of the United States Department of Transportation," Appendix H: WAVE/DSRC Security, 2006.

[WPBG09] R. White, S. Pietrowicz, E. van den Berg, G. Di Crescenzo, D. Mok, R. Ferrer, T. Zhang, and H. Shim: "Privacy and Scalability Analysis of Vehicular Combinatorial Certificate Schemes," IEEE Consumer Communications and Networking Conference 2009 (CCNC 2009), Las Vegas, Nevada, USA, 2009.

[WVZM07] R. White, E. Vandenberg, T. Zhang, D. Mok, R. Ferrer, and G. D. Crescenzo: "Vehicle Segment Certificate Management Scalability Analysis, VII Security Work Order Part I Deliverables 1.2," 2007.

[ZCPV07] T. Zhang, G. D. Crescenzo, S. Pietrowicz, E. Vandenberg, and R. White: "Vehicle Segment Certificate Management, VII Security Work Order Part I Deliverables 2.1 Anonymous Key and Certificate Management Process," 2007.

17

PRIVACY PROTECTION WITH SHORT-LIVED UNIQUE CERTIFICATES

17.1 SHORT-LIVED UNIQUE CERTIFICATES

A vehicle can have multiple unique certificates and chooses a different one to use depending on time, location, or the message to send. Privacy is achieved if adversaries cannot link different certificates to the same certificate owner and each certificate does not provide information that an adversary can use to uniquely identify the certificate owner.

Compared with shared certificates, a main advantage of short-lived unique certificates is that revoking a vehicle's certificates does not affect certificates assigned to other vehicles. However, using short-lived certificates requires each vehicle to use a very large number of certificates. This has direct impacts on public key infrastructure (PKI) system design and operation, including on how certificates can be assigned and revoked, the size and processing time of certificate revocation lists (CRLs), the ability to detect misbehaving vehicles, the ability to protect privacy against certificate authority (CA) operators, and system scalability.

This chapter describes short-lived unique certificate schemes, their main characteristics, and how to improve their scalability and performance.

Vehicle Safety Communications: Protocols, Security, and Privacy, First Edition. Luca Delgrossi and Tao Zhang.
© 2012 John Wiley & Sons, Inc. Published 2012 by John Wiley & Sons, Inc.

17.2 THE BASIC SHORT-LIVED CERTIFICATE SCHEME

The basic form of a short-lived unique certificate scheme works as follows [CrVZ09] [PCZW08] [ReKW06] [ZCPV07]:

- *Certificate Generation*: Each vehicle creates its private and public keys and requests the CA to issue certificates for its public keys.
- *Certificate Assignment*: Each vehicle is assigned a large number of unique certificates, each with a short lifetime.
- *Certificate Revocation and Replacement*: When the CA believes that a vehicle has been misbehaving, the CA will revoke all the certificates allocated to the vehicle, place the identifiers of the revoked certificates on the CRL, and make the updated CRL available to all vehicles. When a vehicle is about to run out of valid certificates, the vehicle will request a batch of new certificates from the CA.
- *CRL Processing on a Vehicle*: Each vehicle verifies whether a certificate has been revoked by checking whether the certificate's identifier is on the CRL.

The level of privacy achievable with short-lived unique certificates depends on several factors. The first factor is the level of difficulty for an adversary to associate each individual certificate to a particular vehicle. The second factor is the certificate lifetime. An adversary can use a certificate to track a vehicle as long as the certificate is being used. The longer the adversary can track a vehicle, the easier it will be for the adversary to uniquely identify the vehicle. The third main factor impacting the privacy level of short-lived certificate schemes is the level of difficulty for an adversary to link multiple certificates to the same certificate owner.

Consider a population of V vehicles. Each certificate is unique to one of the V vehicles. If certificates contain no plaintext vehicle-identifying information, then knowing that a certificate belongs to only one vehicle alone does not help an adversary to determine which particular vehicle is the subject of the certificate. Under these assumptions, the anonymity set will be the entire set of V vehicles. That is, the maximum level of anonymity will be:

$$H_{max} = \log_2 V. \tag{17.1}$$

While a vehicle uses the same certificate, the vehicle will be clearly distinguishable from other vehicles and traceable based on the certificate. The certificate lifetime needs to be designed to meet two main requirements:

- If short-term message linkability is allowed for up to τ time units, the certificate lifetime should not exceed τ, except randomization of the start and the end times of the certificate validity time period.

- Tracking a vehicle for the certificate lifetime should not provide significant advantage to an adversary for identifying which particular vehicle is the subject of the certificate.

Now let us consider long-term certificate unlinkability for time threshold τ. Consider two messages carrying certificates C_1 and C_2, respectively, which are captured by an adversary at least τ time units apart. The probability that certificate C_2 belongs to the same vehicle that owns certificate C_1 will be $1/V$. The adversary, however, could do better. Let's say certificates C_1 and C_2 belong to the same vehicle. If the adversary observes a stream of messages carrying certificates C_1 followed immediately a stream of messages carrying certificate C_2, then the adversary can tell with a high level of certainty that the same vehicle has just switched from using certificate C_1 to using certificate C_2. If the adversary can capture such consecutive certificate changes, it will have a high chance to link multiple certificates to the same vehicle. The adversary may, for example, deploy observation points along the roads to capture consecutive certificate changes made by a vehicle.

Next, we derive the probability of capturing g consecutive certificate changes made by a vehicle using g observation points. Let r be the radius of the radio coverage area of each observation point. Assume that certificates assigned to each vehicle has disjoint lifetimes and that the lifetime of every certificate is t_L. Then, the probability p_1 that an observation point can capture a certificate change is the probability that a vehicle will make a certificate change within the radio coverage area of the observation point, which can be estimated as [CrVZ09]:

$$p_1 = \frac{2r}{s \cdot t_L}. \tag{17.2}$$

The probability p_g that every one of the g observation points captures a certificate change is therefore:

$$p_g = \left(\frac{2r}{s \cdot t_L} \right)^g. \tag{17.3}$$

Figure 17.1 shows numerical samples of the certificate linkability probability. The results show that it is highly unlikely that an adversary can use multiple observation points to capture a long sequence of certificate changes made by a vehicle.

Based on the long-term unlinkability definition in the previous chapter, if each vehicle changes its certificate every t_L time units, the level of long-term message unlinkability for time threshold τ can be expressed as:

$$U(t) = \begin{cases} 0 \text{ for } \tau < T \\ \max \left\{ \dfrac{1}{V}, 1 - \left(\dfrac{2r}{s \cdot t_L} \right)^{\frac{\tau}{t_L}} \right\} \text{ for } \tau \ge T \end{cases}. \tag{17.4}$$

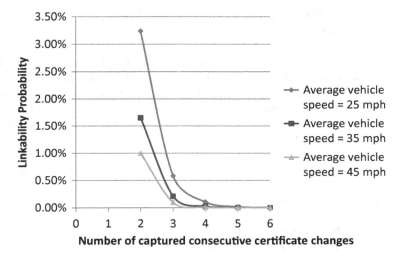

Figure 17.1. Certificate linkability probability

17.3 THE PROBLEM OF LARGE CRL

When each vehicle is assigned a large number of short-lived certificates, revoking a vehicle means that the CRL needs to contain information about all the nonexpired certificates assigned to the vehicle. This can make the CRL prohibitively large for distribution over wireless networks to a large population of vehicles. A large CRL also means that a vehicle has to consume heavy computational resources to search through the CRL to verify whether a certificate is on the CRL. This could result in excessive delays in signature verification.

Each CRL consists of a set of fixed-size data elements and the identifiers of the revoked certificates. The identifier for each certificate is 10 bytes long for the certificates defined in the Institute of Electrical and Electronics Engineers (IEEE) 1609.2 standard [IEEE11]. Therefore, the size of a CRL can be $L_{\text{fixed}} + 10 \sum_{i=1}^{m} n_i$ byte long, where L_{fixed} is the total size of the fixed-length fields in the CRL, m is the number of revoked vehicles, and n_i is the number of nonexpired certificates assigned to the ith revoked vehicle. As the number of revoked vehicles increases, the CRL size will be dominated by $10\sum_{i=1}^{m} n_i$. If Φ is the average number of nonexpired certificates on each revoked vehicle, then the expected size of the CRL is $10\, m \cdot \Phi$.

Let us use the following scenario to illustrate the potentially large size of the CRL: the certificate lifetime is 5 minutes; each vehicle is assigned enough certificates to last for a year each time certificates are loaded onto the vehicle; 500 vehicles have been revoked; and half of the certificates allocated to each revoked vehicle have not expired when these vehicles are revoked. The size of the CRL will be $10\, m \cdot \Phi = 260{,}280{,}000$ bytes or approximately 263 MB.

17.4 ANONYMOUSLY LINKED CERTIFICATES TO REDUCE CRL SIZE

One way to reduce the size of the CRL for short-lived unique certificates is anonymously linked certificates or linked certificates for simplicity [ReKW06]. The idea is to place specially designed tags on the certificates to anonymously link all the certificates assigned to the same vehicle. The certificate tags will be random values to any observer unless the observer knows the secrets used to create the tags. That is, the tags should not help adversaries to link multiple certificates to the same vehicle. The certificate tags for each vehicle are generated using a small amount of secret information referred to as certificate tag seeds or tag seeds for simplicity. Anyone who knows the certificate tag seed for a vehicle can use the seed to reproduce the tags on the certificates allocated to the vehicle and can therefore also tell which certificates have been allocated to the vehicle. For vehicles that have not been revoked, their tag seeds will be kept secret by the CA or other trusted escrows.

To revoke the certificates assigned to a vehicle, the CRL only needs to contain the certificate tag seeds for the vehicle, rather than all the certificates assigned to the vehicle.

17.4.1 Certificate Tags

Here, we describe one way to construct the certificate tags. First, generate a random value s_v and use it as the certificate tag seed for vehicle v. Then, the certificate tag $Tag(v, i)$ for certificate i assigned to vehicle v is computed as in Equation 17.5, where $H()$ is a cryptographic hash function and i can be any form of an identifier of certificate C_i such as the certificate serial number or the hash value of the certificate content without the certificate tag:

$$Tag(v, i) = H(s_v, i). \tag{17.5}$$

Given tag seed s_v and the identifier i of certificate C_i, one can easily compute the tag on the certificate. However, knowing only the value of the tag will not allow one to derive the tag seed because cryptographic hash functions are one-way functions.

The certificate tags must be semantically secure. This means that the certificate tags should not help adversaries to identify the subject of a certificate or to link certificates to the same vehicle.

For the certificate tags to be unlinkable, they cannot form a totally ordered set with respect to the tag seeds or certificates [CrZh10]. The certificate tags for all vehicles form a totally ordered set with respect to the tag seeds when $Tag(s_1, i) \leq Tag(s_2, i)$ holds for every pair of tag seeds s_1 and s_2 that satisfy $s_1 < s_2$. The tags form a totally ordered set with respect to the certificates when $Tag(s_k, i) \leq Tag(s_k, j)$ holds for every pair of certificates C_i and C_j that satisfy $i < j$. As shown in [CrZh10], if the certificate tags form a totally ordered set, an adversary can use polynomial time algorithms to quickly find out the tag seed for each vehicle or to link multiple certificates to the same vehicle.

Therefore, only functions that do not generate totally ordered sets of tags with respect to the tag seeds and the certificates can be used to create anonymous certificate tags.

17.4.2 CRL Processing by Vehicles

Each vehicle uses the CRL to verify whether a certificate has been revoked. Suppose that the CRL contains the tag seeds for m revoked vehicles. To verify whether a certificate C is on the CRL, the vehicle must perform the following steps:

- For each of the m revoked vehicles on the CRL, generate the tags for all the certificates assigned to this vehicle that have not yet expired. These tags are called the revoked tags.
- Search through the revoked tags to check whether certificate C's tag is among them.

Since the tags cannot form a totally ordered set with respect to the tag seeds as discussed previously, each vehicle will have to compute all revoked tags before it can verify whether a particular certificate tag is on the CRL. This means that each vehicle will have to perform $m \cdot \Phi$ hashing operations on average to compute the revoked tags if each revoked vehicle has, on average, Φ nonexpired certificates when the vehicle is revoked.

After the vehicle computes all the revoked tags, it may order the tags in increasing or decreasing order of the tag values. This would take $O(m \cdot \Phi \cdot \log_2(m \cdot \Phi))$ comparison operations. The vehicle can then perform binary searches through this ordered list of the revoked tags to verify whether a given tag is on the list. This binary search would take $O(\log_2(m \cdot \Phi))$ comparisons on average.

As discussed previously, $m \cdot \Phi$ can be a very large number that can easily reach tens of millions in a large consumer vehicle network. Performing such a large number of hashing and comparison operations each time a CRL is received could consume a large amount of the computing resources on a vehicle and causes excessive delays.

A linked certificate scheme reduces the size of the CRL at the price of increasing the processing time each vehicle takes to process each CRL. A comparison of the CRL processing between linked certificates and shared certificate schemes are shown in Table 17.1.

For shared certificates and unlinked short-lived certificates, the identifiers of the revoked certificates on a CRL can be ordered by the CA before they are placed on the CRL. A vehicle therefore does need to order the list of revoked certificates.

Table 17.1. CRL processing on the vehicle by different privacy-preserving certificate schemes

	Shared Certificate	Anonymously Linked Short-Lived Certificates	Unlinked Short-Lived Certificates
Generate certificate tags each time a new CRL is received	0	$O(m \cdot \Phi)$ hashing operations	0
Create ordered list for binary search of CRL membership each time a new CRL is received	0	$O(m \cdot \Phi \log(m \cdot \Phi))$ comparison operations	0
Search to verify whether a certificate is on the CRL for every signature to be verified	$O(\log(\min\{m \cdot n, N\}))$ comparison operations	$O(\log(m \cdot \Phi))$ comparison operations	$O(m \cdot \Phi)$ comparison operations

For shared certificates, there will be no more than the minimum of $(m \cdot n)$ or N revoked certificates on a CRL when m vehicles are revoked, where N is the size of the shared certificate pool and n is the number of certificates allocated to each vehicle.

The extra number of comparison operations a linked certificate scheme would have to perform to search through the ordered list of entries on a CRL, compared with a shared certificate scheme, can be estimated as follows:

$$\begin{cases} O(log_2(m \cdot \Phi)) - O(log_2(m \cdot n)) = O\left(log_2\left(\dfrac{\Phi}{n}\right)\right); \text{if } m \cdot n \ll N \\ O(log_2(m \cdot \Phi)) - O(log_2(N)) = O\left(log_2\left(\dfrac{m\Phi}{N}\right)\right); \text{if } m \cdot n > N \end{cases}. \quad (17.6)$$

More precise estimates can be obtained. However, they do not seem to change the above estimate significantly and will not materially change the following discussions.

Now, let us consider a numerical example:

- For the linked certificate scheme, assume that each revoked vehicle still has enough nonexpired certificates for a 60-day period on average and each certificate is valid for 5 minutes.

Table 17.2. Examples of CRL processing complexities

	Shared Certificate Schemes	Anonymously Linked Short-Lived Certificates	Unlinked Short-Lived Certificates
Generate certificate tags each time a new CRL is received	0	$O(2^{23})$ hashing operations	0
Create ordered list for binary search of CRL membership each time a new CRL is received	0	$O(2^{28})$ comparison operations	0
Search to verify whether a certificate is on the CRL for every signature to be verified	$O(11)$ comparison operations	$O(23)$ comparison operations	$O(2^{23})$ comparison operations

- For the shared certificate scheme, assume that the size of the shared certificate pool is $N = 10,000$ and each vehicle is assigned with $n = 5$ certificates.
- Assume that $m = 500$ misbehaving vehicles are represented on the current CRL.

Table 17.2 summarizes the computational complexities of the shared certificate, linked short-lived certificate, and unlinked short-lived certificate schemes for the above example. In Table 17.2, $O(x)$ indicates that the number of operations is a constant factor times x.

17.4.3 Backward Unlinkability

With the basic linked certificate method, anyone who knows the tag seed for a revoked vehicle will be able to compute the tags on all the certificates assigned to the vehicle, including those used by the vehicle before it is revoked. This poses a privacy risk for vehicles that are wrongly revoked. After a vehicle is wrongly revoked, adversaries could identify all the certificates the vehicle has used and may use this knowledge to picture together the vehicle's historical movement patterns.

It is therefore desirable for the CA to have the ability to release only sufficient information for everyone to identify only the revoked certificates that are still valid when and after a vehicle is revoked. This is referred to as backward unlinkability.

With anonymously linked certificates, backward unlinkability can be achieved using a tag seed $s_{v,i}$ to generate tag $Tag(v, i)$ for certificate C_i assigned to a vehicle v in such a way that $s_{v,i}$ is used to generate $s_{v,i+1}$, but $s_{v,i}$ cannot be derived from $s_{v,i+1}$. An example to construct such a series of tag seeds is shown in Equation 17.7, where $s_{v,1}$ is a randomly selected initial value:

$$s_{v,i} = \begin{cases} s_{v,1}, \, i = 1 \\ H(s_{v,i-1}), \, i \geq 2 \end{cases}. \tag{17.7}$$

The certificate tags for each certificate C_i assigned to vehicle v can be computed as:

$$Tag(v, i) = H(s_{v,i}, i). \tag{17.8}$$

If the CA revokes a vehicle v and wants only its certificates with identifiers i and higher be identifiable, the CA can place the tag seed $s_{v,i}$ on the CRL. This allows anyone with the CRL to reproduce the tags for certificates C_{i+y} for $y \geq 0$; but not for certificates $C_1, C_2 \dots$, and C_{i-1}. When the CA wishes for everyone to reproduce the tags on all the certificates allocated to this vehicle, the CA will place $s_{v,1}$ on the CRL.

17.5 REDUCE CRL SEARCH TIME

A method is proposed in [CrZh10] for improving the CRL search time for linked certificates. With this approach, the CA assigns certificates to the vehicles as follows. It uses a common certificate validity timing plan for all vehicles. Specifically, the CA divides time into intervals defined by their start and end times: (t_0, t_1), (t_1, t_2), (t_2, t_3), $\dots$, (t_k, t_{k+1}), and so on. For each time interval, the CA assigns a different certificate valid only for this time interval to each vehicle. Small randomized overlaps between consecutive validity time intervals can be used to make it difficult for adversaries to capture certificate changes made by each vehicle to increase long-term unlinkability. The length of the validity time periods is configurable to balance short-term linkability and long-term unlinkability.

The tag for vehicle v's certificate for time interval i is $Tag(v, i) = H(s_v, i)$, where s_v is vehicle v's tag seed. When vehicle v is revoked, the CA will place vehicle v's certificate tag seed s_v on the CRL.

To verify whether a certificate C for time interval i is on the CRL, a vehicle takes the following steps:

- *Step 1*: For each revoked vehicle v identified by its certificate tag seed s_v on the CRL, compute the tag $Tag(v, i) = H(s_v, i)$ for vehicle v's certificate for time interval i. This step will generate the tags for all the revoked certificates for time interval i. We refer to these certificate tags as the revoked certificate tags for time interval i.
- *Step 2*: Sort the revoked certificate tags for time interval i in increasing or decreasing order.

- *Step 3:* Perform a binary search through the revoked certificate tags for interval i to see if certificate C's tag is among them. If the answer is YES, certificate has been revoked.

Before the next time interval $i + 1$ starts, each vehicle computes the revoked certificate tags for the next time interval. This will allow the vehicle to check whether a certificate for time interval $i + 1$ is on the CRL.

The m revoked tags for each time interval can be ordered. This can be performed in $O(m \cdot \log_2 m)$ comparisons, where m is the number of revoked vehicles on the CRL. The vehicle can then perform a binary search through this ordered list of revoked tags to verify whether a certificate is on the CRL. This search will take $O(\log_2 m)$ comparisons.

In the ideal case, during time interval i, a vehicle should only receive messages signed with certificates assigned for this time interval. In reality, vehicles' clocks may not be completely synchronized. One vehicle's clock may still be in time interval i while another vehicle's clock has advanced to time interval $i + 1$. Consequently, during time interval i, some vehicles may use certificates allocated for time intervals surrounding the current time interval i. Therefore, to check whether a certificate received during time interval i is revoked, each vehicle may need to compute the revoked certificate tags for not only the current time interval i but also for several surrounding time intervals such as time intervals $(i - 1)$ and $(i + 1)$. Even in these cases, the time a vehicle has to spend to search the ordered list of revoked certificates remains $O(\log_2 m)$.

17.6   UNLINKED SHORT-LIVED CERTIFICATES

An unlinked short-lived certificate scheme has been proposed to address the issue related to distributing and processing the CRL by eliminating the need for the CRL [PCZW08].

The basic principles of unlinked short-lived certificates are:

- Each vehicle is assigned a large number of unique certificates that are not linked with each other in any fashion.
- Each certificate has a very short lifetime and cannot be used after its lifetime expires.
- Certificates are not revoked. Instead, misbehaving vehicles' privileges for receiving new certificates will be revoked. As a result, there will be no need for distributing, processing, and maintaining CRLs.

The advantage of this approach is that it avoids the difficulty of distributing CRLs and the time each vehicle has to spend on processing CRLs. The drawback, however, is that a misbehaving vehicle will be able to mount security attacks until all of its previously assigned certificates expire. This risk can be mitigated by allocating a small number of certificates to each vehicle at a time.

However, doing so will force vehicles to contact the CA more frequently to acquire new certificates.

17.7 REDUCE THE VOLUME OF CERTIFICATE REQUEST AND RESPONSE MESSAGES

Distributing a large number of short-lived certificates to each vehicle can consume an excessive amount of wireless resources and make the vehicle communication systems heavily dependent on the availability and the capacity of roadside networks. Furthermore, when vehicles generate their own private–public key pairs and send their public keys to the CA to obtain certificates, a certificate request message may have to contain a large number of public keys if the vehicle needs a large number of short-lived certificates.

One way to reduce the volume of certificate request and response messages is to load a very large number of certificates, with their associated private–public key pairs, onto each vehicle in encrypted forms. These certificates can be sufficient for the vehicle to use for multiple years until, for example, the driver is due to renew her driver's license or undergo a mandatory vehicle safety inspection, where the vehicle can be loaded with more encrypted certificates.

The vehicle can acquire a decryption key from the CA to unlock one batch of these encrypted certificates at a time. The number of certificates unlocked by each decryption key should be sufficient for the vehicle to use until the next time the vehicle has a chance to contact the CA. A decryption key for unlocking a batch of encrypted certificates will be significantly smaller in size than a large number of certificates.

17.8 DETERMINE THE NUMBER OF CERTIFICATES FOR EACH VEHICLE

The number of certificates allocated to each vehicle directly affects how long the vehicle will have valid certificates to sign safety application messages, and consequently, how frequently the vehicle has to contact the CA to acquire new certificates.

For instance, in a scenario where vehicles communicate to the CA through a cellular network, this impacts how frequently a vehicle needs to use cellular services to obtain new certificates and how much extra cellular traffic is generated. In a scenario where vehicles communicate with the CA through dedicated short-range communications (DSRC), this impacts how many DSRC roadside units (RSUs) are required and at what distance these RSUs should be located.

Let us say each time a vehicle is assigned a batch of n certificates, each certificate has a lifetime of t time units, and the lifetimes of different certificates

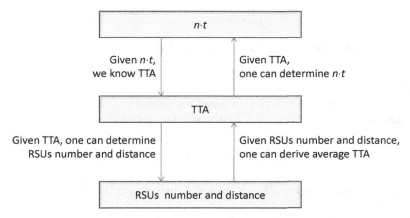

Figure 17.2. TTA, number of certificates for a vehicle, and number of RSUs

are disjoint. Then, the vehicle will need to contact the CA to acquire a new batch of certificates at least once every $n \cdot t$ time units.

To provide a framework for assessing the number of certificates to be loaded to a vehicle each time, we introduce the concept of time to access (TTA). TTA is defined as the time interval during which a vehicle can access the CA at least once. Figure 17.2 illustrates the relations between TTA, the number n of certificates to be assigned to each vehicle each time, the lifetime time t of each certificate, and the number of RSUs required to meet the TTA (assuming that vehicles interact with the CA through DSRC).

Given n and t, we can derive the required TTA value. Given this target TTA value, we can derive the number of required RSUs and at what distance these RSUs should be located so that each vehicle will encounter an RSU at least once every TTA interval with high probability.

On the other hand, if we know how many RSUs we can afford to deploy and where they can be deployed, we can estimate the minimal, maximal, and average TTA values by estimating the time it takes for each vehicle to encounter another RSU after it leaves a previous RSU. We can then determine how many certificates should be given to a vehicle each time so that the vehicle will have enough certificates to use for at least a target TTA interval until it has its next change to contact the CA.

Next, we show a simplified estimate of the TTA based on results in [CrVZ09]. For illustration purpose only, we consider the case where DSRC is used for vehicles to interact with the CA. We assume the RSU locations follow a Poisson process with a density of λ RSUs per mile (more sophisticated models can be found in [CrVZ09]). Under this assumption, the number of RSUs in any stretch of road traversed by a vehicle during time interval $(0, t)$, denoted by $N(t)$, has a Poisson distribution. Let $X_1, X_2, \ldots, X_n$ be the locations of the RSUs. The probability that the distance between two RSUs, $|X_2 - X_1|$, is greater than x is given by:

$$p(|x_2 - x_1| > x) = e^{-\lambda x}. \tag{17.9}$$

On average, the distance between any two RSUs is:

$$E(|X_2 - X_1|) = 1 / \lambda. \tag{17.10}$$

If we assume vehicles travel at a constant velocity v, the distance $|X_2 - X_1|$ translates directly into the TTA between RSU encounters:

$$P(TTA > t) = P\left(\frac{|X_2 - X_1|}{v} > t\right) = P(|X_2 - X_1| > vt) = e^{-\lambda vt}. \tag{17.11}$$

The average time a vehicle has to travel between two RSUs is:

$$E(TTA) = \frac{1}{\lambda v}. \tag{17.12}$$

Figure 17.3 illustrates the average TTA as a function of the RSU density λ for various average speed speeds.

In the state of New Jersey in the United States, there are approximately 39,000 mi of public roads. This means that, on average, approximately 312 RSUs deployed according to a Poisson process along the roads in New Jersey will be able to meet a TTA of 6.25 hours of travel time when the average travel speed is 20 mph. Here, travel time refers to the accumulative time each vehicle is driven on the road. Therefore, if each vehicle is driven for 1 hour on average every day, 6.25 hours of travel time would be equivalent to 6.25 calendar days.

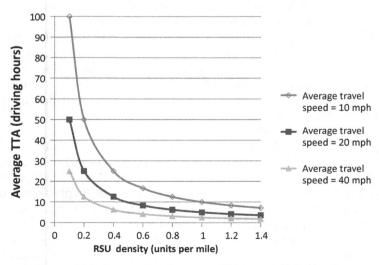

Figure 17.3. Average time to access (TTA) versus RSU density

The TTA will drop to 3.13 hours of travel time when the average travel speed doubles to 40 mph while the number of RSUs remains the same. These results suggest that a relatively small number of RSUs will be sufficient to allow each vehicle to be able to contact the CA once in a few days in New Jersey.

REFERENCES

[CrVZ09] G. D. Crescenzo, E. Vandenberg, and T. Zhang: "Vehicle Safety Communications—Applications VSC-A: Analysis of Infrastructure and Communications Requirements for V2V PKI Security Management," 2009.

[CrZh10] G. D. Crescenzo and T. Zhang: "Efficient CRL Search in Vehicular Network PKIs," the 6th ACM Workshop on Digital Identity Management (DIM 2010), Chicago, IL, 2010.

[IEEE11] IEEE 1609.2/D8: "Draft Standard for Wireless Access in Vehicular Environments—Security Services for Applications and Management Messages," 2011.

[PCZW08] S. Pietrowicz, G. D. Crescenzo, T. Zhang, and R. White: "Vehicle Segment Certificate Management Using Short-Lived, Unlinked Certificate Schemes," United States Patent Application No. 20080232595, 2008.

[ReKW06] E. Rescorla, J. Kelsey, and D. Whiting: "Vehicle Safety Communications Consortium Report to the National Highway Traffic Safety Administration of the US Department of Transportation—Appendix H: WAVE/DSRC Security," 2006.

[ZCPV07] T. Zhang, G. D. Crescenzo, S. Pietrowicz, E. Vandenberg, and R. White: "Vehicle Segment Certificate Management, VII Security Work Order Part I Deliverables 2.1 Anonymous Key and Certificate Management Process," 2007.

18

PRIVACY PROTECTION WITH GROUP SIGNATURES

18.1 GROUP SIGNATURES

In this chapter, we provide an overview of group signature schemes and discuss the properties that impact their suitability for large and dynamic consumer vehicle networks.

The group signature concept was first introduced by David Chaum and Eugene van Heyst in 1991 [ChHe91] to allow a member of a group to sign messages on behalf of the group. Signatures generated by the members of a group are called group signatures. Each group signature can be verified as being generated by a member of the group, but will not reveal which particular member is the signature generator. Each group has a group public key that anyone can use to verify group signatures. A group manager is responsible for establishing and maintaining a group, creating and updating the group public key, and making the group public key available to all potential signature verifiers. The group manager will also be able to "open" a signature generated by any group member to reveal the identity of the message signer. However, no one, not even the group manager, should be able to forge the signature of a group member.

The members in a group form an anonymity set. With a properly designed group signature scheme, each group signature should appear to be generated by any group member equally likely. This leads to the full degree of anonymity for a given group size.

Vehicle Safety Communications: Protocols, Security, and Privacy, First Edition. Luca Delgrossi and Tao Zhang.
© 2012 John Wiley & Sons, Inc. Published 2012 by John Wiley & Sons, Inc.

Group signature schemes seek to meet the following basic requirements:

- *Correctness*: Valid signatures by group members should always verify correctly, and invalid signatures should always fail verification.
- *Unforgeability*: Only members of the group can create valid group signatures.
- *Anonymity*: Given a message and its signature, identifying the actual signer is computationally hard except for the actual signer and the group manager.
- *Unlinkability*: Given two messages and their signatures, deciding whether the signatures were generated by the same signer is computationally hard.
- *Traceability*: The group manager can always open a valid signature to identify the actual signer.
- *Exculpability*: Neither a group member nor the group manager can forge a signature of another group member.

Most early group signature schemes use large group public keys, require heavy computations for signature verification that typically grow linearly with the size of the group, and do not support dynamic addition and revocation of group members. Group signature schemes published since 2000 have made significant progress in reducing the sizes of the keys and signatures, improving signature creation and verification speeds, and supporting dynamic addition and revocation of group members [ACJT00] [AtTS02] [CaGr05] [CaLy02] [CaLy02b] [BeMW03] [BeSZ05] [BoBS04] [BoSh04] [Joye09] [KiYu04].

Most state-of-the-art group signature schemes rely on zero-knowledge proof of knowledge mechanisms. Therefore, we will first provide a brief illustration of the zero-knowledge proof of knowledge concept.

18.2 ZERO-KNOWLEDGE PROOF OF KNOWLEDGE

Zero-knowledge proof of knowledge is to prove possession (i.e., knowledge) of a secret without revealing what the secret is. A zero-knowledge proof of knowledge typically needs to satisfy the following requirements:

- *Completeness*: Any true statement can be proven (i.e., can be verified to be true by a verifier).
- *Soundness*: Any false statement cannot be proven (i.e., will not be verified to be true by a verifier).
- *Zero Knowledge*: The proof process should not reveal the secret the prover wishes to prove she possesses.

One frequently used and easy to understand zero-knowledge proof of knowledge is to prove knowledge of a discrete logarithm, that is, knowledge of a secret $x = \log_g A$ modulo n, where g, A, and n are positive integers. Proofs of

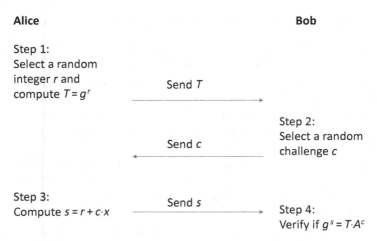

Figure 18.1. Example of zero-knowledge proof of knowledge

knowledge of discrete logarithms are used, for example, in the well-known ACJT group signature scheme [ACJT00] that we will discuss next in this chapter.

A well-known protocol for implementing zero-knowledge proof of discrete logarithms is the Schnorr protocol [Schn91]. We use this protocol to illustrate the concept of zero-knowledge proof of knowledge. The protocol is implemented over a cyclic group with generator g. Recall that a cyclic group is a group in which any member of the group can be generated by raising the generator g to an integer power.

Let us say that Alice wants to prove to Bob that she knows a secret $x = \log_g A \pmod{n}$, where A, g, and n are public knowledge. A zero-knowledge proof based on the Schnorr protocol [Schn91] is shown in Figure 18.1. It works as follows:

- *Step 1*: Alice selects a secret random integer r and computes $T = g^r$ and sends T to Bob.
- *Step 2*: Bob sends a challenge—a random integer c—back to Alice.
- *Step 3*: Alice computes $s = r + c \cdot x$ and sends s to Bob.
- *Step 4*: Bob verifies whether $g^s = T \cdot A^c$ is true. If this verification is positive, then Bob concludes that Alice knows $x = \log_g A$.

If Alice knows $x = \log_g A$ and has used it to compute s, then we have $g^s = g^{r+c \cdot x}$. Since $T \cdot A^c = g^r \cdot (g^x)^c = g^{r+x \cdot c}$, Bob's verification in Step 4 will be positive. On the other hand, if Alice does not know x, she will not be able to compute the right value of s to make g^s and $T \cdot A^c$ equal to each other. Therefore, Bob could verify whether Alice knows the secret $x = \log_g A$ without having to know the value of x.

The Schnorr protocol illustrated above is a zero-knowledge proof of knowledge of $x = \log_g A$ under several conditions. First, the protocol can prove that Alice indeed knows the secret x only when the verifier is honest. That is, the verifier must honestly follow the protocol to randomly select its challenge c from a large enough space so that it is computationally infeasible for Alice to guess the value of c. It is known that if Alice can guess the value of c the verifier will use next, she can compute T in Step 1 in a way that will ensure successful verification in Step 4 without having to know x. This can be achieved, for example, if Alice computes T as $T = g^s/A^c$ in Step 1, where s is an arbitrary value selected by Alice. Then in Step 3, Alice will respond to Bob's challenge by returning s to Bob.

Zero-knowledge proof of knowledge protocols that rely on honest verifiers are said to follow an honest-verifier model. For example, the popular ACJT group signature scheme [ACJT00] follows the honest-verifier model.

Second, Alice must select the secrets r and x randomly from large enough spaces so that it will be computationally infeasible for the verifiers to guess the values of these secrets.

Third, the modulus n must be carefully selected such that it is computationally infeasible to find the xth root of an arbitrary large integer modulo n for any $x > 2$. In other words, given a large integer A computed as $A = g^x$, it should be computationally infeasible for anyone to compute the xth root of A modulo n for any $x > 2$. One way to meet this requirement is to construct the discrete logarithm problem as a Strong Rivest–Shamir–Adleman (RSA) Problem [BaPf97]. Strong RSA Problems are used in ACJT group scheme to implement zero-knowledge proofs of knowledge.

18.3 THE ACJT GROUP SIGNATURE SCHEME AND ITS EXTENSIONS

The ACJT group signature scheme [ACJT00] is named after the four people who published it: Giuseppe **A**teniese, Jan **C**amenisch, Marc **J**oye, and Gene **T**sudik. It is one of the first group signature schemes that gained wide acceptance. It formed the basis for several subsequent state-of-the-art group signature schemes. Here, we will summarize the ACJT group signature scheme at a high level.

The original ACJT group signature scheme does not support revocation of group members. Several subsequent extensions introduced membership revocation abilities to the ACJT scheme. Therefore, we will also discuss representative extensions to the ACJT scheme to support membership revocation.

18.3.1 The ACJT Group Signature Scheme

The ACJT scheme uses statistical zero-knowledge proofs of knowledge to overcome several limitations of previous group signature schemes. First, it eliminates the dependencies of the group public key and the group signature on the size of the group. Second, it eliminates the need to change the group public key when new members join the group.

The ACJT scheme implements zero-knowledge proof of knowledge based primarily on the Strong RSA Assumption [BaPf97]. The Strong RSA Assumption is the assumption that solving the Strong RSA Problem is hard. The Strong RSA Problem is a generalized version of the RSA Problem discussed in Chapter 14 and can be illustrated as follows: Given a composite integer modulus n, which is the product of two secret large prime numbers, and given an arbitrary large integer C, find u that satisfies $u^e = C$ modulo n for *any* integer $e > 1$. The modulus n in the RSA Problem is constructed in the same way as in the RSA signature algorithm. Such a modulus is referred to as an RSA modulus. So far, no polynomial time algorithm has been found for the Strong RSA Problem, although there has been no formal proof on how hard the Strong RSA Problem is.

Figure 18.2 illustrates the high-level interactions among the group manager, group members, and group signature verifiers in the ACJT scheme. First, a set of system-wide security parameters needs to be configured on all entities that

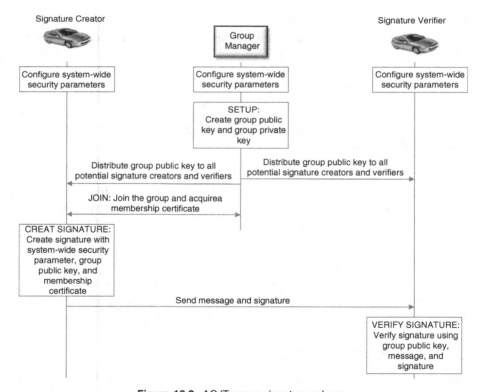

Figure 18.2. ACJT group signature scheme

will use the ACJT group signature scheme. These system-wide parameters include, for example, the size of the hash values used in the signature and parameters that define the ranges of the values in the computation of a group signature. The group manager will then create a group public key and makes it available to all group members and potential signature verifiers. The group public key can be certified by a certificate authority (CA) with a public key certificate signed by a CA just as how regular public keys discussed in previous chapters can be certified by a CA.

Each group member has a unique private key that consists of two parts: a membership secret and a membership certificate. The membership secret is a secret value selected by the member, which will not be revealed to anyone else, not even to the group manager. This ensures that no one, not even the group manager, can forge a member's signature.

The membership certificate is the group manager's signature of the membership secret. The membership certificate is therefore uniquely linked to its corresponding membership secret. This enables the group manager to open any signature generated by a group member to identify which particular member generated a signature.

To join a group, a prospective member provides zero-knowledge proofs of a membership secret to the group manager. Roughly speaking, prospective member U_i selects a membership secret x_i and proves to the group manager in zero knowledge that she knows the discrete logarithm $C = a^{x_i}$ or $x_i = \log_a C$, where a is an integer in the group public key. The prospective member sends C to the group manager. The group manager creates a signature for C as (A_i, e_i) and issues it to the new member as its membership certificate. Here, e_i is a random prime number selected by the group manager and A_i is computed as $A_i = (C \cdot a_0)^{1/e_i} = (a^{x_i} \cdot a_0)^{1/e_i}$, where a_0 is another integer in the group public key.

The prime number e_i in the membership certificate uniquely identifies this membership certificate. Several extensions to the ACJT signature scheme, and several subsequent group signature schemes that are built upon the ACJT scheme, use this prime number to support membership revocation.

Only the member and the group manager know the membership certificate for the member. Signature verifiers do not need to know the membership certificate of a signature generator.

A group member uses its membership certificate, the group public key, and the system-wide security parameters to generate group signatures. It incorporates into each signature a statistical zero-knowledge proof that the signature generator possesses a membership certificate and its associated membership secret.

A signature consists of a challenge c to signature verifiers. This challenge is a hash value of selected elements in the group public key and selected elements in the signature. A signature verifier uses the group public key, the signature, and the received message to construct a value c' that should match the challenge c in the signature if and only if the signature was generated with a valid membership certificate and its associated membership secret.

The information required to verify a group signature will not help a signature verifier to identify the signature generator beyond verifying that the signature generator is a member of the group. This allows a group member to use the same membership certificate to sign multiple messages that will remain unlinkable by adversaries.

Now, let us examine the sizes of the keys used in ACJT and the sizes of the ACJT signatures. For signature schemes based on the Strong RSA Assumption or the regular RSA Assumption, the key and signature sizes are linked closely to the length of the RSA modulus n. Let L_n be the size (bit length) of the RSA modulus n. One can derive the following [ACJT00] [Joye09]:

- *Group Public Key Size*: An ACJT group public key consists of six elements; each element has the same bit length L_n. Therefore, an ACJT group public key is $6L_n$ bits long.
- *Group Private Key Size*: An ACJT group private key consists of three elements, leading to a total bit length of $2L_n$.
- *Membership Certificate Size*: Each membership certificate consists of two values. The first value has a bit length of L_n and the second value has a bit length of $2L_n$. Therefore, the size of a membership certificate is $3L_n$.
- *Signature Size*: An ACJT signature consists of eight elements. Four of these elements have a bit length of approximately $2L_n$ each. Three of these elements have a bit length of L_n each. The remaining element is the challenge c that is hash value with a size that is typically significantly smaller than L_n. Therefore, the bit length of an ACJT signature is approximately $11L_n$.

Table 18.1 summarizes the sizes of the ACJT keys and signatures. The table also shows the key and signature sizes for 112-bit security strength. For 112-bit security strength, the RSA modulus n needs to be 2048 bits long. [Joye09] presented a new implementation of the ACJT scheme that reduces the key sizes and the signature size by half.

Next, we examine the computational complexities (i.e., processing times) of ACJT signature generation and verification. The computational complexities are dominated by multibase modular exponentiations. A k-base modular exponentiation (or k-exponentiation) is to compute the following:

Table 18.1. ACJT key and signature sizes

	Size (Bits) for L_n Bit Long RSA Modulus	Size (Bits) for 112-Bit Security Strength
Group public key	$6L_n$	12,288 (12 Kbits)
Group private key	$2L_n$	4,096 (4 Kbits)
Membership certificate	$3L_n$	6,144 (6 Kbits)
Group signature	$11L_n$	22,528 (22 Kbits)

Table 18.2. Computational complexities of ACJT signature creation and verification

	Signature Creation	Signature Verification
Modular exponentiations	3	0
Two-base modular exponentiations	3	2
Three-base modular exponentiations	1	1
Four-base modular exponentiations	0	1

$$\left(g_1^{e_1}\right)\left(g_2^{e_2}\right)\cdots\left(g_k^{e_k}\right) \text{modulo } n. \tag{18.1}$$

According to [Joye09], the average complexity $C(k,L)$ of computing k-exponentiation with L-bit long exponents can be estimated using Equation 18.2, where S is the computational complexity of a modular squaring operation and M is the computational complexity of a modular multiplication operation, using the same modulus n. A modular squaring operation is a special case of modular multiplication and is typically slightly less computational intensive than a modular multiplication.

$$C(k, L) = (L-1)\left(S + \frac{2^k - 1}{2^k}M\right) \tag{18.2}$$

Equation 18.2 shows that the computational complexity of a multibase exponentiation grows linearly with the size L of the exponents.

The computational complexities of modular squaring and modular multiplication will also grow with the size of the modulus n.

The computational complexities of ACJT signature creation and verification are summarized in Table 18.2 [ACJT00] [Joye09].

For signature creation, two of the three modular exponentiations have exponents the size of approximately L_n and the third exponentiation has an exponent the size of approximately $2L_n$. Each of the three two-base exponentiations has exponents the size of approximately $2L_n$. The three-base exponentiation has exponents the size of approximately $2L_n$. Therefore, based on Equation 18.2, the average complexity of ACJT signature creation can be translated into approximately $12L_nS + 8.25M$, which are $12L_n$ modular squaring operations plus $8.25L_n$ modular multiplications with a modulus n [Joye09].

For signature verification, one of the two-base modular exponentiations has exponents the size of approximately $2L_n$. The other two-base modular exponentiation has one exponent the size of approximately L_n and another exponent the size of the challenge c in the signature. The challenge c is a hash value and its size needs to be twice the required security strength, which is significantly smaller than L_n. For example, for 112-bit security strength, the size of the c needs to be 224 bits, while L_n needs to be 2048 bits long. The three-base modular exponentiation has one exponent the size of approximately $2L_n$, one exponent the size of approximately L_n, and one exponent the size of the challenge c in the signature. The four-base exponentiation has one exponent the size of approximately $3L_n$, two exponents the size of approximately $2L_n$, and one exponent the size of the challenge c in the signature. Therefore, based on Equation 18.2, the average complexity of ACJT signature verification can be estimated as approximately $8L_nS + 7.5M$, which are $8L_n$ modular squaring operations plus $7.5L_n$ modular multiplication operations.

18.3.2 The Challenge of Group Membership Revocation

A straightforward way to revoke a user's group membership is to create a new group that consists of only the unrevoked group members. The new group will be identified by a new group public key. All unrevoked members will need to join the new group and be issued new membership certificates by the group manager. This, however, leads to several issues.

First, each time a member is revoked, a new group public key must be distributed to all unrevoked group members and all potential verifiers. This can create heavy communication burden in a large consumer vehicle network. Since not all vehicles can be expected to have always-on network connectivity with the group manager and some vehicles may be without network connectivity for long periods of time, distributing the new group public key to all unrevoked vehicles can also take an excessive amount of time.

Second, all unrevoked members must join the new group. This can be accomplished by requiring each unrevoked member to repeat the interactive process to join the new group as described in the ACJT scheme. Repeating this interactive process to join a new group can incur heavy processing by the unrevoked members and also generate heavy communication overheads. Alternatively, the new group may be formed by the group manager issuing new membership certificates to the unrevoked members without interacting with these members. This is possible with the ACJT scheme, where the group manager can use the information provided by each member when the member first joined the group to issue subsequent membership certificates without further interactions with the member. However, distributing new membership certificates to all unrevoked members each time a member is revoked can still face the same problem as distributing the new group public key to a large number of vehicles; it can result in heavy communication overhead and experience excessive delays.

Third, with the ACJT scheme, the new membership certificates and the new group public key must be distributed to, or more precisely, committed to be used by, the message signers and verifiers in a synchronized manner (or in a highly coordinated fashion). This is because before a verifier receives the new group public key, it will not be able to successfully verify signatures generated by the new membership certificates. A verifier using the new group public key will not be able to successfully verify the signatures created by the unrevoked members who have not yet received the new group membership certificates. Achieving synchronized distribution of the group public key and the membership certificates, however, will be highly difficult in a large and dynamic consumer vehicle network.

An ideal revocation method for a group signature scheme would employ the revocation paradigm used in public key infrastructure (PKI) as pointed out in [AtTS02]. That is, a verifier checks the signer's certificate against a certificate revocation list (CRL). This is attractive because (1) the signer does not need to be aware of the ever-changing CRL, (2) the revocation checking burden is placed only on the verifier, (3) the unrevoked members do not need to perform any extra work to maintain their group membership, and (4) there is no need to distribute revocation information to both signers and verifiers in a synchronized manner.

However, with the state-of-the-art group signature schemes, including the ACJT scheme, revoking a group member cannot be accomplished simply by publishing on a CRL the private information the group manager has previously issued to the member. For example, with the ACJT group signature scheme, a signature contains an encrypted representation of the signer's group membership certificate and the group manager knows the membership certificate. However, placing the membership certificate on a CRL will not be sufficient to revoke a member. This is because the encrypted version of the membership certificate in the signature must be semantically secure, which means that any signature verifier or adversary should not be able to link a group signature to the membership certificate used to generate the signature. If someone can link a group signature to its corresponding membership certificate, then he can determine whether multiple signatures are created by the same singer, which violates the anonymity requirement for group signatures and also violates the message unlinkability requirement for privacy-preserving vehicle communications. The ability to link signatures to the same signer also allows a verifier to link the signatures generated before the member is revoked, which violates the backward unlinkability requirement.

18.3.3 ACJT Extensions to Support Membership Revocation

Several extensions have been proposed to the ACJT group signature scheme to support group membership revocation. In the succeeding sections, we describe two representative revocation extensions to ACJT presented in [AtTS02] and [CaLy02].

18.3.3.1 *The ATS Revocation Extensions* [AtTS02] presents an extension to the ACJT group signature scheme that uses a CRL to support group membership revocation. We refer to this extension as the ATS extension after the last name initials of the three people who published the scheme: Giuseppe Ateniese, Gene Tsudik, and Dawn Song. With the ATS extension, the signature size and the amount of computation for signature generation are independent of the number of revoked members. No update to the group public key is necessary when new members join the group.

The group manager issues a CRL whenever one or multiple members are revoked. The CRL contains information about the membership certificate previously issued to each revoked member. More specifically, the CRL contains a representation of the prime number e_i in the membership certificate (A_i, e_i) of each revoked member. The CRL also contains a new group public key for the new group that consists of only the unrevoked group members.

The latest CRL must be distributed to all unrevoked group members and all potential signature verifiers. The unrevoked members, and only the unrevoked members, will be able to use the information on the CRL to create valid new membership certificates. Signature verifiers need the CRL to verify whether a signature is created with a revoked membership certificate.

Using its updated membership certificate, an unrevoked member will be able to incorporate in each future signature a zero-knowledge proof that it possesses a membership certificate that is not on the CRL. To accomplish this, two new values T_4 and T_5 are added to the signature structure. If the member has been revoked, a verifier will be able to use T_4 and the information on the CRL to reconstruct T_5. For unrevoked members, the verifier will not be able to reconstruct T_5. The signer will further need to incorporate a zero-knowledge proof in the signature that it has used the same new membership secret and new member certificate to generate the new value T_5 and the other elements in the signature.

The zero-knowledge proofs are implemented using double discrete logarithms. A double discrete logarithm problem is to find x that satisfies the double exponential function illustrated in Equation 18.3, where $y, g, a,$ and n are publicly known integers.

$$y = g^{a^x} \bmod n. \qquad (18.3)$$

Double discrete logarithms are significantly more computationally intensive than regular discrete logarithms. Using the ATS extension, creating or verifying a signature requires up to L double exponentiations and L two-base exponentiations, where L is the bit length of the challenge c in the signature. Recall that the size of challenge c needs to be twice the required security strength. Therefore, to achieve 112-bit security strength, creating or verifying a signature would require approximately 224 double exponentiations plus 224 two-base exponentiations. Although the size of the exponents in these double exponentiations and two-base exponentiations are smaller than the exponents used in the original ACJT scheme, the large number of required double exponentiations and two-base exponentiations would still make the ATS extension significantly more computationally intensive than the original ACJT scheme. For comparison, the RSA algorithm takes only one regular single-base exponentiation to create a signature and one regular single-base exponentiation to verify a signature.

Besides performing zero-knowledge proofs based on double discrete logarithms, the verifier must perform an additional exponentiation for every

Table 18.3. Performance of the ATS revocation extension to the ACJT group signature scheme

	Signature Generation	Signature Verification
Exponentiations		$|CRL|$
Two-base exponentiations	L	L
Double exponentiations	L	L

revoked member on the CRL. Therefore, the verification complexity also grows linearly with the number of revoked members.

Table 18.3 summarizes the computational complexities of signature generation and verification of the ATS membership revocation extension, where $|CRL|$ is the number of revoked members on the CRL and L is the bit length of the challenge c in the signature.

The CRL-based ATS membership revocation approach also requires that message signer and message verifier use the same CRL to create and verify signatures. Before everyone receives the same latest CRL, some vehicles will use an old CRL while others use the new CRL. This can cause messages signed by unrevoked vehicles to fail signature verification because message signers and verifiers are using different versions of the CRL. Therefore, all message signers and verifiers need to receive and use each new CRL in a synchronized manner. Such synchronized information distribution, however, is extremely difficult to accomplish in a large and dynamic consumer vehicle network.

18.3.3.2 The CL Revocation Extensions [CaLy02] proposes another membership revocation extension to the ACJT group signature scheme. We refer to this extension as the CL revocation extension after the last name initials of the two people who published the scheme: Jan **C**amenisch and Anna **L**ysyanskaya.

The CL revocation extension does not have any operation linearly dependent on the number of revoked members. It uses a dynamic accumulator. An accumulator is an algorithm that allows one to hash multiple inputs into one short value called the accumulator, with a unique indicator showing that a given input has been incorporated into the accumulator. A dynamic accumulator allows dynamic addition of elements to and deletion of elements from the accumulator.

When a new user U_i joins the group, the group manager issues a membership certificate to U_i. The current group public key will be incorporated into the membership certificate. The group manager then adds the prime number e_i of U_i's membership certificate to the accumulator. The group manager also updates the group public key to incorporate the prime number e_i of U_i's membership certificate into the group public key.

To generate a group signature, the signature generator incorporates a zero-knowledge proof that it has a membership certificate and its membership certificate is in the accumulator (indicating that the membership certificate has not been revoked). The input parameters to the signature generation algorithm include the group public key.

To revoke a member U_i, the group manager removes U_i's membership certificate from the accumulator. The group manager also updates the group public key accordingly so that U_i's membership certificate is no longer incorporated in the group public key.

Each time a new member joins the group or a current member is revoked, the updated group public key needs to be made known to all the unrevoked group members and all potential signature verifiers in a synchronized manner. Before everyone receives the same version of the group public key, some message signers and verifiers may use different group public keys and, consequently, signatures generated by unrevoked members can fail signature verification. As discussed before, achieving synchronized information distribution to all vehicles in a large and dynamic consumer vehicle network will be highly difficult.

The CL revocation extension uses a public archive to keep track of which members have joined the group and which members have been revoked. Each member has to watch the changes made to the public archive and use the new information in the archive to update its membership certificate.

The amount of data to read from the public archive and the local computation are linear in the number of changes since the last time the user checked the public archive.

The computational complexity of signature creation and verification is approximately twice that of the original ACJT group signature scheme [CaLy02].

18.4 THE CG GROUP SIGNATURE SCHEME WITH REVOCATION

[CaGr05] proposed a new group signature scheme with an integrated membership revocation mechanism. We refer to this group signature scheme as the CG scheme after the last name initials of the two authors of [CaGr05]: Jan Camenishch and Jens Groth.

The CG group signature scheme builds upon the ACJT group signature scheme and the group signature scheme in [CaLy02b] with extensions to support dynamic addition and revocation of membership.

The revocation method is similar to that in [CaLy02]. However, unlike the revocation method in [CaLy02], new members can join a group dynamically without changing the group public key in the CG group signature scheme.

Like the ACJT group signature scheme, when joining the group, each member U_i receives a signature from the group manager on a unique membership secret x_i selected by and known only to the member. This signature by the group manager, referred to as the CL signature, plus the membership secret, serve as the member's private key like in the ACJT group signature scheme.

Table 18.4. Performance of the CG group signature scheme

	Signature Generation	Signature Verification
Exponentiations	7	0
Two-base exponentiations	0	3
Three-base exponentiations	1	1

A signature generator uses its private key and the group public key to generate a signature. It incorporates into the signature a zero-knowledge proof of its knowledge of a membership secret x_i and the CL signature for x_i. This ensures that the signer is a member of the group because nonmembers cannot forge CL signatures.

To revoke a member, the group manager creates a new group public key that reflects the revoked member. The group manager distributes this new group public key and part of the revoked member's private key to all unrevoked members and potential signature verifiers. Unrevoked members, and only unrevoked members, will be able to use such published information to create new valid private keys associated with the new group public key. They will then use their new private keys and the new group public key to generate future signatures.

Table 18.4 summarizes the computational complexities of the signature generation and verification [CaGr05]. For signature creation, the sizes of the exponents for six exponentiations are slightly larger than the size of the hash value used in the signature, which is about twice the security strength. The size of the exponent for the remaining one exponentiation is $L_n/2$, where L_n is the bit length of the RSA modulus. The three-base exponentiation has one exponent of length slightly larger than $L_n/2$ and two exponents of lengths approximately six times the size of the hash value used in the signature. For comparison, the sizes of the exponents of the exponentiations and multiexponentiations in ACJT group signature generation ranges from one to two times the length of the RSA modulus.

For signature verification, each of the three two-base exponentiations has an exponent of length approximately the size of the hash value used in the signature. Since this exponent is significantly smaller than the other exponent in the two-base exponentiations, each of the three two-base exponentiations takes approximately the same time to compute as a regular exponentiation. The sizes of the exponents of the three-base exponentiation are the same as the three-base exponentiation used in signature generation.

Compared with the original ACJT group signature scheme, the CG group signature scheme has faster signature generation and verification speeds while also being able to revoke group members.

However, like the revocation method in [CaLy02], the CG group signature scheme requires the new group public key to be made known to all the unrevoked group members and all potential signature verifiers in a synchronized manner each time a group member is revoked. Signatures generated with the old group public key will not pass verification using the new group public key. Similarly, signatures generated with new private keys will not pass verification using the old group public key. As discussed previously, synchronized information distribution to all vehicles will be difficult to accomplish in a large consumer vehicle network.

18.5   THE SHORT GROUP SIGNATURES SCHEME

18.5.1   The Short Group Signatures Scheme

The Short Group Signatures scheme presented in [BoBS04] also relies on zero-knowledge proofs of knowledge. It is based on the Strong Diffie–Hellman Assumption rather than the Strong RSA Assumption used in the ACJT scheme. That is, the Short Group Signatures scheme uses the Strong Diffie–Hellman Problem to implement zero-knowledge proofs of knowledge. The Strong Diffie–Hellman Problem is a generalized version of the Diffie–Hellman Problem discussed in Chapter 15.

As in previously described group signature schemes, the group public key must be distributed to all members of the group because it will be used to create group signatures. However, the group public key does not need to change when new members join the group.

The group manager will also assign a unique private key to each group member.

A group member uses both its private key and the group public key to create signatures. It incorporates into each signature a zero-knowledge proof that it possesses a private key (and several additional secret values). The size of the group signature is independent of the size of the group.

Each signature includes a challenge c to signature verifiers. To verify a signature, the verifier uses the group public key, the message, and its signature to construct a value c' that will match the challenge c in the signature if and only if the signature was generated with a valid private key issued by the group manager.

The group public key and the signature do not reveal the message signer's private key. The signatures generated with the same private key will not be linkable. Therefore, a group member can use the same private key to sign multiple messages and each signature will only be recognized as being generated equally likely by any member of the group.

The Short Group Signatures scheme uses bilinear pairings to implement zero-knowledge proofs of knowledge. Pairing is a function that maps an element in a group G_1 and an element in a group G_2 to an element in a different group G_T. G_1 can be identical to G_2 depending on the specific pairing implementations. A pairing function e is denoted by $e: G_1 \times G_2 \to G_T$. Informally speaking, a pairing is bilinear when it satisfies Equation 18.4, where $u \in G_1, v \in G_2$, and a and b are positive integers:

$$e\left(u^{a}, v^{b}\right) = (e(u, v))^{ab}. \tag{18.4}$$

For pairings to be used for cryptography, they must be nondegenerate in that if $u \neq 0$, then $e(u, u) \neq 1$. That is, elements cannot all map to unity. Pairings must also be efficiently computable. Today, the two most popular pairing functions are the Weil pairing [Mill04] and the Tate pairing [BKLS02], which are typically implemented over elliptic curves.

The Short Group Signatures scheme uses pairing to implement zero-knowledge proof of knowledge. This leads to smaller keys and signatures than previous group signature schemes such as the ACJT scheme.

The sizes of the keys and the signatures are closely linked to the order p of the groups G_1 and G_2.

The group public key of the Short Group Signatures scheme consists of six separate elements; each has the same bit length as p. Let's denote by L_p the bit length of p. Then, the size of the group public key is $6L_p$. The group private key has two values; each has a bit length of L_p. Therefore, the size of the group private key is $2L_p$. A group member's private key has two values; each has a bit length of L_p. Hence, the size of a member's private key is $2L_p$. The group signature consists of nine separate elements; each has a bit length of L_p.

Table 18.5 shows the sizes of the keys and the group signature [BoBS04] as a function of L_p. The table also shows the sizes of the keys and the group signature for the 112-bit security strength. When bilinear pairing operations are implemented over elliptic curves, achieving an L-bit security strength, L_p needs to be $2L$ or larger. For 112-bit security strength, L_p should be 224.

As shown in Table 18.5, Short Group Signatures scheme requires significantly smaller key sizes and generates much smaller signatures than the ACJT scheme for the same security strength. For 112-bit security, for example, the group public key size is only approximately 10% of the ACJT public key size and the group signature size is less than 10% of the ACJT signature size. Further, the signatures are approximately the same size as standard RSA signatures for comparable security.

The computational complexities of signature creation and verification are illustrated in Table 18.6 based on [BoBS04]. The Short Group Signatures scheme can be implemented over an elliptic curve and as a result the

Table 18.5. Key and signature sizes for group signature scheme in [BoBS04] for 112-bit security

	Size (L_p Bits)	Size (Bits)
Group public key	$6L_p$	1344 (168 bytes)
Group private key	$2L_p$	448 (56 bytes)
Member private key	$2L_p$	448 (56 bytes)
Group signature	$9L_p$	2016 (252 bytes)

Table 18.6. Performance of the Short Group Signatures scheme

	Signature Creation	Signature Verification
Pairing	0	1
Two-base modular exponentiation	2	4
Three-base modular exponentiation	1	1
Exponentiations	5	0

exponentiations will use much smaller exponents than the ACJT scheme. In particular, all the exponents will have the same bit length that is approximately twice the security strength. For 112-bit security strength, for example, the size of each exponent will be approximately 224 bits. In comparison, the sizes of the exponents for ACJT signature generation and verifications range from one to two times the size of an RSA modulus. The size of the RSA modulus needs to be 2048 bits long to achieve the same 112-bit security. Recall that the computational complexity of an exponentiation or a multiexponentiation grows linearly with the lengths of the exponents.

In addition to exponentiations and multiexponentiations, signature generation and verification both require pairing operations. Signature generation requires three pairing operations. But, they can all be precomputed for use in generating future signatures because these pairing operations only use elements in the group public key which does not change for each signature. Signature verification requires the same three precomputable pairing operations as used in signature generation plus one pairing operation that must be dynamically performed for each signature verification.

Pairing is known to be highly computationally intensive [Stog04]. We use the Weil pairing as an example to illustrate the computational complexity of pairing operations. With the Weil pairing, the groups G_1 and G_2 are groups of points on an elliptic curve and the group G_T is a subgroup of the multiplicative group of a related finite field. If $E(F_q)$ is an elliptic curve defined over the finite prime field F_q, then G_1 is a subgroup of $E(F_q)$, G_2 is a subgroup of $E\left(F_{q^k}\right)$, where F_{q^k} is an extension field of F_q. Let L_p denote the common size of the groups G_1, G_2, and G_T. Based on [Shik05], the computational complexity $C_{\text{Weil}}(L_p)$ of the Weil pairing can be estimated using Equation 18.5, where I is the complexity of a modular inversion and M is the complexity of a modular multiplication over field F_q. The Tate pairing can be implemented at approximately half of the computational complexity of the Weil pairing:

$$C_{Weil}\left(L_p\right) = 2\left(4I + 25M\right)log_2\left(L_p\right). \tag{18.5}$$

The size of each group G_1, G_2, and G_T must be large enough so that discrete logarithm problems are computationally infeasible to solve in each group. To achieve L-bit security strength, the size L_p of each group G_1, G_2, and G_T needs to be $2L$ bits long.

The complexities of modular operations also depend closely on the size of the modulus, which is the size of the group over which modular operations are performed. The modular inversions and multiplications mentioned in Equation 18.5 are performed over field F_q. The size of F_q should be $2L$ bit long to achieve L-bit security strength. As a comparison, the modular operations in the ACJT scheme are performed with an RSA modulus, which is significantly larger than the size of field F_q used to implement bilinear pairings over elliptic curves in the Short Group Signatures scheme. As discussed in Chapter 15, the RSA modulus has to be 2048 bits long for 112-bit security strength, while the size of F_q, which is the Weil pairing, can be 224 bit long.

Significant efforts are still underway to develop more efficient algorithms for implementing the Weil pairing and the Tate pairing. Today, however, pairing remains highly computationally intensive.

18.5.2 Membership Revocation

The Short Group Signatures scheme described previously [BoBS04] includes a membership revocation mechanism. This revocation method is similar to the revocation approach in [CaLy02]. To revoke members of a group, the group manager publishes a revocation list (RL) that contains all the private keys of the revoked members.

The RL is then distributed to all signers and verifiers. Everyone can use the information on the RL to construct the new group public key to be used to verify future group signatures. Unrevoked group members will also be able to use the RL to generate a new private key while a revoked group member will not be able to do so.

A signature created with an unrevoked member's new private key will be positively verifiable with the corresponding group public key that a verifier can construct using the information on the RL. However, a signature created with a revoked private key will fail the verification process with the new group public key.

This approach also requires synchronized distribution of the RL to all message signers and verifiers, which will be difficult to implement in large and dynamic vehicle communication systems. Everyone needs the RL to update the group public key. Unrevoked group members also need the RL to update their private keys. Upon receiving an RL, a message signer will use its updated private key and the updated group public key to create future signatures, and a message verifier will use the updated group public key to verify signatures. Suppose a message signer has received the RL and a message verifier has not. Then, the signer will be using its new private key to generate signatures while the verifier does not have the new group public key to verify signatures. Signature verification will fail before both the message signer and the verifier have both received the RL. The message signer could send the RL to the message verifier before it sends signed messages. This, however, is not likely to be feasible when the RL becomes large. Similarly, if the verifier has received the RL while the signer has not, the signer will still be using its old private key and the old public key to sign messages while the message verifier is using the new group public key to verify signatures.

18.6 GROUP SIGNATURE SCHEMES WITH VERIFIER-LOCAL REVOCATION

[BoSh04] and [NaFu05] presented Verifier-Local Revocation (VLR) schemes for group membership revocation, which only need to distribute RLs to signature verifiers rather than both signers and verifiers.

With the approach in [BoSh04], the group manager generates a group public key, a set of private keys for the group members, and a revocation token for each member. To revoke a member, the group manager places the revocation token of the revoked member on an RL and distributes the RL to all potential signature verifiers. The revocation token is the lower half of the private key of the revoked member.

The signing algorithm takes as input the signer's private key, the group public key, and the message. The signer's revocation token will be encoded in its signature.

The verification algorithm uses the group public key and the latest RL to verify a signature. To verify whether a signer has been revoked, the verifier checks whether any revocation token on the RL is encoded in the signature. If no token on the RL is encoded in the signature, then the signer has not been revoked.

The signatures are approximately the same size as standard RSA signatures for comparable security strength.

The computational complexities of signature creation and verification are summarized in Table 18.7. Signature generation requires about eight exponentiations (or multibase exponentiations) and two pairing computations. Signature verification takes six exponentiations and $3 + 2|RL|$ pairing computations, where $|RL|$ is the number of revocation tokens on RL. That is, the number of pairing operations to be performed to verify a signature will grow linearly with the number of revoked group members. This can result in excessively heavy computations in a large consumer vehicle network where an RL can contain a large number of revoked vehicles.

Table 18.7. Performance of group signature with verifier-local revocation [BoSh04]

	Signature Creation	Signature Verification
Pairing	2	$3 + 2|RL|$
Modular exponentiations or multiexponentiations	8	6

Furthermore, [BoSh04] does not support backward unlinkability. [NaFu05] presented a verifier-local revocation method that supports backward unlinkability and also has lower verification complexity than the method in [BoSh04]. However, it requires the use of complicated time intervals that can be difficult to implement in a large and highly dynamic consumer vehicle network.

REFERENCES

[ACJT00] G. Ateniese, J. Camenisch, M. Joye, and G. Tsudik: "A Practical and Provably Secure Coalition-Resistant Group Signature Scheme," Advances in Cryptology—CRYPTO 2000, Springer-Verlag Lecture Notes in Computer Science, Vol. 1880, 2000.

[AtTS02] G. Ateniese, G. Tsudik, and D. Song: "Quasi-Efficient Revocation of Group Signatures," Financial Cryptography, 2002.

[BaPf97] N. Baric and B. Pfitzmann: "Collision-Free Accumulators and Fail-Stop Signature Schemes Without Trees," Advances in Cryptology, EUROCRYPT'97, Springer-Verlag Lecture Notes in Computer Science, Vol. 1233, 1997.

[BKLS02] P. S. L. M. Barreto1, H. Y. Kim, B. Lynn, and M. Scott: "Efficient Algorithms for Pairing-Based Cryptosystems," Advances in Cryptology, CRYPTO 2002, Lecture Notes in Computer Science, Vol. 2442, 2002.

[BeMW03] M. Bellare, D. Micciancio, and B. Warinschi: "Foundations of Group Signatures: Formal Definitions, Simplified Requirements and a Construction Based on General Assumptions," EUROCRYPT'03, Springer-Verlag Lecture Notes in Computer Science, Vol. 2656, 2003.

[BeSZ05] M. Bellare, H. Shi, and C. Zhang: "Foundations of Group Signature: The Case of Dynamic Groups," Springer-Verlag Lecture Notes in Computer Science, Vol. 3376, 2005.

[BoBS04] D. Boneh, X. Boyen, and H. Shacham: "Short Group Signatures," Advances in Cryptology—CRYPTO 2004, Springer-Verlag Lecture Notes in Computer Science, Vol. 3152, 2004.

[BoSh04] D. Boneh and H. Shacham: "Group Signatures with Verifier-Local Revocation," the 11th ACM Conference on Computer and Communications Security, ACM Press, 2004.

[CaGr05] J. Camenishch and J. Groth: "Group Signatures: Better Efficiency and New Theoretical Aspects," Security in Communication Networks, Lecture Notes in Computer Science, Vol. 3352, 120–133, 2005.

[CaLy02] J. Camenisch and A. Lysyanskaya: "Dynamic Accumulators and Application to Efficient Revocation of Anonymous Credentials," CRYPTO 2002, Springer-Verlag Lecture Notes in Computer Science, Vol. 2442, 2002.

[CaLy02b] J. Camenisch and A. Lysyanskaya: "A Signature Scheme with Efficient Protocols," 3rd International Conference on Security in Communication Networks, SCN'02, 2002.

[ChHe91] D. Cham and E. van Heyst: "Group Signatures," 10th Annual International Conference on Theory and Application of Cryptographic Techniques (EUROCRYPT'91), Springer-Verlag Lecture Notes on Computer Science, 1991.

[Joye09] M. Joye: "On Cryptographic Schemes based on Discrete Logarithms and Factoring," 8th International Conference on Cryptography and Network Security (CANS), Kanazawa, Japan, 2009.

[KiYu04] A. Kiayias and M. Yung: "Group Signatures: Provable Security, Efficient Constructions, and Anonymity from Trapdoor-Holders," Cryptology ePrint Archive, Report 2004/076, 2004.

[Mill04] V. S. Miller: "The Weil Pairing, and Its Efficient Calculation," Journal of Cryptology, vol. 17, no. 4, pp. 235–261, 2004.

[NaFu05] T. Nakanishi and N. Funabiki: "Verifier-Local Revocation Group Signature Schemes with Backward Unlinkability from Bilinear Maps," Advances in Cryptology —ASIACRYPT 2005, Springer-Verlag Lecture Notes in Computer Science, Vol. 3788, 2005.

[Schn91] C. Schnorr: "Efficient Signature Generation by Smart Cards," Journal of Cryptology, vol. 4, no. 3, pp. 161–174, 1991.

[Shik05] A. Shikfa: "Bilinear Pairing over Elliptic Curves," Master Thesis, Universite de Nice—Sophia Antipolis, 2005.

[Stog04] M. Stogbauer: "Efficient Algorithms for Pairing-Based Cryptosystems," Diploma Thesis, Darmstadt University of Technology, 2004.

19

PRIVACY PROTECTION AGAINST CERTIFICATE AUTHORITIES

19.1 INTRODUCTION

Using conventional public key infrastructure (PKI) designs, each certificate authority (CA) has full knowledge about which certificates it has assigned to which vehicles. A CA operator can use such information to identify and track vehicles. CA operators may provide the information to others who may use it to breach vehicle privacy.

In this chapter, we describe new PKI architectures that prevent any individual CA from having sufficient information to track vehicles based on certificates. We first describe the basic idea and the baseline architecture, with its enabling protocol message flow and message processing procedure. We then describe enhancements to the baseline architecture and protocol to support shared certificates, unlinked short-lived certificates, anonymously linked short-lived certificates, and to allocate batches of unlinked or anonymously linked short-lived certificates.

19.2 BASIC IDEA

The idea is to split the functionality performed by a CA in the traditional PKI design into *Authorizing CA* functions and *Assigning CA* functions that can be

Vehicle Safety Communications: Protocols, Security, and Privacy, First Edition. Luca Delgrossi and Tao Zhang.
© 2012 John Wiley & Sons, Inc. Published 2012 by John Wiley & Sons, Inc.

operated by independent certificate management entities so that no single certificate management entity has sufficient information to track vehicles based on certificates. We refer to PKI architectures built upon this idea as *Split CA Architectures*.

A first such split CA architecture was described in [WVZM07]. Independently, a different approach was presented in [KCKL07]. Using the approach in [WVZM07], a vehicle only needs to communicate directly with one CA while the approach in [KCKL07] requires a vehicle to communicate with two separate CAs.

Building upon the basic idea in [WVZM07], this chapter presents enhanced split CA architectures and protocols that can support shared certificates, unlinked short-lived certificates, anonymously linked short-lived certificates, and allocation of batches of certificates.

Here, we first describe the basic idea in [WVZM07]. The Authorizing CA decides whether each certificate request should be granted. It maintains information required to authorize certificate requests. Such information can identify individual vehicles and may include, for example, vehicle identifiers such as vehicle identification numbers (VINs), vehicle makes and models, vehicle communication software and hardware configurations, and security credentials for authenticating certificate requests from vehicles.

The Authorizing CA can issue non-privacy-preserving keys and certificates. These may include security management keys and certificates that vehicles use to authenticate with the CA, acquire privacy-preserving certificates, and as backup certificates to sign vehicle safety messages if its privacy-preserving certificates have all expired or been revoked.

The Authorizing CA, however, cannot create privacy-preserving certificates. Instead, it is responsible for creating privacy-preserving certificates upon authorization by the Authorizing CA.

Neither the Authorizing CA nor the Assigning CA alone will have complete knowledge of the association between certificates and vehicles. The Authorizing CA knows how many certificates have been authorized for each vehicle; but it will not know which certificates are assigned to which particular vehicles. The Assigning CA knows the certificates it has created for each certificate request; but it will not know which particular vehicle will be assigned these certificates. Furthermore, the Assigning CA will not know whether the certificates it has created for different certificate requests are assigned to the same vehicle. Consequently, no individual CA will be able to link different certificates to the same vehicle.

If the Authorizing CA and the Assigning CA collude, they collectively will have sufficient information to identify and track vehicles based on certificates. To prevent such collusion, the Authorizing CA and the Assigning CA should be operated by independent certificate management entities. Collusion can be prevented or made difficult if at least one of these entities is motivated to protect vehicle privacy and refuses to collude.

19.3 BASELINE SPLIT CA ARCHITECTURE, PROTOCOL, AND MESSAGE PROCESSING

To acquire privacy-preserving certificates, vehicles can interact with the Authorizing CA or the Assigning CA, but not necessarily both. This is illustrated in Figure 19.1. If vehicles acquire certificates from the Authorizing CA, the Authorizing CA receives, authenticates, and authorizes the certificate requests from the vehicles; obtains certificates from the Assigning CA; and then replies to the vehicles.

If vehicles acquire certificates from the Assigning CA, the Assigning CA receives certificate requests from the vehicles, obtains authorization from the Authorizing CA to issue the requested certificates, and then creates and issues certificates to the vehicles.

In the rest of this chapter, we assume for illustration purpose that vehicles interact with the Assigning CA to acquire certificates. The mechanisms described here, however, can be readily adapted to allow vehicles to interact with the Authorizing CA to acquire their certificates.

Figure 19.2 shows the Baseline Protocol Flow for vehicles to acquire privacy-preserving certificates, assuming that the vehicles interact with the Assigning CA.

A vehicle sends a Certificate Request message to the Assigning CA. These messages should not contain any information that will allow the Assigning CA to identify the vehicle or to link different Certificate Request messages to the same vehicle. Figure 19.3 shows a baseline Certificate Request message format for achieving this goal. It consists of three main parts:

- *Part I*: Information elements that should only be seen by the Authorizing CA.

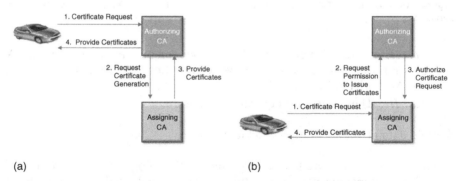

(a)   (b)

Figure 19.1. Baseline Split CA architectures. (a) Vehicles interact with the Authorizing CA to acquire privacy-preserving certificates. (b) Vehicles interact with the Assigning CA to acquire privacy-preserving certificates.

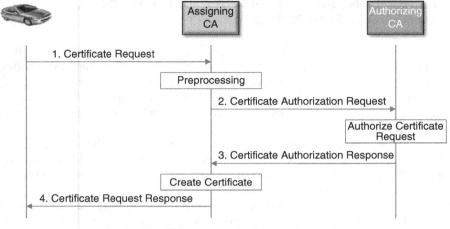

Figure 19.2. Baseline Split CA protocol

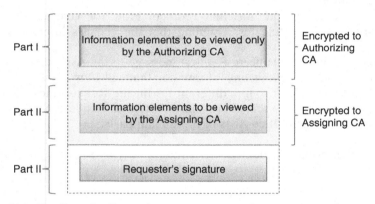

Figure 19.3. Baseline Certificate Request message for acquiring privacy-preserving certificates

- *Part II*: Information elements that can be seen only by the Assigning CA plus information items that can be seen by both the Assigning CA and the Authorizing CA.
- *Part III*: The certificate requester's signature.

Sample data elements in Part I include:

- Information for authenticating and authorizing certificate requests. This may include vehicle identifier, security credentials that have been previously assigned to the vehicle, the makes and models of the vehicles, and the hardware and onboard unit (OBU) software configuration parameters.

- A symmetric encryption key and its associated encryption algorithm for the Authorizing CA to encrypt its subsequent messages to the certificate requester directly without revealing the information to the Assigning CA.

To ensure that the Assigning CA does not know which vehicle originated a Certificate Request message, the vehicle encrypts Part I of the Certificate Request message to the Authorizing CA. This can be done using, for example, the Authorizing CA's public encryption key or a secret symmetric key that has been previously established between the vehicle and the Authorizing CA. The encryption results of Part I in different Certificate Request messages should be randomized so that the Assigning CA cannot link different Certificate Request messages to the same vehicle. This can be achieved by adding a random value to Part I before it is encrypted.

Part II of the Certificate Request message can be encrypted to the Assigning CA. Sample data elements in Part II include:

- The types of certificates requested. Main types of certificates include security management certificates, which are typically not privacy preserving, and privacy-preserving safety application certificates.
- The number of certificates requested.
- The reason for the certificate request. Main reasons would include new certificates, replacing expired certificates, or replacing revoked certificates.
- A symmetric encryption key and its associated encryption algorithm for the Assigning CA to encrypt its responses to the certificate requester.

The types of certificates, the number of certificates requested, and the reason for the certificate request should typically be known to both the Assigning CA and the Authorizing CA. The Assigning CA can decrypt these data elements and send them in the Certificate Authorization Request message to the Authorizing CA.

The symmetric encryption key in Part II, to be used by the Assigning CA to encrypt its response to the certificate requester, should not be revealed to the Authorizing CA. Different Certificate Request messages should carry randomly selected and different encryption keys for the Assigning CA so that it cannot link different Certificate Request messages to the same vehicle.

The same field in Part II of different Certificate Request messages can contain a repetitive value only when the value is identical in Certificate Requests from all users, and hence cannot be used to link multiple Certificate Requests to the same vehicle. For example, since any vehicle can request privacy-preserving certificates, different Certificate Request messages can carry the same value for the type of requested certificates.

The certificate requester should sign the Certificate Request to protect message integrity and to help the Assigning CA determine whether the message is from a vehicle. To preserve privacy, the certificate requester's signatures, the verification keys for these signatures, and the certificates for these verification keys should not be repetitive. Therefore, a vehicle should use different privacy-preserving certificates to sign different Certificate Request messages.

Upon receiving a Certificate Request, the Assigning CA keeps Part II of the message in its local database. It then sends a Certificate Authorization Request message, which contains Part I of the Certificate Request received from the certificate requester, to the Authorizing CA to request permission to issue certificates. The Certificate Authorization Request message should also contain any information elements in Part II of the Certificate Request that are needed by the Authorizing CA to authorize the certificate request.

The Authorizing CA decides whether the certificate request should be granted based on information in the Certificate Authorization Request and additional information it has collected from other sources such as the misbehaving detection system (MDS).

Collectively, the Authorizing CA and the Assigning CA should have the ability to revoke all the certificates that have been assigned to a vehicle. This requires the Authorizing CA and the Assigning CA to be able to collectively determine the set of certificates that have been assigned to the same vehicle. To accomplish this goal, the Authorizing CA creates an *authorization number* for each certificate or each batch of certificates it authorizes for each vehicle. Each authorization number is a random but unique number. The Authorizing CA stores the authorization numbers for each vehicle.

The Authorizing CA sends its authorization decision back to the Assigning CA in a Certificate Authorization Response message. For each authorized certificate request, the Certificate Authorization Response message will carry a new authorization number and any additional information that the Assigning CA will need to create certificates.

Upon authorization by the Authorizing CA, the Assigning CA creates the requested certificates and sends them in a Certificate Request Response message to the certificate requester. The Assigning CA stores each newly issued certificate with its authorization number received from the Authorizing CA.

The Assigning CA encrypts the Certificate Request Response using the symmetric encryption key provided by the certificate requester in its Certificate Request.

To ensure that the Assigning CA cannot link different Certificate Request messages to the same vehicle, the vehicle should use a random return address for each certificate request.

Next, we describe the Baseline Revocation Procedure. To support certificate revocation, the Authorizing CA and the Assigning CA can maintain mappings shown in Figure 19.4 so that they can collaborate to find all the certificates that have been assigned to a revoked vehicle.

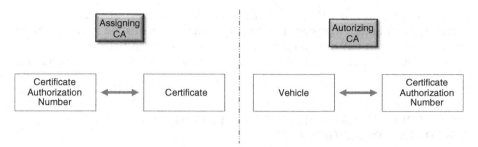

Figure 19.4. Mappings for assigning and revoking certificates in Baseline Split CA architecture

If the vehicle to be revoked is already known, the Authorizing CA can find in its local database the authorization numbers for all the certificates that have authorized for this vehicle and sends the authorization numbers to the Assigning CA. The Assigning CA can locate the actual certificate that corresponds to each authorization number. Once located, the certificates can be placed on the certificate revocation list (CRL) by either the Assigning CA or the Authorizing CA.

When a certificate is detected to be misused and the vehicle that misused this certificate is to be revoked, the Assigning CA and the Authorizing CA can collectively find out which vehicle owns the misused certificate and the other certificates that have been assigned to this vehicle. First, the Assigning CA can use the misused certificate to locate the authorization number for the misused certificate and sends the authorization number to the Authorizing CA. The Authorizing CA can identify the vehicle associated with this authorization number to locate all the other authorization numbers it has issued for this vehicle. These authorization numbers can then be sent back to the Assigning CA, which can locate the actual certificates corresponding to these authorization numbers.

Allocating and revoking anonymously linked certificates or batches of certificates will require additional processing in order to prevent each individual CA from knowing which certificates have been assigned to which vehicles.

19.4 SPLIT CA ARCHITECTURE FOR SHARED CERTIFICATES

The Baseline Split CA Architecture and the Baseline Protocol Flow are a natural fit for supporting shared certificates. The Assigning CA can be responsible for creating and managing the Shared Certificate Pool. No further extension will be necessary.

The Baseline Protocol Flow can be used by a vehicle to acquire a shared certificate. Since each vehicle typically only needs to have a small number of shared certificates, a vehicle can acquire one certificate at a time until it reaches the required number of certificates. This eliminates the need to extend the baseline split CA architecture to assign certificates in batches.

Certificate revocation can also be carried out using the Baseline Revocation Procedure.

19.5 SPLIT CA ARCHITECTURE FOR UNLINKED SHORT-LIVED CERTIFICATES

We first describe how the split CA architecture can be extended to allow a vehicle to acquire one unlinked certificate at a time. We then discuss how to allocate batches of unlinked certificates.

19.5.1 Acquire One Unlinked Certificate at a Time

The Baseline Split CA Architecture and Protocol Flow can be used, with minor extensions, to enable a vehicle to acquire one unlinked certificate at a time.

The extensions are to address unique requirements that can arise when unlinked short-lived certificates are used. For example, with short-lived unlinked certificates, a vehicle can generate its private–public key pairs and request the CA to issue certificates for its public keys. Alternatively, the key pairs and the certificates can all be created by the CAs. Therefore, the message formats and message processing can be extended to support both cases.

If a vehicle creates its private–public key pair, Part II of the Certificate Request should contain the public key to be certified by the CAs. The vehicle must also include proofs in the Certificate Request that it possesses the private key associated with each public key to be certified. This can be achieved, for example, if the vehicle signs each public key with its corresponding private key. The CAs can then use the public key in the Certificate Request to verify the signature to determine whether the certificate requester has the correct private key.

Figure 19.5 shows a sample Certificate Request message for supporting unlinked short-lived certificates and how its data elements are prepared.

Part I of the Certificate Request can be the same as the baseline Certificate Request message format in the Baseline Split CA Architecture. The rest of the Certificate Request message differs from the baseline message format in the following ways:

- The type of certificates can be extended to indicate whether a request is for certificates of vehicle-generated public keys.
- Part II of the message can contain an optional field for vehicle-generated public keys. These public keys should not be known to the Authorizing CA. Otherwise, the Authorizing CA will know which specific certificates

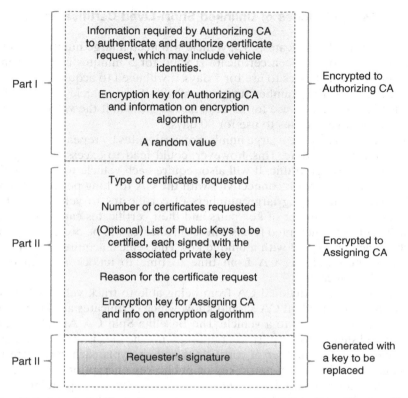

Figure 19.5. Sample Certificate Request message format for unlinked short-lived certificates

will be assigned to each vehicle. Each public key can be signed with its corresponding private key to prove that the certificate requester possesses the corresponding private key.

- The certificate requester can sign the Certificate Request message with the private key that corresponds to a public key to be certified by the CAs. This way, the signature does not leak more private information than the other information already contained in the Certificate Request. The certificate requester may also sign the certificate request with a privacy-preserving certificate assigned to it previously.

Upon receiving the Certificate Request, the Assigning CA verifies that the certificate requester possesses the private key associated with each public key listed in the Certificate Request. Then, the Assigning CA and the Authorizing CA can follow the Baseline Protocol Flow to authorize the certificate request and to issue a certificate to the vehicle.

Certificate revocation can also be carried out using the Baseline Revocation Procedure.

19.5.2   Assign Batches of Unlinked Short-Lived Certificates

Using short-lived certificates, each vehicle will need a large number of certificates. For example, if each certificate is valid for 5 minutes and the vehicle wants enough certificates to use for 7 days, it will need to acquire 2016 certificates each time. This number jumps to 60,480 if the vehicle wants to have sufficient certificates to use for 30 days and to 181,440 if the vehicle wants to have sufficient certificates to use for 90 days.

A vehicle can acquire a large number of certificates by repeatedly acquiring one certificate at a time. This, however, could lead to excessive delays and over-the-air message traffic. It will also require each vehicle to have continuous or frequent network connectivity with the CA for long periods of time.

To avoid transporting large numbers of certificates to vehicles over the network, a large number of key pairs and their certificates can be preloaded to each vehicle in encrypted forms. These certificates will be divided into small batches, each encrypted with a unique key. The vehicle acquires a decryption key from the Assigning CA from time to time to unlock a batch of the encrypted certificates.

To prevent any individual CA from being able to track vehicles based on certificates, no individual CA should know which certificates are in any batch of certificates allocated to a vehicle. The Baseline Split CA Architecture can be modified slightly to support this requirement. The idea is that the Assigning CA creates a large number of certificates and encrypts each certificate with the same key using a symmetric key or public key encryption algorithm. We refer to this key as the *certificate encryption key* and its corresponding decryption key as the *certificate decryption key*. The Assigning CA keeps the certificate decryption key to itself and sends the individually encrypted certificates to the Authorizing CA.

The Authorizing CA divides these individually encrypted certificates into batches and encrypts each batch with a unique *batch encryption key* using symmetric key or public key algorithm. The Authorizing CA keeps the *batch decryption key* for each certificate batch to itself.

The Authorizing CA creates a random but unique authorization number for each authorized certificate batch. This *certificate batch authorization number* can be sent with the certificate batch to the Assigning CA and to the certificate requester so that everyone can use the same identifier to refer to the same certificate batch. This will allow the Authorizing CA and the Assigning CA to collectively determine all the certificate batches that have been allocated to a revoked vehicle. The Authorizing CA sends the encrypted certificate batches to the Assigning CA, which will, in turn, give them to the certificate requester.

The Authorizing CA and the Assigning CA maintain mappings as illustrated in Figure 19.6.

Neither the Assigning CA nor the Authorizing CA will know which certificates are in each certificate batch.

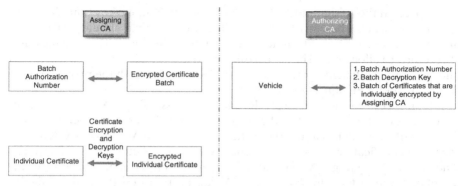

Figure 19.6. Mappings used for assigning and revoking batches of unlinked certificates

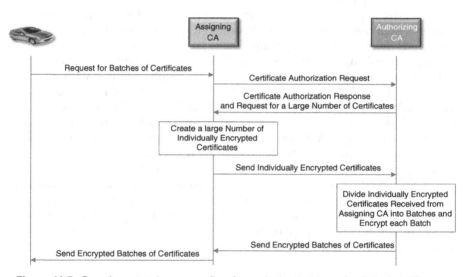

Figure 19.7. Sample protocol message flow for assigning batches of unlinked certificates

Before a vehicle has the necessary security credentials to directly acquire certificates from the CAs, the creation and assignment of the initial batches of certificates can be requested by an entity that is authorized to install certificates on vehicle. Once a vehicle is equipped with its certificate management keys and certificates, it can acquire new batches of certificates by sending Certificate Request messages to the Assigning CA. Regardless who originates the certificate requests, the Certificate Request messages can use the same message format shown in Figure 19.5.

A simplified protocol message flow is illustrated in Figure 19.7 for the case in which a vehicle requests batches of certificates from the Assigning CA directly.

To prevent the Assigning CA from knowing which public key (and therefore which certificates) belongs to which vehicle, the vehicle should not include multiple vehicle-generated public keys in the same Certificate Request message. Therefore, to allocate batches of certificates to the vehicles, it will be easier and more resource efficient for the Assigning CA to create both the private–public key pairs and their certificates for vehicles.

Upon receiving the Certificate Request, the Assigning CA sends a Certificate Authorization Request to the Authorization CA to request permission to issue certificate batches to the requesting vehicle. If the Authorizing CA authorizes the certificate request, it will send a Certificate Authorization Response back to the Assigning CA to request for a large number of certificates for this certificate requester. The Assigning CA can be required to create a significantly larger number of certificates than the number of certificates to be in a single certificate batch.

The Assigning CA creates the certificates requested by the Authorizing CA, encrypts each certificate individually using the same certificate encryption key, and sends them to the Authorizing CA.

The Authorizing CA divides these individually encrypted certificates into batches and encrypts each batch with a unique batch encryption key. It generates a random but unique *batch authorization number* for each certificate batch and stores each encrypted certificate batch with its batch decryption key and batch authorization number. This batch authorization number will be used by the Authorizing CA to locate the batch decryption key for a certificate batch when a batch of certificates should be revoked.

The Authorizing CA sends the encrypted certificate batches, with their batch authorization numbers, back to the Assigning CA. The Assigning CA in turn sends these encrypted certificate batches to the requesting vehicle.

If the Authorizing CA still has unused certificate batches for a vehicle after servicing a vehicle's current certificate request, it will use them for future certificate requests from this vehicle.

19.5.3 Revoke Batches of Unlinked Certificates

To revoke a batch of certificates with a known certificate batch authorization number, the Authorizing CA can locate the batch decryption key from its local database, decrypt the batch to recover individual certificates encrypted by the Assigning CA, and send the individually encrypted certificates to the Assigning CA. The Assigning CA decrypts each of these certificates. It can then place these certificates on the CRL or send them to the Authorizing CA if the Authorizing CA is responsible for issuing CRLs.

To revoke all batches of certificates that have been previously assigned to a given vehicle, the Authorizing CA can locate the batch authorization numbers for all the certificate batches that have been assigned to a vehicle, decrypt each batch to recover the individually encrypted certificates, and send them to the Assigning CA. The Assigning CA decrypts each certificate. It can then place these certificates on the CRL or send them to the Authorizing CA if the Authorizing CA is responsible for issuing CRLs.

If all the certificate batches that have been assigned to the vehicle that owns a misused certificate are to be revoked, the Assigning CA and the Authorizing CA can collaborate to use this misused certificate to find out which certificate batches have been assigned to this vehicle. This can be achieved as follows: The Assigning CA encrypts the misused certificate to generate an individually encrypted certificate and sends it to the Authorizing CA. The Authorizing CA searches through the certificate batches to locate the certificate batch that contains this encrypted certificate. The Authorizing CA sends all the individually encrypted certificates in this batch to the Assigning CA. The Assigning CA decrypts each individual certificate in the batch. Now, these certificates can be placed on the CRL.

The Authorizing CA further determines which vehicle has been assigned this batch of certificates and locates all the certificate batches assigned to this vehicle. It can open all these certificate batches and send the individually encrypted certificates to the Assigning CA. The Assigning CA can decrypt each individual certificate and put them on the CRL to be revoked.

19.5.4 Request for Decryption Keys for Certificate Batches

To request for decryption keys to unlock a batch of encrypted certificates that have been preloaded on a vehicle, the vehicle sends a Certificate Request to the Assigning CA using the sample message format shown in Figure 19.5 and sets the type of request to "decryption keys."

The Certificate Request message can contain the batch authorization number for a certificate batch that is on the vehicle and ask the CAs to provide the decryption keys for this batch. Alternatively, the vehicle can ask the Authorizing CA to select a batch to be unlocked.

Upon receiving a Certificate Request, the Assigning CA sends a Certificate Authorization Request to the Authorizing CA to request the batch decryption key. Upon positive authorization of the request, the Authorizing CA will select a certificate batch previously assigned to the vehicle if the vehicle does not supply a certificate batch authorization number. The Authorizing CA locates the batch decryption key for the selected certificate batch and sends it in a Certificate Authorization Response to the Assigning CA. This batch decryption key cannot be revealed to the Assigning CA because it will allow the Assigning CA to know which certificates are in this certificate batch. Therefore, the Authorizing CA encrypts the batch decryption key to the requesting vehicle using either the vehicle's public encryption key or a secret symmetric encryption key received from the vehicle in Part I of the Certificate Request.

The Assigning CA sends the encrypted batch decryption key from the Authorizing CA and its certificate decryption key to the vehicle.

The vehicle uses the batch decryption to open the batch and then uses the certificate decryption key to decrypt each certificate in the batch. Only the vehicle will know which certificates are in the certificate batch.

19.6   SPLIT CA ARCHITECTURE FOR ANONYMOUSLY LINKED SHORT-LIVED CERTIFICATES

Anonymously linked short-lived certificates pose unique challenges in preventing individual CAs from knowing which certificates are assigned to which vehicles. For example, using the certificate tag generation method described in Chapter 17, the CA that knows the tag seeds can determine which tags (and therefore which certificates) belong to the same vehicle.

To ensure that no individual certificate management entity can identify or track vehicles based on certificates, we need to ensure that no individual certificate management entity knows all the secret materials used to create certificate tags. Here, we will first describe how the Baseline Split CA Architecture can be extended to achieve this goal when assigning one certificate at a time. We then describe ways to allocate batches of anonymously linked short-lived certificates.

19.6.1   Assign One Anonymously Linked Short-Lived Certificate at a Time

The tag for each certificate can be generated using two *partial tags* supplied by different and independent entities that do not know the partial tags created by each other.

The Authorizing CA provides one partial tag for each certificate. A separate Independent Tag Provider is introduced into the split CA architecture to provide a second partial tag for each certificate.

To prevent any partial tag provider from knowing the partial tags provided by the other partial tag provider, each partial tag provider encrypts its partial tags to the Assigning CA. The Assigning CA decrypts the partial tags and uses them to create a final tag for each certificate.

Revoking a vehicle can be done by publishing the seeds used by the Authorizing CA and the Independent Tag Provider to create the partial tags for the revoked vehicle.

Figure 19.8 illustrates a sample protocol message flow to support anonymously linked short-lived certificates. To request an anonymously linked certificate, a vehicle v sends a Certificate Request to the Assigning CA. The Assigning CA requests permission from the Authorizing CA to issue the requested certificate and also requests the Authorizing CA to provide the partial tags for the certificate.

Upon authorizing the certificate request, the Authorizing CA creates a first partial tag for the certificate. It can use the same tag generation method described in Chapter 17 to generate the partial tag. Specifically, it first creates

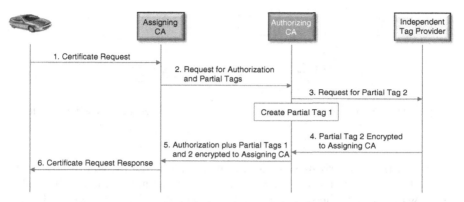

Figure 19.8. Sample protocol message flow for supporting anonymously linked certificates

a tag seed s_1 (v) for vehicle v if it has not already done so. It then creates a partial tag for the ith certificate for vehicle v as: Partial_Tag$_1$ $(v, i) = H (s_1(v), i)$. The Authorizing CA encrypts this partial tag to the Assigning CA

In the meantime, the Authorizing CA requests a second partial tag from the Independent Tag Provider. To prevent the Independent Tag Provider from knowing which particular vehicle will receive a partial tag, the Authorizing CA creates a pseudonym v' for vehicle v in such a way the Independent Tag Provider cannot link the pseudonyms to the vehicle. The Authorizing CA then requests the Independent Tag Provider to issue a new partial tag for pseudo vehicle v'.

The Independent Tag Provider can use the same method used by the Authorizing CA to create its partial tags. That is, the Independent Tag Provider first creates a tag seed s_2 (v') for pseudo vehicle v' if it has not done so already. The Independent Tag Provider then creates a partial tag for the ith certificate for pseudo vehicle v' as: Partial_Tag$_2$ $(v', i) = H (s_2(v'), i)$. The Independent Tag Provider encrypts this partial tag to the Assigning CA.

The partial tag provided by the Authorizing CA can be used as the certificate authorization number for each certificate authorized by the Authorizing CA. Therefore, the Authorizing CA does not have to create separate certificate authorization numbers.

The Authorizing CA sends the following to the Assigning CA:

- The partial tag created by the Authorizing CA, which is encrypted by the Authorizing CA to the Assigning CA, and
- The partial tag received from the Independent Tag Provider, which is encrypted by the Independent Tag Provider to the Assigning CA.

The Assigning CA decrypts the partial tags and uses them to create the final tag for the certificate. The final tag can be created, for example, by hashing the two partial tags together as $H(\text{Partial\_Tag}_1(v, i), \text{Partial\_Tag}_2(v', i))$.

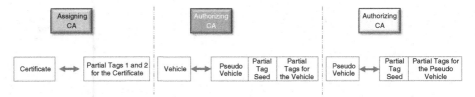

Figure 19.9. Mappings used for assigning and revoking anonymously linked certificates

Neither the Authorizing CA nor the Independent Tag Provider knows both partial tags used to create a final tag. Therefore, neither of them will be able to know or produce the final tag on any certificate. Neither of them will know which final tag is assigned on which certificate or vehicle. Therefore, neither of them will have sufficient information to link certificates to the same vehicle.

The Assigning CA knows both partial tags used to create the final tag for each certificate. But it cannot link different partial tags or different final tags to the same vehicle. This is because (1) the Assigning CA does not know which vehicle will be assigned with any particular certificate and (2) different partial tags cannot be linked to the same vehicle without knowing the seeds used to create them.

The Authorizing CA, the Assigning CA, and the Independent Tag Provider maintain mappings illustrated in Figure 19.9.

To revoke all the certificates that have been previously assigned to a vehicle v, the Authorizing CA locates the seed it used to generate its partial tags for this vehicle. It sends the pseudonym v' of vehicle v to the Independent Tag Provider to request the seed used by the Independent Tag Provider for the vehicle. The vehicle v can then be revoked by placing the two seeds on the CRL.

If all the certificates previously assigned to the vehicle that owns a misused certificate may need to be revoked, the Assigning CA can use the misused certificate to locate the pair of partial tags for the certificate and sends them to their respective providers—the Authorizing CA and the Independent Tag Provider. Each partial provider locates the seed it used to create its partial tag. The vehicle can be revoked by placing the two partial tag seeds on the CRL.

A vehicle, knowing the procedure to create final tags with partial tags, can reproduce the final tags generated by each pair of partial seeds on the CRL and use these final tags to determine whether a received certificate is on the CRL.

The Assigning CA could cheat. It could, for example, ignore the partial tags provided by the Authorizing CA or the Independent Tag Provider and create the final tags in its own way so that it can link the certificates to the same vehicle. This, however, can be readily detected by an auditing agency. An auditing agency could sample past certificates to verify whether the tags on these certificates are created correctly. The auditing agency can randomly select past

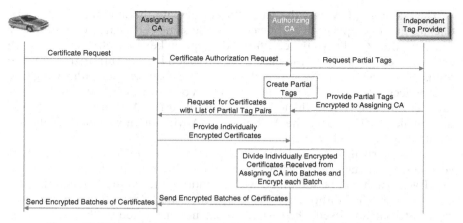

Figure 19.10. Creating encrypted batches of anonymously linked certificates

certificates, obtain the corresponding partial tags from the Authorizing CA and the Independent Tag Provider, and reproduce the final tags for the selected certificates to verify the correctness of the actual tags on the certificates.

19.6.2 Assign Batches of Anonymously Linked Short-Lived Certificates

We describe how to extend the Baseline Split CA Architecture to assign batches of anonymously linked certificates vehicles while preventing any individual CA from having sufficient information to track vehicles based on certificates.

To prevent any individual CA from having sufficient information to use certificates to track vehicles, no individual CA should know which certificates are in each batch of certificates. A sample protocol message flow to achieve this goal is illustrated in Figure 19.10.

The vehicle requests batches of certificates from the Assigning CA. The Assigning CA requests permissions from the Authorizing CA to issue the requested certificate batches. When sending its certificate authorization to the Assigning CA, the Authorizing CA will request the Assigning CA to create a large number of certificates as in the case of unlinked short-lived certificates. The Authorizing CA creates for each vehicle a list of partial tags and requests the Independent Tag Provider to provide a second list of independent partial tags for the same vehicle. As before, the partial tags will be encrypted by their creators to the Assigning CA so partial tag creators will not know the partial tags created by each other. The Authorizing CA sends the pairs of partial tags $<\text{Partial\_Tag}_1(v, i), \text{Partial\_Tag}_2(v', i)>$ to the Assigning CA in random orders. The Authorizing CA can mix the partial tag pairs for different vehicles so that the Assigning CA will not know which pairs are for which vehicles.

The Assigning CA uses the partial tags to create the final tags for the certificates and encrypts each certificate with the same certificate encryption key. The Assigning CA keeps the certificate decryption key to itself. It indexes each encrypted certificate with the partial tag provided by the Authorizing CA and sends the encrypted certificates with their indices back to the Authorizing CA. The indices allow the Authorizing CA to recognize which encrypted certificates are for the same vehicle but will not allow the Authorizing CA to see the certificates. Therefore, the Authorizing CA will not know which certificates are assigned to which vehicles.

The Authorizing CA divides the individually encrypted certificates for each vehicle into batches and encrypts each batch with a unique batch encryption key. Each batch will be identified with a random but unique batch authorization number. The Authorizing CA keeps the batch decryption key for each certificate batch to itself. The certificate batches encrypted by the Authorizing CA can then be loaded onto to vehicles by either the Assigning CA or the Authorizing CA in the same way batches of unlinked short-lived certificates can be assigned to vehicles.

Neither the Assigning CA nor the Authorizing CA will know which certificates are in each certificate batch.

Figure 19.11 summarizes the mappings maintained by the Assigning CA, the Authorizing CA, and the Independent Tag Provider.

19.6.3 Revoke Batches of Anonymously Linked Short-Lived Certificates

With the mappings shown in Figure 19.11, the Assigning CA, the Authorizing CA, and the Independent Tag Provider can collectively identify the partial tag seeds used for a vehicle to revoke all the certificates that have been assigned to the vehicle. Once identified, the partial seeds can be placed on the CRL to revoke the corresponding vehicle.

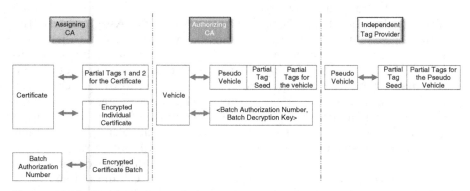

Figure 19.11. Mappings for assigning and revoking batches of anonymously linked certificates

If the vehicle to be revoked is known, the Authorizing CA can locate its tag seed for this vehicle. The Authorizing CA can send the vehicle's pseudonym to the Independent Tag Provider to allow the Independent Tag Provider to locate its seed for the vehicle. The vehicle can now be revoked by placing the two tag seeds on the CRL.

To locate the tag seeds for the vehicle that owns a misused certificate, the Assigning CA uses the misused certificate to locate the two partial tags used to create the final tag for the misused certificate and then sends the partial tags to their respective providers. The Authorizing CA and the Independent Tag Provider can each locate their tag seeds for generating their partial tags. The vehicle can now be revoked by placing the two tag seeds on the CRL.

The Assigning CA, the Authorizing CA, and the Independent Tag Provider can also collaborate to revoke an individual batch of certificates, such as a batch that contains a misused certificate. This can be accomplished as follows: The Assigning CA encrypts the misused certificate with the same certificate encryption key it used to encrypt the certificates when the certificates were initially created. The Assigning CA sends the encrypted misused certificate to the Authorizing CA, which can use it to locate the batch that contains this certificate. The Authorizing CA opens this batch with its batch decryption key and sends the individually encrypted certificates in the batch to the Assigning CA. The Assigning CA decrypts each individually encrypted certificate. These certificates can be placed on the CRL to be revoked. If the tags on the certificates in each batch are created with a pair of batch-specific tag seeds, this pair of tag seeds can be located by the Authorizing CA and the Independent Tag Provider so that only this pair of tag seeds needs to be placed on the CRL.

19.6.4 Request for Decryption Keys for Certificate Batches

To request for the decryption keys to unlock a batch of encrypted certificates, a vehicle sends a Certificate Request to the Assigning CA using the same sample format shown in Figure 19.5 and sets the type of request in the message to "decryption keys."

As in the case of requesting decryption keys for batches of unlinked short-lived certificates, the Certificate Request can contain the batch authorization number for a certificate batch that is already on the vehicle to ask the CAs to provide decryption keys for this certificate batch. Alternatively, the vehicle can ask the Authorizing CA to select a certificate batch to be unlocked.

Upon receiving the Certificate Request, the Assigning CA sends a Certificate Authorization Request to the Authorizing CA to request the batch decryption keys. Upon positive authorization of the request, the Authorizing CA selects a certificate batch to be unlocked if the Certificate Request message does not carry a batch authorization number. The Authorizing CA sends a Certificate Authorization Response to the Assigning CA with the batch decryption key. This batch decryption key cannot be revealed to the Assigning CA because it will allow the Assigning CA to know which certificates are in this batch of certificates. Therefore, the Authorizing CA encrypts the batch decryption to the vehicle.

The Assigning CA sends its certificate encryption key together with the batch decryption key received from the Authorizing CA to the vehicle.

The vehicle uses the batch decryption key to open the certificate batch and then uses the certificate decryption key to decrypt each certificate in the batch. Only the vehicle will know which certificates are in each certificate batch.

REFERENCES

[KCKL07] T. Kwon, J. H. Cheon, Y. Kim, and J.-I. Lee: "Privacy Protection in PKIs: A Separation-of-Authority Approach," Information Security Applications, Lecture Notes in Computer Science, Vol. 4298, 2007.

[WVZM07] T. Zhang, R. White, D. Mok, R. Ferrer, G. D. Crescenzo, and E. van den Berg: "Vehicle Segment Certificate Management Scalability Analysis, VII Security Work Order Part I Deliverables 1.2," 2007.

20

COMPARISON OF PRIVACY-PRESERVING CERTIFICATE MANAGEMENT SCHEMES

20.1 INTRODUCTION

This chapter summarizes and compares the main characteristics of the three categories of privacy-preserving certificate management schemes: shared certificates, short-lived unique certificates, and group signatures. First, we will provide a summary of their main characteristics. Then, we will compare the schemes in several areas: misbehavior detection, misbehavior reporting, and ability to prevent certificate authority (CA) operators from abusing vehicle privacy. Finally, we will summarize the remaining technical challenges that should be addressed to make each category of privacy-preserving certificate scheme deployable in a large-scale consumer vehicle network.

The characterization and comparison consider the following aspects:

- Privacy levels
- Signature verification speeds
- Signature generation speeds
- Certificate revocation list (CRL) size

Vehicle Safety Communications: Protocols, Security, and Privacy, First Edition. Luca Delgrossi and Tao Zhang.
© 2012 John Wiley & Sons, Inc. Published 2012 by John Wiley & Sons, Inc.

- Difficulty to support misbehavior detection
- Difficulty to prevent CA operators to breach vehicle privacy
- Impact on other protocols layers

We use the following notations:

- For all certificate schemes:
  - V is the number of vehicles in the network
  - m is the number of revoked vehicles on a CRL
  - Standard signature algorithms means: Elliptic Curve Digital Signature Algorithm (ECDSA), Rivest–Shamir–Adleman (RSA), or Digital Signature Algorithm (DSA)
- For shared certificates:
  - N is the number of certificates in the Shared Certificate Pool
  - n is the number of certificates assigned to each vehicle
  - R_1 is the number of revoked certificates on a CRL
- For short-lived certificate schemes:
  - R_2 is the total number of nonexpired certificates on all the m revoked vehicles identified on a CRL. R_2 can be estimated as $m \cdot \Phi$, where Φ is the expected number of nonexpired certificates on each of the m revoked vehicles identified on a CRL.

20.2   COMPARISON OF MAIN CHARACTERISTICS

Table 20.1 summarizes the characteristics of shared certificates, short-lived certificates, and digital signatures.

Table 20.1. Comparison of privacy-preserving certificate schemes

	Shared Certificates	Short-Lived Certificates	Group Signatures
Privacy levels	Average anonymity size: $(n/N)V$. Privacy reduces in low vehicle density areas. But reasonable solutions exist.	Anonymity size: V assuming information on certificates is completely randomized and each certificate has very short lifetime. Messages are linkable when a vehicle uses the same certificate.	Anonymity size: size of a group. Privacy reduces when a vehicle travels to areas with few members of its group.
Signature verification speed	Standard signature algorithms. No special processing required for CRL. $O(\log_2 R_1)$ comparisons to search CRL to determine if a certificate is revoked. Fastest signature verification among the three categories of schemes.	Standard signature algorithms. $O(m \cdot \Phi)$ hashing plus $O(m \cdot \Phi \log_2(m \cdot \Phi))$ comparisons to process a CRL. $O(\log_2(m \cdot \Phi))$ comparisons to search CRL to determine if a certificate is revoked.	Group signature schemes. Significantly more time consuming than standard signature algorithms. Typically $O(m)$ exponentiations to verify if a group member has been revoked.
Signature generation speed	Standard signature algorithms.	Standard signature algorithms.	Group signature algorithms. Significantly slower than standard signature generation algorithms in general.
CRL size	$O(m)$	$O(m)$: linked certificates. $O(m \cdot \Phi)$: unlinked certificates	$O(m)$

(Continued)

Table 20.1. (*Continued*)

	Shared Certificates	Short-Lived Certificates	Group Signatures
Vehicle local misbehavior detection	Can use message contents and message headers to detect suspicious messages	Can use message contents and message headers to detect suspicious messages	Can use message contents and message headers to detect suspicious messages
Misbehavior reporting	Can report certificates of suspicious messages	Can report certificates of suspicious messages	Have to report suspicious messages with their signatures. This can result in significantly heavier overheads.
Global misbehavior detection	Can support anonymous detection: MDS does not need to identify vehicles or link certificates to vehicles for detecting misused certificates or misbehaving vehicles. This helps prevent MDS operators from breaching vehicle privacy.	MDS needs to be able to identify vehicles and links certificates to vehicles to enable misbehavior detection. MDS operators need to have the ability to breach vehicle privacy.	MDS needs to be able to identify vehicles and links certificates to vehicles to enable misbehavior detection. MDS operators need to have the ability to breach vehicle privacy.
Revocation	Revoking a certificate impacts all innocent vehicles that share the certificate. This negative impact grows with the level of privacy and the number of revoked vehicles.	Revoking a vehicle can significantly reduce signature verification speed on all vehicles. This negative impact grows with the number of revoked vehicles.	Revoking a vehicle can significantly reduce signature verification speed on all vehicles. This negative impact grows with the number of revoked vehicles.

Privacy protection against CA and MDS operators	Can be accomplished with a simple split CA architecture. Misbehaving vehicles may be detected and revoked anonymously.	Significantly more difficult to accomplish when anonymously linked short-lived certificates are used, as compared with using shared certificates or unlinked short-lived certificates. Likely requires more independent certificate management entities, such as a separate certificate tag server, when anonymously linked certificates are used. MDS will likely need the ability to breach vehicle privacy, that is, to link certificates to the same vehicle, in order to detect misbehaving vehicles.	Infeasible with existing group signature schemes because a group manager, which is a necessary component of group signature schemes, can determine who generated each signature. MDS needs the ability to link multiple certificates to the same vehicle to detect misbehaving vehicles. That is, the MDS needs the ability to open signatures.
Impact on networking protocols	Requires media access layer and Internet Protocol (IP) layer addresses to change each time a vehicle uses a different certificate. This, however, does not occur frequently	Requires media access layer and IP layer addresses to change each time a vehicle uses a different certificate. This occurs highly frequently.	Does not cause media access layer and IP layer addresses to change.

20.3   MISBEHAVIOR DETECTION

With every privacy-preserving certificate scheme discussed in this book, vehicles can use their local misbehavior detection capabilities to detect erroneous or malicious vehicle safety messages based on the plaintext information in message contents and headers. These local misbehavior detection capabilities will be largely independent of specific privacy-preserving certificate schemes.

The privacy-preserving certificate schemes, however, have a direct and profound impact on the following:

- how vehicles can report locally detected misbehaviors to global misbehavior detection systems (MDSs)
- how an MDS can detect misused certificates and misbehaving vehicles
- whether an MDS needs to have the ability to breach vehicle privacy in order to detect misbehavior

With shared certificate schemes, vehicles can report to an MDS the certificates associated with malicious messages they have detected. The MDS can then decide whether a certificate has been misused based on the misbehavior reports and the statistical differences vehicles have demonstrated in the process of requesting for replacement certificates (Chapter 17). The MDS may do so without having to know which particular vehicle misused a certificate. The CA revokes misused certificates and does not have to explicitly revoke vehicles. When a misbehaving vehicle exhausted its certificate quota, the misbehaving vehicle is effectively revoked. This anonymous revocation can prevent a global misbehavior detection system from having the ability to breach vehicle privacy.

With short-lived certificates, each certificate is only used for a very short period of time. Therefore, messages signed with each individual certificate may not be sufficient to decide whether the message originator is misbehaving. Messages signed by a vehicle with different certificates may be detected as malicious and these certificates may be reported to the MDS. The MDS will need to link the suspected certificates to their originating vehicles to determine whether a vehicle is misbehaving. Furthermore, with short-lived certificates, it no longer makes sense to revoke just one individual certificate because a suspected certificate will likely have expired by the time it is detected to be misused. The CA needs to revoke all the short-lived certificates previously assigned to a misbehaving vehicle and/or the misbehaving vehicle's privilege to receive future certificates, Therefore, the MDS and the CA needs the ability to identify and track vehicles, that is, the ability to breach vehicle privacy.

Today's leading group signature mechanisms do not use public certificates for individual group members. Instead, the proof of group membership is embedded in each signature and in the signature verification process using zero-knowledge proof of knowledge techniques. Only the group manager can open a signature to determine which group member signed a message. Opening a signature, however, will require the group manager to have the signed message and the signature. Therefore, when a vehicle reports a suspicious message to the MDS, the vehicle will have to send the signed message and its signature to the MDS. The MDS will need to have the functionality of the group manager to determine which vehicle generated the signature on the suspected message. This means that the MDS will need the ability to identify and track vehicles, that is, the ability to breach vehicle privacy. Furthermore, sending messages and their signatures to the MDS can consume heavy wireless bandwidth.

20.4   ABILITIES TO PREVENT PRIVACY ABUSE BY CA AND MDS OPERATORS

As discussed in Chapter 19, a split CA architecture can be used to prevent any individual CA from having enough information to breach vehicle privacy when shared certificates are used. Furthermore, when shared certificates are used, a misused certificate can be detected and revoked without having to know which vehicles misused the certificate. That is, a misbehaving vehicle can be revoked anonymously. This means that an MDS does not need to have the ability to identify or track vehicles and hence does not need to have sufficient information to breach vehicle privacy.

When unlinked short-lived certificates are used, a vehicle requests for a batch of new certificates at a time. A split CA architecture can still be used to ensure that any individual CA will have complete knowledge of which certificates have been assigned to which vehicles. However, more sophisticated cryptographic mechanisms will be required to anonymously provide batches of certificates to a vehicle without giving any individual CA complete knowledge of certificate-to-vehicle mappings.

When anonymously linked short-lived certificates are used to reduce CRL size, however, multiple independent entities would be required to create independent tag seeds or partial tags that can be integrated by a CA to create the final tags for the certificates. This, however, will significantly increase system complexity and reduce system operability because more independent operators will be required to operate these tag-generating entities. More sophisticated cryptographic mechanisms and protocols will be required to anonymously provide batches of anonymously linked certificates to vehicles to prevent each individual CA from acquiring sufficient information that can be used to identify or track vehicles. This, however, will further increase the complexity of the CA system. Furthermore, the MDS will need the ability to link certificates to their subjects to detect misbehaving vehicles.

With existing group signature schemes, a group manager by design will be able to determine who created a signature and will therefore have sufficient information to identify and track vehicles. This makes it infeasible to protect privacy against CA or MDS operators.

20.5 SUMMARY

All existing privacy-preserving certificate schemes still have significant technical issues that must be addressed before they can become suitable for supporting a nationwide consumer vehicle network.

- Shared certificate schemes require more effective ways to reduce collateral damages of certificate revocation.
- Short-lived certificate schemes require ways to significantly reduce the time required for processing a CRL to determine whether a certificate is revoked. More sophisticated public key infrastructure (PKI) architecture and certificate management mechanisms will be required to protect privacy against CA operators. With existing anonymously linked short-lived certificates, the global misbehavior detection system will need to have the ability to link messages to the message-originating vehicles.
- Group signature schemes need to be improved to significantly reduce signature creation and signature verification delays, especially when a large number of vehicles are revoked. Furthermore, with existing group signature schemes, a group manager by design will be able to identify and track vehicles.
- For all privacy-preserving certificate schemes, significantly more effective ways are required to detect misused certificates and misbehaving vehicles and to revoke misbehaving vehicles without causing significant inconvenience to innocent drivers.

21

IEEE 1609.2
SECURITY SERVICES

21.1 INTRODUCTION

The Institute of Electrical and Electronics Engineers (IEEE) 1609.2 standard specifies a set of security services for supporting vehicular communications. It defines message formats and message processing for authenticating wireless access in vehicular environments (WAVE) management messages and vehicle application messages, and for encrypting messages to known recipients.

This chapter provides an overview of the IEEE 1609.2 standard based on its latest version published in April 2011 [IEEE11]. We focus on the message formats and message processing for supporting certificate management.

We will also discuss several capabilities that are required to support certificate management for consumer vehicle networks but are not covered in the current IEEE 1609.2 standard. These capabilities need to be addressed either by future extensions to the IEEE 1609.2 standard or by mechanisms external to the IEEE 1609.2 standard.

21.2 THE IEEE 1609.2 STANDARD

To use IEEE 1609.2 security services, each communicating entity must have an IEEE 1609.2 security service subsystem. Figure 21.1 illustrates the

Vehicle Safety Communications: Protocols, Security, and Privacy, First Edition. Luca Delgrossi and Tao Zhang.
© 2012 John Wiley & Sons, Inc. Published 2012 by John Wiley & Sons, Inc.

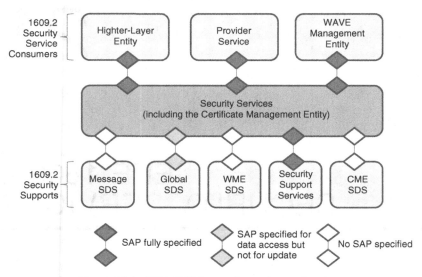

Figure 21.1. IEEE 1609.2 security services subsystem

functional architecture of the IEEE 1609.2 security service subsystem in the context of a WAVE device.

The IEEE 1609.2 security services are *used* by security consumers. Security consumers include (1) higher-layer applications and protocols, (2) WAVE provider services that provide services to end users and applications, and (3) WAVE management entities that manage WAVE devices.

The IEEE 1609.2 security services *use* security data stores (SDSs) to maintain security-related information. Message SDSs are used to store security-related data for each local security consumer on the device. A Global SDS is used to store security-related data that are relevant to the entire security services subsystem.

The IEEE 1609.2 security services also *use* the security support services provided by the device on which the security services run. These security support services include providing the current time, current location, and a source of random numbers that the security services need to use to perform security operations.

The functional entities showed in Figure 21.1 access each other's services through service access points (SAPs). Figure 21.1 shows which SAPs related to security services are defined in the current version of IEEE 1609.2.

The security services defined in IEEE 1609.2 rely on elliptic curve cryptography (ECC), public key certificates and the public key infrastructure (PKI).

Each entity, such as a vehicle or a certificate authority (CA) server, must have IEEE 1609.2 certificates and a *Certificate Management Entity* (CME) that manages these certificates. A CME is a set of functions for (1) requesting certificates and certificate revocation lists (CRLs) from the CA; (2) processing received certificates, CRLs, and other security management messages; and (3) managing security-related information. CME is part of the IEEE 1609.2 security services subsystem.

In addition to the certificate management functions provided by the CME, the IEEE 1609.2 security services subsystem provides cryptographic services such as creating digital signatures, verifying signatures, and encrypting and decrypting messages.

A security consumer, such as a vehicle safety application, must obtain a certificate before it can send signed or encrypted messages. Once a certificate is acquired, a security consumer can call the IEEE 1609.2 security services to prepare signed or encrypted messages to send.

21.3   CERTIFICATES AND CERTIFICATE AUTHORITY HIERARCHY

The IEEE 1609.2 standard classifies all the entities that provide or use IEEE 1609.2 security services into two categories: certificate authority entities (CA entities) and end entities. CA entities issue certificates and CRLs. All other entities that use IEEE 1609.2 certificates, but cannot issue certificates or CRLs, are end entities. End entities include vehicles, roadside units (RSUs), application servers, and applications.

The IEEE 1609.2 standard defines the following types of CA entities:

- *Root CAs*: Root CAs are trusted to issue certificates to all other CA entities and all end entities. The public keys of a Root CA are trusted by end entities and no certificates for these public keys will be required. A Root CA may issue certificates to other CA entities to authorize them to issue certificates or CRLs to end entities.
- *Message CAs*: Message CAs issue certificates to end entities that send application messages secured with IEEE 1609.2.
- *WAVE Service Advertisements (WSA) CAs*: WSA CAs issue certificates to end entities that send WSA. An end entity uses WSAs to tell other end entities what WAVE services it provides.
- *CRL Signers*: CRL Signers are CA entities that are authorized to issue CRLs, but cannot issue certificates.

The CA hierarchy defined in IEEE 1609.2 is illustrated in Figure 21.2.

IEEE 1609.2 defines three types of end entities: Identified, Identified Not Localized, and WSA Signers. The Identified and the Identified Not Localized end entities are entities that send application messages secured with IEEE 1609.2 security services. These end entities obtain their certificates from the Message CAs. WSA Signers are end entities that send signed WSAs. WSA Signers obtain their certificates from the WSA CAs. All end entities obtain CRLs from the CRL Signer.

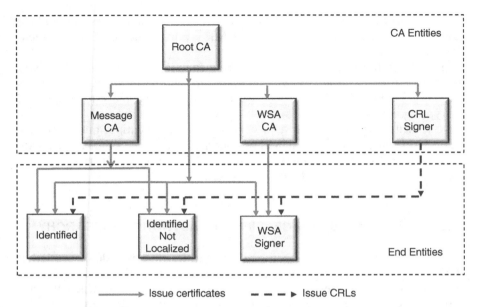

Figure 21.2. CA hierarchy defined in IEEE 1609.2

The IEEE 1609.2 standard classifies messages into two basic categories: certificate management messages and application messages. Certificate management messages are messages sent between end entities (e.g., vehicles) and CA entities to support certificate management functions such as for vehicles to acquire certificates and CRLs from the CA. Application messages are messages sent by the applications, such as vehicle safety applications, that run on a vehicle or other WAVE devices.

Each end entity uses separate sets of certificates to process certificate management messages and application messages. The certificates used to process certificate management messages are called *security management certificates*. The certificates used to process application messages are called *communications certificates*.

Communications between an end entity and a CA requires mutual authentication. This mutual authentication is achieved using two types of security management certificates: (1) a Certificate Signing Request (CSR) certificate used by the end entity to authenticate to the CA and (2) a CA certificate used by the CA to authenticate to the end entity.

The Identified and the Identified Not Localized end entities use Message CSR certificates to authenticate to the Message CAs. That is, they use Message CSR certificates to sign the Message Certificate Signer Request (CSR) messages they send to the Message CAs to request certificates. The WSA Signers use WSA CSR certificates to authenticate to the WSA CAs.

A certificate contains, implicitly or explicitly, at least one public key for a public key cryptosystem, and a list of the permissions associated with that public key. The permissions specify what the private–public key pair associated with this certificate can be used for.

21.4 FORMATS FOR PUBLIC KEY, SIGNATURE, CERTIFICATE, AND CRL

This section describes the data structures defined in the IEEE 1609.2 standard for public keys, digital signatures, certificates, and CRLs.

21.4.1 Public Key Formats

The IEEE 1609.2 standard uses Elliptic Curve Digital Signature Algorithm (ECDSA) for digital signatures and Elliptic Curve Integrated Encryption Scheme (ECIES) for public key encryption. Using ECDSA and ECIES, a public key is a point on an elliptic curve that can be represented by the x- and y-coordinates of this point on the elliptic curve. The IEEE 1609.2 standard defines a public key format illustrated in Figure 21.3, which can be used to encode an ECDSA or ECIES public key.

The Algorithm field indicates which public key algorithm this public key should be used with. The current IEEE 1609.2 standard supports the following public key algorithms:

- ECDSA over two elliptic curves defined by National Institute of Standards and Technology (NIST) over prime fields: the P244 curve for 112-bit security strength and the P256 curve for 128-bit security strength.
- ECIES over the P256 elliptic curve defined by NIST.

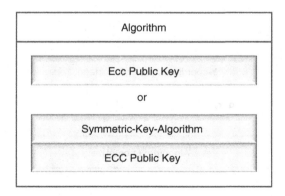

Figure 21.3. The Public Key format

For ECDSA, an ECDSA Public Key structure follows the Algorithm field. For ECIES, a Symmetric-Key-Algorithm flag and an ECC Public Key structure will follow the Algorithm field. The Symmetric-Key-Algorithm flag indicates which symmetric key algorithm should be used with ECIES for symmetric key encryption of subsequent messages. The only symmetric key algorithm currently supported in IEEE 1609.2 is Advanced Encryption Standard (AES).

The Public Key structure in Figure 21.3 can also be used to encode a public key for an arbitrary public key algorithm. In this case, the Algorithm field will be set to UNKNOWN and an opaque public key will follow the Algorithm field. The communicating parties must use other information to determine which public key algorithm to use.

The ECC public key format defined in IEEE 1609.2 is illustrated in Figure 21.4. The Field-Size field contains the length of the integers representing the x- and y-coordinates of a point on the selected elliptic curve. The Field Size will be 28 octets (224 bits) for the NIST P224 elliptic curve and 32 octets (256 bits) for the NIST P256 curve. The Type field indicates whether the elliptic curve point is compressed. If it is not compressed, the y-coordinate field will contain the y-coordinate of the elliptic curve point. If the point is compressed, the y-coordinate field will not be present.

21.4.2 Signature Formats

The digital signature structure is shown in Figure 21.5. The Algorithm field indicates the public key algorithm that was used to create the signature. For ECDSA over elliptic curve P224 or P256, the Signature field will contain an ECDSA signature structure. If the Algorithm field indicates an unknown algorithm, the Signature field will contain a variable-length opaque signature.

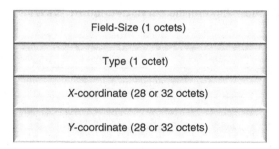

Figure 21.4. The ECC Public Key format

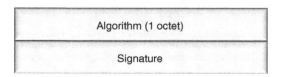

Figure 21.5. The digital signature format

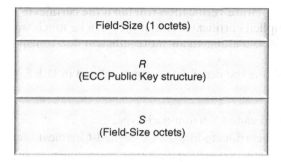

Figure 21.6. The ECDSA Signature format

Figure 21.6 illustrates the ECDSA signature structure defined in the IEEE 1609.2 standard. This ECDSA signature structure can be encapsulated in the signature format in Figure 21.5. Recall that an ECDSA signature is a pair of integers (r, s). An ECDSA signature can also be represented as (R, s), where R is a temporary point on the elliptic curve that is used as part of the signature by an accelerated ECDSA signature verification algorithm proposed in [AnGV05], and s is an integer. The IEEE 1609.2 standard supports both of these ECDSA signature representations. To achieve this flexibility, the ECDSA signature structure consists of an ECC Public Key structure—the same as shown in Figure 21.6—that can be used to encode either r or R. The ECDSA signature structure further consists of an integer field for encoding integer s. The Field-Size field contains the bit length of the x-coordinate and the y-coordinate of the elliptic curve, which will be the bit length of r and s. The value of the Field-Size field should be 28 octets (224 bits) for elliptic curve P224 and 32 octets (256 bits) for elliptic curve P256.

When the ECDSA signature is encoded as an integer pair (r, s), r shall be encoded as an ECC Public Key structure shown in Figure 21.6 with Type set to x-coordinate only. If the ECDSA signature is encoded as (R, s), the elliptic curve point R shall be encoded as an ECC Public Key structure with Type set to compressed_lsb_y_0, compressed_lsb_y_1, or uncompressed at the sender's discretion.

21.4.3 Certificate Format

The IEEE 1609.2 standard supports both explicit certificates and implicit certificates. An explicit certificate includes the public key certified by the certificate and the digital signature of the certificate issuer. A user verifies the certificate by verifying the certificate issuer's signature. An implicit certificate is a variant of public key certificate. It does not explicitly include the public key certified by the certificate. Any user can reconstruct the public key from the information on the certificate. An implicit certificate does not include the signature of the certificate issuer. Upon reconstructing the public key, the user simply uses the public key as input to the ECC signature verification algorithm. The signature verification will fail if the certificate is invalid. A main advantage of implicit certificates is that they can be much smaller in size than explicit certificates, making them more efficient for transport over wireless networks.

Figure 21.7 shows the certificate format defined in IEEE 1609.2. It consists of three parts:

- a header field called Version-and-Type,
- the unsigned certificate in a To-Be-Signed-Certificate format, and
- the signature of the certificate issuer for explicit certificate or a reconstruction value for reconstructing the public key for an implicit certificate.

The header field Version-and-Type contains the version of the certificate format and indicates whether the certificate is explicit or implicit. An explicit

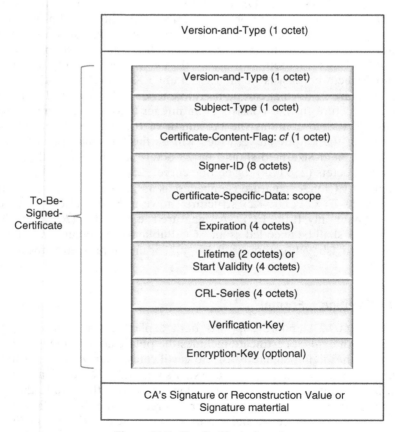

Figure 21.7. The Certificate format

certificate explicitly contains the certificate holder's public key and the signature of the CA that issued this certificate. The CA's signature covers the To-Be-Signed-Certificate. The CA that signed this certificate is identified in the Signer ID field inside the To-Be-Signed-Certificate.

If this is an implicit certificate, it will not explicitly contain the certificate holder's public key or the CA's signature. Instead, it will contain a Reconstruction Value provided by the CA that can be used by any verifier to recover the certificate holder's public key.

The To-Be-Signed-Certificate contains the certificate contents. This structure has the following main fields:

- The Subject-Type field indicates the type of the certificate. The following types of certificate are defined in the current version of the IEEE 1609.2:
  - Message Anonymous (this field is for future use)
  - Message Identified Not Localized
  - Message Identified Localized
  - Message CSR
  - WSA
  - WSA CSR
  - Message CA
  - WSA CA
  - CRL Signer
  - Root CA.
- The Certificate-Content-Flag *cf* indicates whether additional optional fields are present in the certificate.
- The Signer-ID field is present only on an explicit certificate and it will contain the identifier of the certificate of the CA that issued this certificate. The identifier of a certificate is the low-order 8 octets of the SHA-256 hash value of the certificate.
- The Certificate-Specific Data field scope contains information that is unique to this certificate.
- The Expiration field contains the last time the certificate is valid.
- The Lifetime or Start Validity field will be present depending on the value of the Certificate-Content-Flag. Lifetime means that the certificate is valid from (Expiration–Lifetime) to Expiration. Start Validity means that the certificate is valid from the Start Validity time to the Expiration time.
- The CRL-Series field indicates which CRL this certificate will appear on if it is revoked.
- The Verification-Key field contains the public key that should be used to verify signatures generated by holder of this certificate. The public key is formatted in the Public Key structure shown in Figure 21.7.
- The optional Encryption Key field contains a public key for encryption.

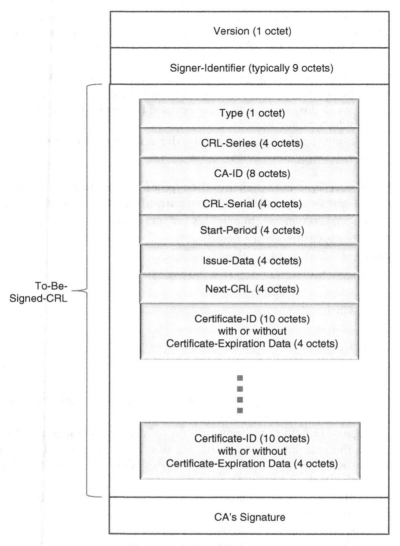

Figure 21.8. The CRL format

21.4.4 CRL Format

The CRL format is shown in Figure 21.8. It consists of three parts: (1) two header fields, (2) the contents of the CRL in the To-Be-Signed-CRL format, and (3) the signature of the CA that issued the CRL.

The first header field Version contains the version number of the CRL format, which is 1 in the current version of IEEE 1609.2. The second header field Signer-Identifier identifies the certificate for the public key used to generate the signature in the Signature field. This field can contain the following:

- The identifier of the certificate, which is the low-order 8 octets of the SHA-256 hash of the certificate.
- The certificate itself.
- A certificate chain.
- The identifier of the certificate and the indication of the public key algorithm when ECDSA is not the public key signature algorithm used to generate the signature in the Signature field.
- The Signature field contains the signature of the CA that issued this CRL and the signature covers the To-Be-Signed-CRL field.

The To-Be-Signed-CRL contains information about the revoked certificates and has the following main fields:

- The Type field indicates the type of the entries in the CRL.
- The CRL-Series field indicates the CRL series to which this CRL belongs.
- The CA-ID field contains the identifier of the CA that issued this CRL. The identifier of the CA is represented by the low-order 8 octets of the hash value of the CA's certificate. The CRL-Serial field contains the serial number of this CRL. Each time a new CRL is issued, its serial number will increment by one.
- The Start-Period and the Issue-Date fields define the time period that this CRL covers. That is, this CRL will contain all the certificates belonging to the CRL-Series that were revoked within this time period.
- The Next-CRL field contains the time when the next CRL is expected to be issued.
- The Certificate-Identifiers are the identifiers of the revoked certificates. The identifier of a revoked certificate is the low-order 10 octets of the hash value of the revoked certificate. Optionally, the expiration date of each revoked certificate can also be included.

21.5 MESSAGE FORMATS AND PROCESSING FOR GENERATING ENCRYPTED MESSAGES

IEEE 1609.2 allows a message to be encrypted to one or multiple known recipients. The message sender encrypts each message using a symmetric key algorithm with a freshly generated symmetric key. This symmetric key is then encrypted for each recipient using the recipient's public key. The current IEEE 1609.2 only supports ECIES algorithm over the NIST P256 curve as the public key encryption algorithm.

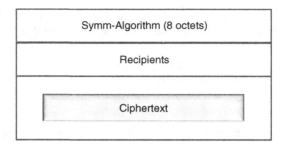

Figure 21.9. The Encrypted-Message format

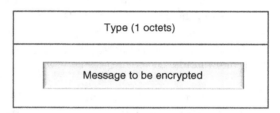

Figure 21.10. The To-Be-Encrypted message format

Each encrypted message is packaged in the Encrypted-Message format shown in Figure 21.9. The Symm-Algorithm field contains an identifier for the symmetric key algorithm used to encrypt the message. The Recipients field contains the following information about each recipient of the encrypted message:

- identifier of the recipient's certificate that contains the recipient's public key;
- public key algorithm used to encrypt the symmetric key; and
- symmetric key, which was used by the message sender to encrypt this message, encrypted with the recipient's public key.

The last field of the Encrypted-Message format contains the ciphertext or the encryption result of the message. Using IEEE 1609.2, a message is not encrypted directly. Instead, the message is first packaged into a To-Be-Encrypted data structure, which is then encrypted to generate the ciphertext to be included in the Encrypted-Message format.

The To-Be-Encrypted structure is shown in Figure 21.10. The Type field contains the type of the message to be encrypted. The current version of IEEE 1609.2 specifies the following message types to be encrypted: unsecured (or plaintext), signed, signed external payload, signed partial payload, certificate request, certificate response, anonymous certificate response, certificate request error, CRL request, CRL, certificate response acknowledgement, and unknown.

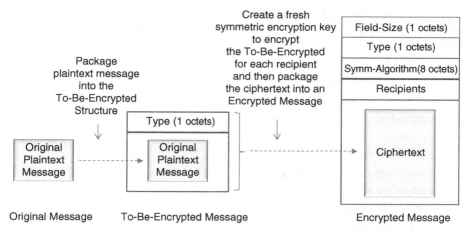

Figure 21.11. Process for generating an encrypted message

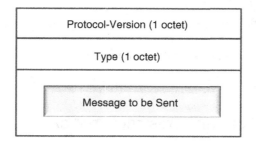

Figure 21.12. The 1609Dot2Message format

The process for generating an encrypted message is summarized in Figure 21.11.

21.6 SENDING MESSAGES

Messages that use IEEE 1609.2 security services, and messages that are used to support IEEE 1609.2 security operations, are required to be encapsulated in the 1609Dot2Message format. Messages that use IEEE 1609.2 security services include the application messages that are signed or encrypted using IEEE 1609.2 security services. Messages for supporting IEEE 1609.2 security operations include messages sent between the vehicles and the CAs for the vehicles to acquire certificates and CRLs.

The 1609Dot2Message format is shown in Figure 21.12. It consists of two header fields followed by the message to be sent.

The first header field, Protocol-Version, contains the version of the IEEE 1609.2 protocol. The current IEEE 1609.2 standard is version 2 [IEEE11].

The second header field Type contains the type of the message to be sent. The current version of the IEEE 1609.2 defines the following message types [IEEE11]:

- unsecured (plaintext)
- signed
- encrypted
- certificate request
- certificate response
- anonymous certificate response
- certificate request response
- certificate request error
- CRL request
- CRL
- signed partial payload
- signed external payload
- signed WSA
- certificate response acknowledgement
- unknown

21.7 REQUEST CERTIFICATES FROM THE CA

Figure 21.13 illustrates the message flows and processing for a messaging entity to request certificates from a CA. A messaging entity is a higher-layer entity (HLE) that sends application messages.

The HLE sends a WaveCertificateManagement-RequestCertificate.req message, which is also known as a WCM-RC.req message, to the CME on the vehicle to trigger it to initiate an interaction with the CA to acquire the requested certificate. The CME on the vehicle acknowledges this request by returning a message WCM-RC.cfm to the HLE. The CME then checks to see if it already has the requested certificate in the SDSs on the vehicle. If it does not have the requested certificate and network access to the CA is currently available, the CME will send a Certificate Request message to the CA.

The Certificate Request must be signed by the certificate requester and encrypted to the CA.

If network access to the CA is not currently available, the CME may cache the Certificate Request message and send it later when network access becomes available.

The CA authenticates the Certificate Request first and then decides whether to authorize the requested certificate. The CA then sends back a Certificate Response message carrying the requested certificate to the requesting CME

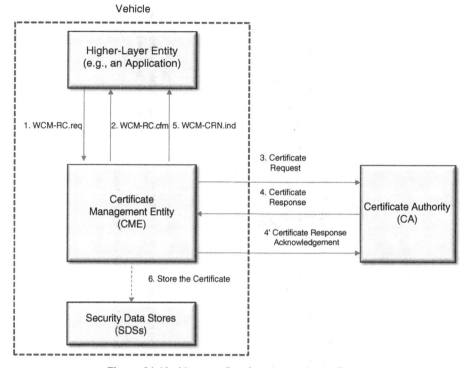

Figure 21.13. Message flow for requesting certificates

if the certificate request is authorized or a Certificate Request Error message if the request is denied.

The Certificate Response message will be encrypted to the certificate requester.

The CME decrypts the Certificate Response message and checks its content for errors. If the decryption and all subsequent checks pass positively, the CME stores the certificate in a local Message SDS. The CME informs the HLE of the status of the certificate request by sending a WaveCertificateManagement-CertificateResponseNotification message, also known as a WCM-CRN message, to the HLE.

The CME on the vehicle will return an acknowledgement message to the CA if so requested by the Certificate Response message.

The process for generating a Certificate Request message is shown in Figure 21.14. It consists of the following main steps:

- *Step 1*: Create a Certificate-Request message that contains the certificate request signed by the certificate requester.
- *Step 2*: Create an encrypted version of the above message and place it in an Encrypted-Message message format.

338

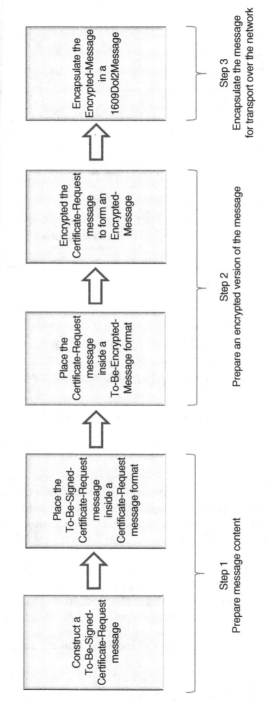

Figure 21.14. Process for constructing and sending a Certificate Request message

Step 1
Prepare message content

Step 2
Prepare an encrypted version of the message

Step 3
Encapsulate the message
for transport over the network

Construct a
To-Be-Signed-
Certificate-Request
message

Place the
To-Be-Signed-
Certificate-Request
message
inside a
Certificate-Request
message format

Place the
Certificate-Request
message
inside a
To-Be-Encrypted-
Message format

Encrypted the
Certificate-Request
message
to form an
Encrypted-
Message

Encapsulate the
Encrypted-Message
in a
1609Dol2Message

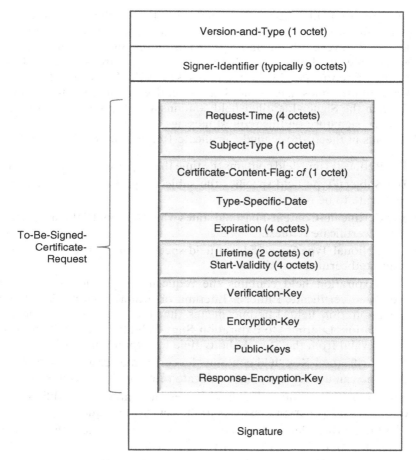

Figure 21.15. The Certificate Request format

- *Step 3*: Encapsulate the encrypted message in the 1609Dot2Message format to be transported over the network to the CA.

The Certificate-Request message format is shown in Figure 21.15. It consists of the following three parts: (1) two header fields, (2) the body of the certificate request message in the To-Be-Signed-Certificate format, and (3) the signature of the certificate requester.

The first header field Version-and-Type contains the version of certificate that the requester is requesting. This field is also used to indicate whether the certificate requester is requesting an implicit or explicit certificate.

The second header field Signer-Identifier contains information about the certificate requester's public key and the certificate for this public key.

The Signature field contains the signature of the certificate requester. This signature will cover the To-Be-Signed-Certificate-Request part of the message. The key used to verify the signature depends on the Signer-Identifier field. If Signer-Identifier field is equal to SELF, then the signature is generated using the private key corresponding to the public key in the Public-Key field of this certificate request message. If the Signer-Identifier field is set to CERTIFICATE, then the Signer-Identifier field contains a certificate that can be used to verify the signature on this request message.

The fields in the To-Be-Signed-Certificate-Request are explained below.

- The Request-Time field contains the time the request message was formed.
- The Subject-Type field specifies the subject type to be contained in the certificate to be issued.
- The Certificate-Content-Flag field states which optional fields are included in this certificate request.
- The optional Type-Specific-Data field specifies the desired scope of the requested certificate.
- The Expiration field contains the requested expiration time for the requested certificate. Either a Lifetime or a Start Validity field follows the Expiration field. Lifetime means that the certificate is valid from (Expiration–Lifetime) to Expiration. Start-Validity means that the certificate is valid from the Start-Validity time to Expiration time.
- The Verification-Key field contains the certificate requester's public key that one can use to verify the certificate requester's signatures.
- The optional Encryption-Key field, if present, contains the public key that can be used to encrypt messages to the certificate requester.
- The Public-Keys field contains one or multiple public keys the certificate requester wants the CA to certify.
- The Response-Encryption-Key field contains the public key to be used by the certificate requester to encrypt its response messages to the CA.

The Certificate Response message is constructed using the process illustrated in Figure 21.16. IEEE 1609.2 requires that the Request Response message be encrypted to the certificate requester by the CA. The steps for constructing the Certificate Response message are:

- *Step 1*: Creates the main body of the Certificate Response message in the To-Be-Encrypted-Certificate-Response format that contains the certificates to be assigned to the certificate requester.
- *Step 2*: Prepares an encrypted version of the Certificate Response message.
- *Step 3*: Encapsulates the encrypted version of the Certificate Message into a 1609Dot2Message format for transport over the network to the certificate requester.

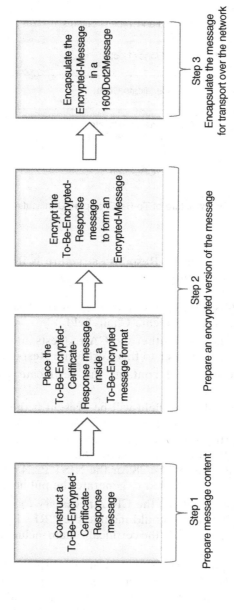

Figure 21.16. Process for constructing and sending a Certificate Response message

341

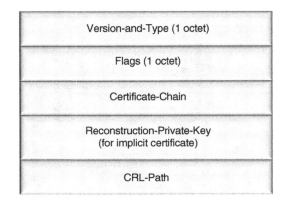

Figure 21.17. The structure of To-Be-Encrypted-Certificate-Response

The To-Be-Encrypted-Certificate-Response structure is illustrated in Figure 21.17. The fields in this structure are explained below:

- The Version-and-Type field is obtained from the last certificate in the certificate chain in the Certificate-Chain field.
- The Flags field indicates whether the CA requests an acknowledgement from the certificate requester to this Certificate Response message.
- The Certificate-Chain field contains the certificate path of the new certificate.
- The Reconstruction-Private field will be present if the issued certificate is an implicit certificate. It will contain the reconstruction private value to be used by the certificate requester to derive the private key corresponding to the public key that can be recovered from the implicit certificate. The size of this field should be 28 for the NIST P224 curve or 32 for the NIST P256 curve that was used to generate the public key.
- The CRL-Path field contains the CRLs necessary to validate the issued certificate. This CRL-Path should include the CRLs issued by the other CAs that issued certificates in the certificate chain included in this message.

After the To-Be-Encrypted-Certificate-Response message is constructed, it will be placed inside a To-Be-Encrypted structure. The resulting To-Be-Encrypted message is then encrypted and packaged into the Encrypted-Message format. The Encrypted-Message is encapsulated in the 1609Dot2Message format to be sent to the certificate requester.

If the certificate requester is requested by the CA to send a Certificate Response Acknowledgement message back to the CA, the certificate requester first creates a To-Be-Encrypted-Certificate-Response-Acknowledgement message. This message will contain the first 10 octets of the SHA-256 hash of the received To-Be-Encrypted-Certificate-Response message. This message is then encrypted and packaged into an Encrypted-Message format, which will in turn be encapsulated in a 1609Dot2Message format to be sent back to the CA.

21.8 REQUEST AND PROCESSING CRL

The CME on each vehicle can request CRLs from the CA by sending to the CA a CrlRequest message, encapsulated in the 1609Dot2Message format. The current IEEE 1609.2 standard states that the CrlRequest message shall not be signed or encrypted, and the response shall not be encrypted either (Figure 21.18).

The format for the CrlRequest message is shown in Figure 21.19. The Issuer field identifies the CA that issues the CRL. The CRL-Series field indicates which CRL series the requester wants to receive and the time duration for the revoked certificates. The Issue-Date field contains the issue date of the last

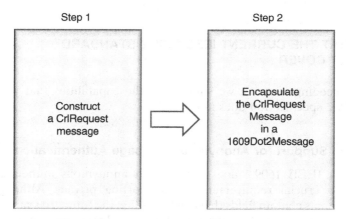

Figure 21.18. Process for creating a CRL Request message to be sent to the CA

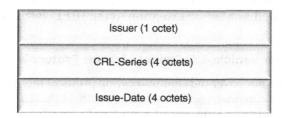

Figure 21.19. The CrlRequest message format

CRL in the series for which no subsequent CRLs are missing. If this field is set to 0, it indicates that the requester wants only the most recent CRL.

The CA will respond by sending all the CRLs in the requested CRL series with issue date later than the Issue-Date field in the CrlRequest message. The current version of the IEEE 1609.2 does not specify the CRL Response message.

Upon receiving a CRL from the CA, the CME on the vehicle processes the CRL as follows. It first authenticates the CRL. If the CRL carries its issuing CA's certificate or a certificate chain, the CME verifies the CA's certificate or the certificate chain. It then uses the public key on issuing the CA's certificate to verify the CA's signature on the CRL. If the signature is verified to be authentic, the CME accepts the CRL. If the CRL does not explicitly carry its issuing CA's certificate, it will carry a certificate digest which the CME on the vehicle can use to locate the corresponding certificate it already has in its local data stores.

For each accepted CRL, the CME on the vehicle stores the revoked certificates listed on the CRL in the Global SDS on the vehicle. If any of the vehicle's own certificates is listed on the CRL, the CME will mark it as REVOKED so the vehicle will not use this certificate later to send messages.

21.9 WHAT THE CURRENT IEEE 1609.2 STANDARD DOES NOT COVER

In the succeeding sections, we summarize the capabilities that the current IEEE 1609.2 specification does not cover.

21.9.1 No Support for Anonymous Message Authentication

The current IEEE 1609.2 does not support anonymous authentication of messages—a crucial requirement for supporting privacy. Although some message formats contain fields that can be used in the future to support anonymous message authentication, the currently defined message formats are insufficient for supporting anonymous message authentication. Message processing for supporting anonymous message authentication will also be needed.

Support for anonymous certificates and for anonymous message authentication could be added as a future extension to the IEEE 1609.2 specification.

21.9.2 Separate Vehicle-CA Communication Protocols Are Required

IEEE 1609.2 does not cover the communication protocol required for a vehicle to exchange certificate management messages with a CA. IEEE 1609.2 defines messages for a vehicle to request certificates and CRLs from a CA. However, these messages are defined in a way that requires them to be carried by a separate communication protocol between the vehicle and the CA. A separate communication protocol will be needed to handle the following:

- The naming and addressing required for two communicating parties to exchange messages.
- Intermittent availability of network access from the vehicle to the CA.
- Errors that can occur during message exchange between the vehicle and the CA.
- Buffering of the certificate management messages that the vehicle cannot send out immediately.
- Potential needs for message retransmissions.
- Security protections for transmissions over networks.

Several protocols are available for transporting the certificate management messages between a vehicle and a CA. These include:

- *TCP (Transport Control Protocol)*: It provides reliable packet transport over IP (Internet Protocol) networks.
- *HTTP (Hypertext Transfer Protocol)*: It is used to support Web access.
- *HTTPS (HTTP Secure)*: It supports encrypted transport over TCP/IP.
- *UDP (User Datagram Protocol)*: It provides unreliable message transport over IP networks.
- *DTLS (Datagram Transport Layer Security)*: It provides encrypted message transport over UDP/IP.
- *V-DTLS (Vehicular DTLS)* [PSCZ08]: It provides secured message transport over UDP/IP networks. V-DTLS is designed to address several unique requirements in a vehicular communication environment. For example, it has significantly lower message overheads than DTLS for establishing secure connections.
- *SIP (Session Initiation Protocol)*: It is an application-layer protocol for controlling end-to-end real-time sessions. It can run over TCP/IP or UDP/IP.

TCP and TCP-based protocols such as HTTP and HTTPS are, in general, better suited for long message flows that require reliable transport due to their overheads for setting up TCP sessions and for realizing reliable packet transmissions. UDP and UDP-based protocols such as DTLS and V-DTLS are, in general, better suited for short message exchanges over not overly lossy networks.

It typically takes only two to three messages for a vehicle to acquire a set of certificates from a CA (a certificate request from the vehicle to the CA, a response from the CA to the vehicle, and an optional acknowledgement from the vehicle to the CA). Therefore, UDP-based transport protocols tend to fit more naturally for transporting certificate management messages between the vehicle and the CA.

21.9.3   Interactions and Interfaces between CA Entities Not Addressed

The current IEEE 1609.2 does not specify interactions and interfaces between different types of CA entities.

The current IEEE 1609.2 does not define the interface or a protocol between the root CA and the subordinate CAs. Such an interface will be needed to allow a root CA to issue certificates to other CAs automatically.

The current IEEE 1609.2 does not define the interface between the RA and the CA. Such an interface will be needed to support privacy-preserving certificates and to automate the certificate management.

Furthermore, to protect privacy against CA operators, split CA architectures may be needed. The current IEEE 1609.2 does not support split CA architectures.

REFERENCES

[AnGV05] A. Antipa, R. Gallant, and S. Vanstone: "Accelerated verification of ECDSA signatures," Selected Areas in 21 Cryptography, 12th International Workshop, SAC 2005, Kingston, ON, Canada, August 2005.

[IEEE11] IEEE 1609.2/D8: "Draft Standard for Wireless Access in Vehicular Environments—Security Services for Applications and Management Messages," 2011.

[PSCZ08] S. Pietrowicz, H. Shim, G. D. Crescenzo, and T. Zhang: "VDTLS—Providing Secure Communications in Vehicle Networks," *Infocom MOVE (MObile Networking for Vehicular Environments) Workshop*, 2008.

22

4G FOR VEHICLE SAFETY COMMUNICATIONS

22.1 INTRODUCTION

In this chapter, we will first provide a brief overview of the leading fourth-generation (4G) cellular technology—the Long-Term Evolution or LTE. We will then use examples to illustrate the technical feasibility and potential issues to be addressed when using LTE to support vehicle-to-infrastructure (V2I) and vehicle-to-vehicle (V2V) safety communications.

22.2 LONG-TERM REVOLUTION (LTE)

The leading 4G cellular network standard is the LTE developed by the Third-Generation Partnership Project (3GPP). LTE offers significantly higher capacities and lower delays than third-generation (3G) cellular networks. It has the potential to meet the stringent requirements of many vehicle safety communications.

LTE was first defined in 3GPP Release 8 specifications published in 2008 [3GPP08a] [3GPP08b] [3GPP09] [StTB11]. A first fundamental difference between LTE and previous generations of cellular networks is that LTE uses an end-to-end all-Internet Protocol (IP) network architecture. Second-generation (2G) cellular networks use circuit-switched technologies. Third-generation (3G) cellular networks use IP protocols to support data services but continued to use circuit-switched technologies to support voice services.

Vehicle Safety Communications: Protocols, Security, and Privacy, First Edition. Luca Delgrossi and Tao Zhang.
© 2012 John Wiley & Sons, Inc. Published 2012 by John Wiley & Sons, Inc.

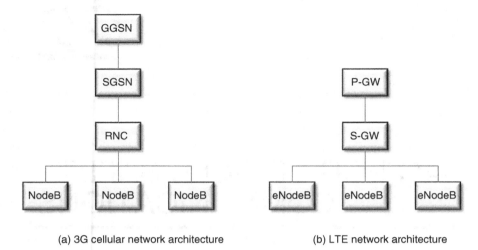

(a) 3G cellular network architecture   (b) LTE network architecture

Figure 22.1. Comparison of (a) LTE and (b) 3G network architectures.

The IP network components and the overall control protocols in 3G networks depend heavily on legacy circuit-switched networks and protocols. Adopting an end-to-end all-IP network architecture makes LTE significantly more suitable for supporting advanced wireless services and for integrating with the Internet.

LTE uses a flatter network architecture than previous generations of cellular networks, which reduces packet delay and system complexity. Figure 22.1 shows a simplified view of the user planes of LTE and 3G network architectures defined by the 3GPP:

- *3G Networks*: Multiple radio base stations (NodeB) are connected to a radio network controller (RNC) that manages the radio interface. Multiple RNCs are connected to a serving GPRS support node (SGSN) that serves as the interface between radio access networks and the core network. Multiple SGSNs are, in turn, connected to a gateway GPRS support node (GGSN) that serves as an anchor point for supporting IP routing and packet forwarding among mobile nodes and also serves as the gateway with external networks.
- *LTE Networks*: Multiple enhanced radio base stations (eNodeB) are connected directly to a core network gateway—a serving gateway (S-GW). RNCs are eliminated. Multiple S-GWs are connected to a packet data network gateway (P-GW), which serves as an anchor point for IP routing and as the gateway with external networks.

LTE uses several proven technologies to increase capacity, enhance spectrum efficiency, and to reduce delay. The air interface uses orthogonal frequency division multiplexing (OFDM) technology on the downlink (from base station

Table 22.1. LTE peak downlink data rates (Mbps)

	1.4 MHz	3 MHz	5 MHz	10 MHz	15 MHz	20 MHz
2 × 2 MIMO (64-QAM)	12.1	25.9	43.2	86.4	129.6	172.8
4 × 4 MIMO (64-QAM)	22.9	49.0	81.6	163.2	244.8	326.4

to mobile device) but uses single-carrier frequency division multiple access (SC-FDMA) for the uplink. It uses multiple-input and multiple-output (MIMO) antenna technologies to achieve higher total capacity. LTE also supports flexible channel bandwidths ranging from 1.4 to 20 MHz to support a wide range of data rates.

LTE supports peak downlink data rate up to 326 Mbps. It uses various antenna configurations and modulation schemes to provide a range of peak data rates to fit varying real-world needs.

Table 22.1 shows the peak downlink data rates for different bandwidths and MIMO antenna configurations [3GPP08a]. It reaches a maximum data rate of 326 Mbps using 20 MHz bandwidth, 4 × 4 MIMO antenna, and 64-point quadrature amplitude modulation (64-QAM) modulation. This means that an LTE cell using a single 4 × 4 MIMO antenna can provide up to 326 Mbps downlink capacity. LTE networks can use antenna diversity to further increase cell capacity. It allows each cell to be divided into multiple sectors each served by a separate antenna array. A three-sector cell using 4 × 4 MIMO antennas, for example, will triple the cell capacity to approximately 1 Gbps.

As a comparison, 3G cellular systems were designed to support peak downlink data rates up to 2 Mbps for stationary users, 384 Kbps for users traveling below 120 km (75 mi) per hour, and 144 kbps for users traveling over 120 km/h. In other words, LTE increases cellular network peak data rates by tens to hundreds times over 3G.

More significantly for supporting vehicle safety communications, LTE offers significantly lower end-to-end delays compared with previous-generation cellular networks. LTE specifications require the user packet delay through the LTE radio access network to be below 10 ms and the call setup latency to be less than 100 ms [3GPP08b]. It has been shown that a well-designed LTE network can support end-to-end delays well below 100 ms. Most recent trials show delays ranging from 30 to 50 ms, which satisfy the requirements of many vehicle safety applications such as distributing traffic controller signal phase and timing (SPAT) information to vehicles [UDOT05].

LTE supports nine differentiated quality of service (QoS) classes. Table 22.2 shows the upper bounds on the delays and packet loss ratios for each QoS class [3GPP10]. The Packet Delay Budget is the upper bound on the delay between the mobile device and the gateway inside the LTE core network that controls the user packet flows and interfaces with external networks such as another LTE network domain or the Internet.

Table 22.2. LTE quality of service classes

QoS Class Identifier	Type of Resources	Priority	Packet Delay Budget (ms)	Packet Error Loss Rate	Target Applications
1	Guaranteed	2	100	10^{-2}	Conversational voice
2	bit rate	4	150	10^{-3}	Conversational video
3	(GBR)	3	50	10^{-3}	Real-time gaming
4	services	5	300	10^{-6}	Nonconversational video
5	Non-GBR	1	100	10^{-6}	IMS signaling
6	services	6	300	10^{-6}	Video, TCP-based applications (e.g., WWW, e-mail, chat)
7		7	100	10^{-3}	Voice, video live streaming, interactive gaming
8		8	300	10^{-6}	Video, TCP-based applications (e.g., WWW, e-mail, chat)
9		9			

IMS, IP multimedia subsystem; TCP, Transport Control Protocol; WWW, World Wide Web.

Four service classes provide guaranteed delays and packet loss rations designed to support highly delay-sensitive applications. For example, one service class guarantees delays below 50 ms and another service class guarantees delays below 100 ms. Four additional services classes support stringent but not guaranteed delay requirements. For example, one service class targets delays below 100 ms for applications like video gaming.

To put the performance of LTE in perspective, Figure 22.2 shows the delays one can expect from the following 2G and 3G cellular networks:

- *General Packet Radio Service (GPRS)*: GPRS is 2G cellular technology and it was the first cellular packet data radio system that received global adoption. GPRS was integrated with the most widely adopted 2G cellular voice standards Global Systems for Mobile Communications (GSM) to provide cellular packet data services. The GPRS architecture and protocols were later enhanced to form a foundation for the 3G cellular system defined by 3GPP.
- *Enhanced Data Rates for GSM Evolution (EDGE)*: EDGE was an enhancement to GPRS to provide higher data rates and lower delays.
- *Wideband Code Division Multiple Access (WCDMA)*: WCDMA is the most widely used 3G cellular standard [ChZh04]. It was defined by the 3GPP.
- *High-Speed Data Access (HSPA)*: HSPA is an enhancement to WCDMA.

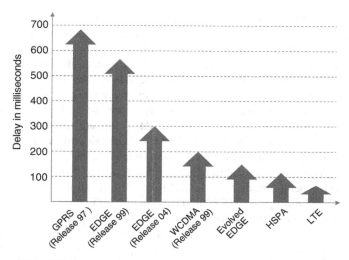

Figure 22.2. LTE delays compared with previous generations of cellular systems

LTE has received overwhelming support worldwide. It is the 4G technology of choice for most major wireless network operators around the world. Since the first commercial LTE networks were deployed by TeliaSonera in Stockholm, Sweden, and Oslo, Norway, in 2009, commercial LTE networks have been rapidly deployed in many countries. In the United States, Verizon Wireless LTE coverage reached over 190 cities by the end of year 2011, covering nearly two-thirds of the American population. Verizon LTE networks offer data rates from 5 to 12 Mbps. According to statistics collected by TeleGeography [Tele12], there were 5.6 million LTE subscribers already in the United States by the end of year 2011, which represent a nearly 8,000% growth over the same time year ago.

Most current commercial deployments in the United States are in the 700 MHz bandwidths. However, LTE can be implemented in a number of spectrum bands defined in [3GPP09], including four 700 MHz bands.

In the United States, the Federal Communications Commission (FCC) has adopted LTE as the technology standard for public safety broadband communications [FCC11]. In January 2011, the FCC mandated that all networks deployed in the 700 MHz public safety broadband spectrum adopt LTE. The Association of Public Safety Officials (APCO) and the APCO Global Alliance have adopted LTE as the worldwide standard for broadband emergency communications.

Commercial deployments have confirmed LTE's ability to meet its designed capacity and delay targets. For example, measurements taken from TeliaSonera's commercial LTE networks in Finland in March 2011 showed an average of 36.1 Mbps download speed and an average of 23-ms round-trip delay through the commercial LTE networks [Epit11]. The measurements are taken to assess real-world user experience of running a range of Internet applications over the LTE network. This LTE network was designed to support a data downlink data rates between 20 and 80 Mbps. The measured user experiences matched well with the LTE network design goals. Furthermore, the performances of these commercial LTE networks are comparable to the broadband wireline network performances in many leading countries.

Numerous trials have been conducted to evaluate LTE performance for vehicle safety communications. With 20 MHz spectrum and 2 × 2 MIMO antennas, the results have repeatedly shown average downlink data rates from tens to over 100 Mbps and end-to-end data delays below 50 ms. For example, Nokia Siemens Networks' trial in Munich in 2009 used commercial LTE equipment, 20 MHz spectrum, and 2 × 2 MIMO antennas supporting a maximum data rate of 172.8 mbps [NOKI09]. They tested LTE performance for both urban and open-road environments at various vehicle speeds. The results show that, with a 300-m cell diameter in a crowded urban environment where vehicles travel at 10–15 km/h, the average layer-1 (L1) throughput ranged from 60 to 100 Mbps with the maximum measured data rate reaching 140 Mbps. Along a more open but still heavily traveled road where the vehicles travel between 30 and 40 km/h, the average L1 throughput ranged from 14 to 83 Mbps depending on the shadowing conditions along the route. Even at a distance of about 1000 m from the base station, the average L1 throughput reaches 30 Mbps.

As LTE deployment progresses worldwide, 3GPP has been developing LTE Advanced [StTB11]. LTE Advanced specifications are in 3GPP Release 10, which passed stage 3 functional freeze in March 2011[3GPP12a]. LTE Advanced offers significantly greater network capacity and spectral efficiency than LTE as shown in Table 22.3. It also offers enhanced multimedia broadcast multicast services (MBMS) capabilities.

Table 22.3. LTE Advanced versus LTE

	LTE	LTE Advanced
Peak data rate (downlink)	326.4 Mbps	1 Gbps for low mobility, 100 Mbps for high mobility
Peak data rate (uplink)	75 Mbps	500 Mbps
Maximum channel bandwidth (downlink)	20 MHz	100 MHz
Maximum channel bandwidth (uplink)	20 MHz	40 MHz
Spectral efficiency	16.3 bps/Hz	30 bps/Hz
Scalable channel bandwidth	1.4, 3, 5, 10, 15, 20	Up to 100 MHz
Simultaneous active sessions	200 for each 5 MHz	Three times higher than LTE

22.3 LTE FOR VEHICLE SAFETY COMMUNICATIONS

In this section, we discuss how commercially available point-to-point LTE connections can be used to support V2I and V2V communication applications and some of the technical issues to be addressed.

22.3.1 Issues to Be Addressed

Using cellular networks for supporting the communications of a large number of vehicles can have a profound impact on the cellular networks and requires an in-depth examination of several fundamental issues.

First, vehicles can generate a significant volume of extra cellular calls in addition to the calls made by human users. The number of simultaneous calls or transmissions each radio base station can support is limited. For example, this limit depends on the number of time slots in each transmission frame, or the number of orthogonal frequencies in a frequency division multiplexing radio system, or the number of orthogonal codes in a code division multiplexing system. Today, an LTE cell can typically support 200 active sessions at the same time in every 5 MHz bandwidth. This means that a three-sector LTE cell site using 10 MHz bandwidth in each sector could support up to 1200 simultaneous sessions. This number will triple with LTE Advanced. A question becomes: can LTE networks support the extra volume of calls made by vehicles?

Second, today's cellular networks are engineered under the assumption that only a small percentage of all subscribers will make phone calls at any time. The networks are typically optimized to take advantage of the statistical characteristics of human conversions, which have long and predictable silent periods, to multiplex multiple calls to increase network utilization. Cellular calls made by vehicles have significantly different characteristics than human conversations. A heavy volume of vehicle-generated calls can fundamentally change the characteristics of the traffic in cellular networks just as Internet traffic has changed the traffic characteristics over the telecommunications networks that used to be dominated by voice traffic. An in-depth study of the statistical characteristics of vehicle-made cellular calls will be necessary to assess their impact on cellular network designs and configurations.

22.3.2 LTE for V2I Safety Communications

We illustrate how a cellular network, such as an LTE network, can be used to support intersection collision avoidance applications. Specifically, we consider applications based on SPAT information.

Figure 22.3 illustrates how to use LTE (or other cellular networks) to support SPAT-based applications. As a vehicle approaches an intersection, it uses LTE to connect to an LTE SPAT server. The LTE SPAT server sends

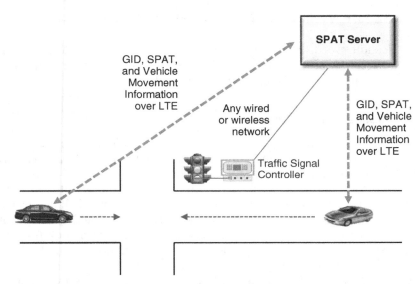

Figure 22.3. Using cellular networks to support SPAT-based applications. GID, geometric intersection description

SPAT information to vehicles approaching the intersection over point-to-point LTE connections.

In low vehicle density areas, each vehicle makes a cellular call to the LTE SPAT server as it approaches an intersection and keeps this connection alive until it exits the intersection. To reduce the delay for establishing a cellular connection with the SPAT server, a vehicle can initiate its cellular connection to the LTE SPAT server a few seconds before it needs the SPAT messages. In high vehicle density areas, smart solutions may be required to keep the volume of vehicle calls below a cellular cell's capacity without jeopardizing the performance of SPAT applications.

The LTE SPAT server acquires SPAT information in several ways. The signal phases and timings of many signaled road intersections are controlled by centralized traffic management centers (TMCs). For these intersections, the LTE SPAT server can be located at, or connected to, the TMC to obtain SPAT information directly from the TMC. Some intersections are equipped with controllers that can adjust signal phase and timing autonomously in response to local traffic conditions. These traffic controllers typically already have network connectivity with the TMC and can use it to send their locally computed signal phase and timing to the LTE SPAT Server.

Next, we use an example to examine whether an LTE cell has sufficient capacity to support SPAT-based applications. We consider the volume of cellular calls made by vehicles and people inside an LTE cell during the rush hours in the busy midtown Manhattan of New York City. We assume that the

LTE cell has a 500-m radius. This cell would cover approximately 7 avenues, 13 streets, and 65 intersections in Manhattan. Let us assume that the cell uses a three-sector configuration and each sector supports a maximum data rate of 173 Mbps using 20 MHz bandwidth and 2 × 2 MIMO antennas. Furthermore, let's assume that each sector can support up to 400 simultaneous cell phone connections based on the specifications of typical currently available LTE base stations.

A recent study reported traffic counts at major intersections between the 34th Street and the 57th Street in Manhattan during different time periods of a day [NYC10]. It shows that the number of vehicles traversing an intersection in each direction ranges approximately from 750 to 2500 and averages around 1500. This means that approximately 0.833 vehicles would cross each intersection every second on average, or approximately 0.833 × 65 ≅ 55 vehicles on average would pass through the intersections inside the LTE cell every second. If each vehicle makes one 5-second long cellular call to retrieve SPAT information for each intersection it traverses, then approximately 550 calls would be made from the vehicles every second. This accounts for about 22.5% of the total number of simultaneous calls the cell can support. This number would reduce to about 6% if the cell radius is reduced to 0.25 km. These results are summarized in Table 22.4.

Now let us consider the total number of calls made by both humans and vehicles. According to the 2010 U.S. census data, the population of Manhattan is 1,585,873 and the area of Manhattan is about 59 km^2 or about 23 mi^2. Up to half of the work force in Manhattan comes from outside Manhattan. There are also many tourists in Manhattan. Assuming the number of mobile phones in Manhattan is approximately 1.5 times the local population, there could be about 31,667 mobile phone users inside the cell under our consideration at any time. Table 22.5 shows the total volume of calls made by both people and vehicles and the maximum data rate for each call.

Although the data in Tables 22.4 and 22.5 are rough estimates, they reveal an interesting observation. Even in densely populated urban areas such as

Table 22.4. Cellular calls generated by vehicles to support SPAT applications

	Cell Radius = 0.5 km	Cell Radius = 0.25 km
Number of vehicles crossing each intersection per second	0.83	0.83
Number of vehicles crossing intersections inside a cell	0.83 × 65 = 54	0.83 × 17 = 14
Average duration of a call	5 seconds	5 seconds
Maximum number of simultaneous vehicle-made calls in a cell	54 × 5 = 270	14 × 5 = 70
Cell capacity	1200 simultaneous calls	1200 simultaneous calls
Utilization	270/1200 = 22.5%	70/1200 = 5.8%

Table 22.5. Cellular calls generated by both people and vehicles inside a LTE cell

	Cell Radius = 0.5 km	Cell Radius = 0.25 km
Number of cellular phone users in a cell	31,667	7917
Number of simultaneous calls by people (active radio: 2%)	633	158
Total number of calls by people and vehicles	903	238
Average bandwidth per call	0.6 Mbps	2.2 Gbps
Cell capacity	1200 simultaneous calls	1200 simultaneous calls
Utilization by people and vehicles	75%	20%

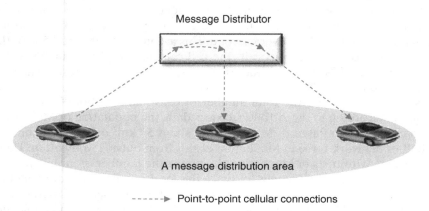

Figure 22.4. Using commercial point-to-point cellular calls to support V2V communications

New York City, a LTE network with approximately a 500-m cell radius could be sufficient to support both regular mobile phones and vehicle-made calls for supporting SPAT applications.

22.3.3 LTE for V2V Safety Communications

Figure 22.4 illustrates how commercial point-to-point cellular connections and a message distributor (reflector) can be used to support V2V communications. A message distributor is a functional entity that relays messages from one vehicle to other vehicles.

Vehicles can use their onboard digital maps and their current positions to determine when they should send messages through a message distributor to other vehicles or to receive messages from other vehicles via the message distributor. For example, vehicles in an accident-prone area can detect dangerous driving conditions and send messages to the message distributor to be distributed over cellular connections to other vehicles inside this accident-prone area. Each vehicle entering this accident-prone area can contact the message distributor to receive messages from other vehicles in real time. A vehicle may also send its cellular phone number to the message distributor so that the message distributor can proactively connect to the vehicle to send messages to it when certain events occur—such as when new messages are received from other vehicles indicating a hazardous driving condition.

Recently, the CoCar project and the subsequent CoCarX project in Germany conducted trials of using cellular networks to send V2V hazard warning and other messages [BBCD09] [PhRS11]. The CoCar and CoCarX programs used a reflector, a functional entity connected to infrastructure networks, to reflect messages from one vehicle to other vehicles using point-to-point cellular connections. The trials show that it is feasible to use LTE to support a range of event-driven time-sensitive V2V message dissemination applications such as event-driven V2V hazard warning messages and other environmental notification messages. The trials used a 5 MHz frequency division duplex (FDD) LTE system (5 MHz bandwidth downlink capacity and 5 MHz uplink capacity). The LTE network consisted of seven sites of three-sector cells for a total of 21 radio cell sectors. The distances between the radio sites ranged from 500 m to 6 km to reflect both urban and suburban network configurations. The results show that, in an urban environment, this LTE network can support lower than 150 ms average end-to-end V2V messaging delays when 10 vehicles are reporting incidents, at one message per second, to 150 vehicles in each cell. This average delay increase to below 200 ms when 40 vehicles are reporting incidents in each cell.

Since a point-to-point cellular connection is used to deliver messages to each vehicle, the more vehicles receiving messages, the more connections must be established and more copies of the same message must be delivered over the air. That is, the traffic generated by each reporting vehicle is multiplied by the number of receiving vehicles. This is why the delay performance degrades as the number of reporting or receiving vehicles increases.

22.3.4   LTE Broadcast and Multicast Services

Native broadcast and multicast capabilities are defined in LTE standards, which can be used to deliver delay-sensitive I2V and V2V messages to a large population of users in a more resource-efficient manner. Native broadcast capability will also avoid the constraint of limited simultaneous point-to-point calls or connections that each cellular cell can support.

Third-generation (3G) cellular networks support cell broadcast service (CBS) that broadcasts low-rate text messages to all receivers inside one cellular cell. CBS has been used for broadcasting public safety alerts, location-based news, and weather information. National-scale public safety alert systems are being developed in the United States. The FCC recently announced its new

Personal Localized Alerting Network (PLAN) designed to send emergency warnings to mobile devices. The system will send out three types of commercial mobile alert system (CMAS) alerts: Presidential, Imminent Threats and AMBER (America's Missing: Broadcast Emergency Response). Mobile phone subscribers will receive these alerts free of charge. Today, however, not all mobile network operators have activated the CBS function activated in their networks due primarily to the lower revenues from broadcast services than from other cellular services.

LTE and LTE Advanced support evolved multimedia broadcast and multicast service (E-MBMS) [3GPP11] [3GPP12b] [3GPP12c]. MBMS was first introduced into 3GPP Release 6, a pre-LTE specification. LTE (3GPP Release 8) and LTE Advanced (3GPP Release 10) have enhanced MBMS to specify evolved MBMS or E-MBMS.

E-MBMS supports two modes of operation: broadcast and multicast. E-MBMS broadcast mode delivers multimedia traffic including text, audio, and video to all receivers in one or multiple cellular cells. Users do not need to explicitly activate or subscribe to MBMS broadcast mode.

E-MBMS multicast mode sends multimedia traffic to a group of mobile stations in one or multiple cellular cells. Unlike the broadcast mode, the multicast mode requires a user to subscribe to a multicast subscription group and then join the corresponding multicast group.

E-MBMS traffic can be sent over a dedicated frequency to reduce interference between E-MBMS traffic and other cellular traffic. The same frequency will be used to broadcast or multicast to multiple cells. This eliminates the need for a vehicle to switch channels to receive broadcast or multicast messages each time it enters a new LTE cell. The cells that participate in an LTE broadcast or multicast service over a single frequency is called the multicast/broadcast single frequency network or MBSFN. E-MBMS user services can also be delivered to a user at different bit rates and quality of service depending on specific radio network conditions.

REFERENCES

[3GPP08a] 3rd Generation Partnership Project: "3GPP TR 25.912 V8.0.0, Technical Specification Group Radio Access Network, Feasibility Study for Evolved Universal Terrestrial Radio Access (UTRA) and Universal Terrestrial Radio Access Network (UTRAN) (Release 8)," 2008.

[3GPP08b] 3rd Generation Partnership Project: "3GPP TR 25.913 V8.0.0, Technical Specification Group Radio Access Network, Requirements for Evolved UTRA (EUTRA) and Evolved UTRAN (E-UTRAN) (Release 8)," 2008.

[3GPP09] 3rd Generation Partnership Project: "3GPP TS 36.141 V8.2.0 (2009-03), Technical Specification, Technical Specification Group Radio Access Network, Evolved Universal Terrestrial Radio Access (E-UTRA), Base Station (BS) Conformance Testing (Release 8)," 2009.

[3GPP10] 3rd Generation Partnership Project: "3GPP TS 23.203 V8.9.0 (2010-03), Technical Specification Group Services and System Aspects, Policy and charging control architecture (Release 8)," 2010.

[3GPP11] 3rd Generation Partnership Project: "3GPP TS 22.146 V10.1.0 (2011-12), Technical Specification Group Services and System Aspects; Multimedia Broadcast/ Multicast Service (MBMS); Stage 1 (Release 10)," 2011.

[3GPP12a] 3rd Generation Partnership Project: "Overview of 3GPP Release 10 V0.1.3 (2012-01)," 2012.

[3GPP12b] 3rd Generation Partnership Project: "3GPP TS 23.246 V10.3.0 (2012-03), Technical Specification Group Services and Architecture, Multimedia Broadcast/ Multicast Service (MBMS), Architecture and Functional Description (Release 10)," 2012.

[3GPP12c] 3rd Generation Partnership Project: "3GPP TS 26.346 V10.3.0 (2012-03), Technical Specification Group Services and System Aspects, Multimedia Broadcast/ Multicast Service (MBMS), Protocols and Codecs (Release 10)," 2012.

[BBCD09] C. Birle, B. Borsetzky, Y. Chen, U. Dietz, G. Gehlen, S. Gläser, G. Jodlauk, J. Kahle, G. Nöcker, A. Schmidt, C. Sommer, S. Sories, and U. Dietz, ed.: AKTIV-CoCar Adaptive and Cooperative Technologies for Intelligent Traffic—Cooperative Cars: CoCar Feasibility Study Technology, Business and Dissemination, CoCar Consortium, 2009.

[ChZh04] J.-C. Chen and T. Zhang: IP-Based Next-Generation Wireless Networks: Systems, Architectures, and Protocols, John Wiley& Sons, 2004.

[Epit11] Epitiro Ltd: "LTE Real World Performance Study: Broadband and Voice over LTE (VoLTE) Quality Analysis: TeliaSonera, Turku, Finland," Epitiro Whitepaper, 2011.

[FCC11] Federal Communications Commission (FCC) FCC 11-6: "Third Report and Order and Fourth Further Notice of Proposed Rulemaking," 2011.

[NOKI09] Nokia Siemens Networks: White Paper, "LTE Performance for Initial Deployments," 2009.

[NYC10] New York City Department of Transportation: "Green Light for Midtown Evaluation Report," 2010.

[PhRS11] M.-A. Phan, R. Rembarz1, and S. Sories: "A Capacity Analysis for the Transmission of Event and Cooperative Awareness Messages in LTE Networks," ITS World Congress, Orlando, Florida, USA 2011.

[StTB11] S. Sesia, I. Toufik, and M. Baker: LTE—The UMTS Long Term Evolution—From Theory to Practice, Second edition, John Wiley & Sons, 2011.

[Tele12] TeleGeography: "US Remains at Forefront of LTE Service Adoption," [online]. Available at: <http://www.telegeography.com/products/commsupdate/ articles/2012/03/15/us-remains-at-forefront-of-lte-service-adoption> (accessed July 29, 2012), 2012.

[UDOT05] United States Department of Transportation, National Highway Safety Administration (NHTSA): "Vehicle Safety Communications Project Task 3 Final Report Identify Intelligent Vehicle Safety Applications Enabled by DSRC," 2005.

GLOSSARY

2G	Second generation
3G	Third generation
3GPP	Third-generation partnership project
4G	Fourth generation
ABS	Antilock braking system
ACC	Adaptive cruise control
ACJT	Ateniese–Camenisch–Joye–Tsudik
ACK	Acknowledgment
AES	Advanced Encryption Standard
AIAD	Additive increase/additive decrease
AIFS	Arbitration interframe spacing
AIMD	Additive increase/multiplicative decrease
AMP	Arbitration on message priority
APCO	Association of Public Safety Officials
ARIB	Association of Radio Industries and Businesses
ASTM	American Society for Testing and Materials
BAS	Brake assist system
BCM	Body control module
BLIS	Blind Spot Information System
BPSK	Binary phase-shift keying
BSM	Basic safety message
BSS	Basic service set
BSSID	Basic service set identifier
CA	Certificate authority
CALTRANS	California Department of Transportation

Vehicle Safety Communications: Protocols, Security, and Privacy, First Edition. Luca Delgrossi and Tao Zhang.
© 2012 John Wiley & Sons, Inc. Published 2012 by John Wiley & Sons, Inc.

CAMP	Crash Avoidance Metrics Partnership
CAN	Controller area network
CBR	Channel busy ratio
CBS	Cell broadcast service
CCA	Clear channel assessment
CCH	Control channel
CD	Communication density
CDMA	Code division multiple access
CDP	Certificate revocation list distribution points
CEN	European Committee for Standardization
CICAS-SLTA	Cooperative Intersection Collision Avoidance System–Signalized Left Turn Assistance
CICAS-V	Cooperative Intersection Collision Avoidance System for Violations
CLW	Control loss warning
CMAS	Commercial mobile alert system
CMDI	Channel monitoring and decision interval
CME	Certificate management entity
CMRS	Commercial mobile radio services
CRC	Cyclic redundancy check
CRL	Certificate revocation list
CSMA	Carrier sensing multiple access
CSMA/CA	Carrier sensing multiple access with collision avoidance
CSMA/CD	Carrier sensing multiple access with collision detection
CSR	Certificate Signing Request
CST	Carrier sensing threshold
CTS	Clear to send
CVN	Consumer vehicle network
CWAB	Collision Warning with Auto Brake
DARPA	Defense Advanced Research Projects Agency
DAS	Data acquisition system
DCF	Distributed coordination function
D-FPAV	Distributed Fair Transmission Power Adjustment for Vehicular Ad hoc Networks
DGNSS	Differential Global Navigation Satellite Systems
DGPS	Differential Global Positioning System
DIFS	DCF interframe spacing
DNPW	Do not pass warning
DoS	Denial of service

DSA	Digital Signature Algorithm
DSRC	Dedicated short-range communications
DSS	Digital Signature Standard
DTLS	Datagram Transport Layer Security
DVI	Driver–vehicle interface
ECC	Elliptic curve cryptography
ECDSA	Elliptic Curve Digital Signature Algorithm
ECIES	Elliptic Curve Integrated Encryption Scheme
ECM	Engine control module
ECU	Electronic control unit
EDCA	Enhanced distributed channel access
EDGE	Enhanced Data Rates for GSM Evolution
EEBL	Emergency electronic brake light
EIFS	Extended interframe spacing
E-MBMS	Evolved multimedia broadcast and multicast service
ESA	European Space Agency
ESC	Electronic stability control
ETC	Electronic toll collection
ETSI	European Telecommunications Standards Institute
FCC	Federal Communications Commission
FCW	Forward collision warning
FDD	Frequency division duplex
FHWA	Federal Highway Administration
FIPS	Federal Information Processing Standard
FOC	Full operational capability
FOT	Field operational trial
GDP	Gross domestic product
GGSN	Gateway GPRS support node
GID	Geometric intersection description
GNSS	Global navigation satellite system
GPRS	General Packet Radio Service
GPS	Global Positioning System
GSM	Global Systems for Mobile Communications
GUI	Graphical user interface
HALL	High availability, low latency
HDOP	Horizontal dilution of precision
HLE	Higher-layer entity
HPLR	High power, long range
HSPA	High Speed Packet Access

HTTP	Hypertext Transfer Protocol
HTTPS	HTTP Secure
HV	Host vehicle
IEEE	Institute of Electrical and Electronics Engineers
IETF	Internet Engineering Task Force
IMA	Intersection movement assist
IOC	Initial operational capability
IOV	In-orbit validation
IP	Internet Protocol
IPv6	Internet Protocol version 6
IRP	Intersection reference point
ISI	Intersymbol interference
ISP	Internet Service Provider
ITS	Intelligent transportation systems
ITSA	Intelligent Transportation Society of America
ITU-T	International Telecommunications Union, Telecommunication Standardization Sector
Kbps	Kilobits per second
km	Kilometer
LCW	Lane change warning
LIN	Local Interconnect Network
LLC	Logical link control
LoS	Line of sight
LTE	Long-Term Evolution
MAC	Media access control layer (in communications contexts)
MAC	Message authentication code (in security contexts)
MANET	Mobile ad hoc network
MBL	Maximum beaconing load
Mbps	Megabits per second
MBSFN	Multicast/broadcast single frequency network
MDS	Misbehavior detection system
MIAD	Multiplicative increase/additive decrease
MIB	Management information base
MIMD	Multiplicative increase/multiplicative decrease
MIMO	Multiple input and multiple output
MLIT	Ministry of Land, Infrastructure, Transport and Tourism
MLME	MAC Layer Management Entity
MOST	Media Oriented Systems Transport
NAV	Network allocation vector

NHTSA	National Highway Traffic Safety Administration
NIST	National Institute of Standards and Technology
OBD	Onboard diagnostics
OCB	Outside the context of a basic service set
OEM	Original equipment manufacturer
OFDM	Orthogonal frequency division multiplexing
OTP	Objective test procedures
P-GW	Packet data network gateway
PHY	Physical layer
PKCS	Public Key Cryptography Standard
PKI	Public key infrastructure
PLAN	Personal Localized Alerting Network
PLCP	Physical layer convergence procedure
PLME	Physical Layer Management Entity
PMD	Physical media dependent
PPS	Pulse per second
PSID	Provider Service Identifier
PULSAR	Periodically Updated Load Sensitive Adaptive Rate Control
PWR	Insufficient power
PXB	Prereception busy
QAM	Quadrature amplitude modulation
QoS	Quality of service
QPSK	Quadrature phase-shift keying
RA	Registration authority
RDU	Radar decision unit
RF	Radiofrequency
RFC	Request for Comments
RITA	Research and Innovative Technology Administration
RL	Revocation list
RNC	Radio network controller
RSA	Rivest–Shamir–Adleman
RSS	Received signal strength
RSU	Roadside unit
RTCM	Radio Technical Commission for Maritime Services
RTK	Real-Time Kinematics
RTS	Request to send
RV	Remote vehicle
RXB	Reception busy

SAP	Service access point
SC-FDMA	Single-carrier frequency division multiple access
SCH	Service channel
SDARS	Satellite digital audio radio systems
SDS	Security data store
SG	Study group
SGSN	Serving GPRS support node
S-GW	Serving gateway
SHA	Secure Hash Algorithm
SINR	Signal-to-interference-noise ratio
SIP	Session Initiation Protocol
SP	Single Point
SPAT	Signal phase and timing
TCP	Transport Control Protocol
TDEA	Triple Data Encryption Algorithm
TMC	Traffic Management Center
TOA	Time of arrival
TPMS	Tire pressure monitoring system
TTA	Time to access
TXB	Transmission busy
UDP	User Datagram Protocol
URE	User range error
USDOT	U.S. Department of Transportation
UTM	Universal Transverse Mercator
V2CA	Vehicle-to-certificate authority
V2I	Vehicle-to-infrastructure
V2MDS	Vehicle-to-misbehavior detection system
V2V	Vehicle-to-vehicle
VCM	Vehicle certificate manager
V-DTLS	Vehicular datagram transport layer security
VIN	Vehicle identification number
VLR	Verifier-Local Revocation
VMT	Vehicle miles traveled
VSC2	Vehicle Safety Communications 2
VSC-A	Vehicle Safety Communications–Applications
VTTI	Virginia Tech Transportation Institute
WAAS	Wide Area Augmentation System
WAVE	Wireless access in vehicular environments

WCDMA	Wideband Code Division Multiple Access
WGS	World Geodetic System
WME	WAVE Management Entity
WSA	WAVE Service Advertisement
WSMP	WAVE Short Message Protocol
WSU	Wireless safety unit

INDEX

Vehicle Safety Communications: Protocols, Security, and Privacy, First Edition. Luca Delgrossi and Tao Zhang.
© 2012 John Wiley & Sons, Inc. Published 2012 by John Wiley & Sons, Inc.

Printed in the United States
By Bookmasters